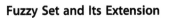

Fuzzy Set and Its Extension

Fuzzy Set and Its Extension

The Intuitionistic Fuzzy Set

Tamalika Chaira
Midnapore (West)
West Bengal, India

This edition first published 2019
© 2019 John Wiley & Sons, Inc.

The right of Tamalika Chaira to be identified as the author of this work has been asserted in accordance with law.

Registered Office(s)
John Wiley & Sons, Inc., 111 River Street, Hoboken, NJ 07030, USA

Editorial Office
111 River Street, Hoboken, NJ 07030, USA

For details of our global editorial offices, customer services, and more information about Wiley products visit us at www.wiley.com.

Wiley also publishes its books in a variety of electronic formats and by print-on-demand. Some content that appears in standard print versions of this book may not be available in other formats.

Library of Congress Cataloging-in-Publication data applied for

ISBN: 9781119544197

Cover Design: Wiley
Cover Image: © kentoh / Shutterstock

Set in 10/12pt Warnock by SPi Global, Pondicherry, India

Printed in the United States of America

V10008901_031919

To my parents

Barid Baran Chaira
and
Puspa Chaira

Contents

Preface

Since Lofti A. Zadeh introduced fuzzy set theory about 50 years ago, i.e. in 1965, theory of fuzzy sets has evolved in many directions and has received more attention from many researchers. Applications of the theory can be found ranging from pattern recognition, control system, image processing, decision making, operations research, robotics, and management.

This book discusses on connections between fuzzy set and crisp set, fuzzy relations, operations on fuzzy sets, various aggregation operators using fuzzy sets, fuzzy numbers, arithmetic operations on fuzzy numbers, fuzzy integrals, fuzzy matrices and determinants, and fuzzy groups. Applications on decision making and image processing is also given.

Apart from fuzzy set, intuitionistic fuzzy set is also discussed in this book. Since its inception by K. Atanassov in 1985, intuitionistic fuzzy set theory has also received attention but to limited number of researchers as compared to fuzzy set. Though its use in application is not as comparable as that of fuzzy set, but still research studies are carried out in the areas that use fuzzy set. In intuitionistic fuzzy set, computational complexity is more as two types of uncertainties are used. But, for obtaining better result, where uncertainty present is more, especially in diagnosis of medical images, accurate result is very much important compromising the computational complexity. So, researchers try to use it on real-time application.

The book discusses the basics of intuitionistic fuzzy set, intuitionistic fuzzy relations, operations on intuitionistic fuzzy sets, various intuitionistic fuzzy aggregation operators, intuitionistic fuzzy numbers, arithmetic operations on intuitionistic fuzzy numbers, intuitionistic fuzzy integrals, and intuitionistic fuzzy matrices. Also, application in decision making and image processing using intuitionistic fuzzy set is also given.

This book is an attempt to unify both fuzzy/intuitionistic fuzzy set and their existing work in application. The primary goal of this book is to help the readers to know the mathematics of both fuzzy set and intuitionistic fuzzy set so that with both these concepts, they can use either fuzzy/intuitionistic fuzzy set in their applications.

Finally, I would like to acknowledge the authors of the papers that have been referred in the book. I acknowledge my beloved daughter, Shruti De, for giving the title of the book. I acknowledge my parents for their continuous support while writing this book. I am also indebted to John Wiley & Sons, Inc. for making the publication of this book possible.

Tamalika Chaira

Organization of the Book

The book contains 10 chapters. Each chapter begins with an introduction, theory, and also several examples that will help the readers to understand the chapters in a better way. *Chapter 1* starts with preliminaries of fuzzy sets and relations. Different types of membership function, composition of fuzzy relation, and fuzzy binary relation that includes symmetric, reflexive, transitive, and equivalent relations are explained with examples. Similar to fuzzy set, intuitionistic fuzzy sets, operations, relations, and compositions are also explained with examples.

Chapter 2 deals with fuzzy numbers. Zadeh's extension principle is explained that states how an image of a fuzzy subset is formed using a function. Using the extension principle, arithmetic operations on fuzzy numbers are explained. Fuzzy numbers with α-cut, operations on fuzzy numbers, and L–R representation of fuzzy numbers are explained with examples. Intuitionistic fuzzy numbers such as triangular and trapezoidal fuzzy numbers, along with operations, are also explained with examples.

Chapter 3 details fuzzy similarity measures and measures of fuzziness. Similarity measures on fuzzy sets and fuzzy numbers are discussed. More emphasis is given on similarity measure on fuzzy numbers. Different types of similarity measures based on the center of gravity, area, perimeter, and graded mean integration of fuzzy numbers are discussed in detail. Measures of fuzziness and different types of entropy are also explained. Intuitionistic fuzzy similarity measures, distance measures, and entropy are also discussed.

Chapter 4 outlines fuzzy measures and fuzzy integrals. Definition and properties of fuzzy measures are discussed. Sugeno measure is a special type of fuzzy measure is discussed with examples. Other types of fuzzy measures such as belief measure, possibility measure, plausibility measure, and necessity measure are discussed. Fuzzy integrals such as Choquet and Sugeno integrals are explained with figures and example on decision making problem is also provided. Intuitionistic fuzzy Choquet integral is also discussed.

Chapter 5 discusses on fuzzy operators where different types of fuzzy operators are used. Fuzzy algebraic operations such as complement, sum, difference, bounded sum, bounded difference, union, and intersection are explained with examples. Fuzzy set theoretic operations that include fuzzy triangular norms (t-norms) and triangular conorms (t-conorms) are explained. Triangular norms suggested by different authors are discussed. Fuzzy/intuitionistic fuzzy aggregation operators that combine different pieces of information into a single object in a same set are explained. Different types of aggregation operators such as fuzzy/intuitionistic fuzzy generalized ordered weighted averaging, hybrid averaging operator, quasi-arithmetic weighted averaging operator, fuzzy/intuitionistic fuzzy-induced generalized averaging operator, fuzzy/intuitionistic fuzzy Choquet and induced Choquet ordered aggregation operator are explained with examples on decision making.

Chapter 6 examines matrices and determinants of a fuzzy matrix. Fuzzy matrix/determinant operations are explained with properties and examples. Adjoint and determinants of a fuzzy matrix, and inverse of a fuzzy matrix are discussed with examples. Intuitionistic fuzzy determinants and matrices are also discussed.

Chapter 7 outlines fuzzy linear equation. It is a continuation of Chapter 6 where an unknown vector is computed using general equation method and also using Cramer's rule. Finding inverse of a fuzzy matrix is discussed with examples. Fuzzy linear equation using L–R-type fuzzy numbers is also discussed with examples where left and right spread of a fuzzy number are considered.

Chapter 8 is dedicated to fuzzy subgroups. Definition of fuzzy subgroup is provided along with properties. Many examples of fuzzy subgroups are mentioned. Other types of fuzzy subgroups such as fuzzy-level subgroup and fuzzy normal subgroup are also discussed with examples. Definition of fuzzy subgroup with respect to fuzzy t-norm is also included. Product of fuzzy subgroups with respect to t-norm with propositions are explained.

Chapter 9 is based on the application on image processing. Introduction on image processing along with image enhancement, segmentation, clustering, edge detection, and morphology are explained with examples using both fuzzy and intuitionistic fuzzy set. Results on medical images for detection of abnormal lesions/clot/hemorrhage in CT scan brain image are shown.

Lastly, the book ends up with *Chapter 10* where Type-2 fuzzy set is explained. Introduction and representation of Type-2 fuzzy set is discussed. Operations on Type-2 fuzzy set along with examples are provided.

1

Fuzzy/Intuitionistic Fuzzy Set Theory

1.1 Introduction to Fuzzy Set

A classical set is normally defined as a collection of objects or elements x in $X = \{x_1, x_2, x_3, \ldots, x_n\}$ that are finite. Each element or object either belongs or does not belong to a set. Most of our traditional tools, modeling, and methods are based on crisp set theory where elements are deterministic and precise. By crisp we mean, answer is "yes" or "no" rather "a little bit less or a little bit more" type. This means that the statement is either "true or false" and in mathematics it may be defined as either "0 or 1." Elements have a Boolean state of nature that means either belongs to the set or does not belong to the set. This belongingness to a set may be termed as "membership value" or the degree of belongingness. So, if an element in a set is present, then its membership value is "1" else its membership value is "0." The membership function or the degree of belongingness of an element "x" in the set is denoted by $\mu(x)$.

But in reality, this crisp theory does not follow. This deterministic and precise theory does not work. It may happen many times in day-to-day situation, when the terms like "less," "more or less," or "very high" are required. To deal with such type of situations, fuzzy set theory is used. In fuzzy set theory, instead of having precise or sharp values, gradually varying values are used. Prof. L.A. Zadeh in 1965 [1] introduced the concept of fuzzy set theory on the basis of principles of uncertainty, ambiguity, and vagueness. He suggested that in real world, classes of objects do not always have precisely defined membership values. The objects do not have a rigid demarcating boundary, i.e. either present or do not present and thus there is a gradual transition from zero to unity membership. These sets are named as fuzzy sets and the elements in these sets have membership values lying between 0 and 1, i.e. $0 \leq \mu(x) \leq 1$. These fuzzy sets have been extensively used in many application areas such as image processing,

Fuzzy Set and Its Extension: The Intuitionistic Fuzzy Set, First Edition. Tamalika Chaira.
© 2019 John Wiley & Sons, Inc. Published 2019 by John Wiley & Sons, Inc.

pattern recognition, decision-support systems, artificial intelligence, control system, and so on to model uncertainties, imprecision, and vagueness inherent in them.

Example 1 Consider a following set X and its subset A as:

$$X = \{x_1, x_2, x_3, x_4, x_5, x_6, x_7\} \text{ and } A = \{x_1, x_3, x_5, x_6, x_7\}.$$

Considering the set to be a crisp set, subset A may be represented using membership function (either 0 or 1):

$$\mu_A(x_1) = 1, \ \mu_A(x_3) = 1, \ \mu_A(x_5) = 1, \ \mu_A(x_6) = 1, \ \mu_A(x_7) = 1.$$

Thus, A may be written as

$$A = \{(x_1, 1), (x_2, 0), (x_3, 1), (x_4, 0), (x_5, 1), (x_6, 1), (x_7, 1)\}.$$

Now, let us imagine a situation where the membership degrees of the elements in a set take any value in the interval [0,1]. That means each element in the set has a fractional membership, depending on the degree of its presence in the set that may be partially, moderately, or fully present. Elements having partial membership in the set have membership values that lie between $0 < \mu_A(x_i) < 1$ and elements with full membership have membership value $\mu_A(x_i) = 1$. This membership concept may also be represented as:

$$A = \{(x_1/0.7), (x_2/0.8), (x_3/0.9), (x_4/1), (x_5/0.7), (x_6/0.6), (x_7/0.3)\},$$

where x_i is an element of the set A, followed by the membership value of the element x_i that lies between 0 and 1. It is a measure of the degree of belongingness of the element in the set.

An example relating to person's height is shown.

It is observed that in crisp set theory, there is a sharp transition of height. If the membership degree is 0, i.e. $\mu = 0$, then the person is not tall and if the membership degree is $\mu = 1$, then the person is tall. So, membership degree does not have any role, if a person's height is 6 ft or 7 ft. They are simply both tall. But there is a significant difference in the heights. So, crisp set works better in binary mathematics but not in real-world situation.

Fuzzy approach to the set leads to a better approximation of a person's height as shown in Figure 1.1b. The figure shows a continuous function. Persons with different heights do not have same membership degree. So, if a person whose membership degree is 0.4 whereas a person whose membership degree of 0.55, then their heights are different. That means the second person is little bit taller than the first person. If another person has membership degree 0.95, then this person is considered to be significantly tall.

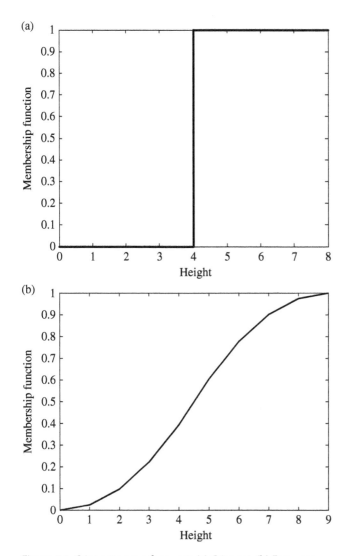

Figure 1.1 Crisp set versus fuzzy set. (a) Crisp set. (b) Fuzzy set.

1.2 Mathematical Representation of Fuzzy Sets

If X be a collection of objects denoted by x, then a fuzzy set A in X is defined as:

$$A = \{(x, \mu_A(x)) \mid x \in X\},$$

where $\mu_A(x)$ is the degree of membership of x in A. The degree of membership lies between $[0,1]$. Membership degree "zero" means no presence and membership degree "1" means full presence of that element in a set. In between values means partial presence of that element in a set.

Example 2   Suppose there are 10 agricultural fields and they are classified based on their fertility. Let $X = \{1,2,3,4,5,6,7,8,9,10\}$ be the agricultural lands. Now a fuzzy set for "fertile land" may be described as:

$$A = \{(1,0.3),(2,0.4),(3,0.5),(4,0.7),(6,1.0),(7,0.8),(8,0.6),(10,0.4)\}.$$

It is observed that the fertility of lands 5 and 9 is 0, i.e. infertile as their membership is "0" and land 6 is the most fertile as the membership degree is 1. The fertility of other lands depends on the membership degree. The higher the membership value, the more is the fertility.

Example 3   An integer close to "8" is written as:

$$A = \{(5,0.2),(6,0.6),(7,0.8),(8,1),(9,0.8),(10,0.7),(11,0.5)\}.$$

So, the elements around "8" has more membership value than those elements are away from "8" and "8" has membership degree 1.

Here, we present a definition of convex fuzzy set.
Convex fuzzy set – A fuzzy set A is said to be convex if [2]

$$\mu_A[\lambda x_1 + (1-\lambda)x_2] \geq \min\{\mu_A(x_1),\mu_A(x_2)\}, x_1,x_2 \in A, \lambda \in [0,1].$$

Figure 1.2 shows an example of a convex and non-convex fuzzy set.

Suppose for a particular value of $\lambda \in [0,1]$, let $\lambda x_1 + (1-\lambda)x_2 = x_3$, which lies between x_1 and x_2. Then the condition $\mu_A(x_3) \geq \min\{\mu_A(x_1), \mu_A(x_2)\}$ fulfils for the convex fuzzy set but the condition is not fulfilled for the non-convex fuzzy set.

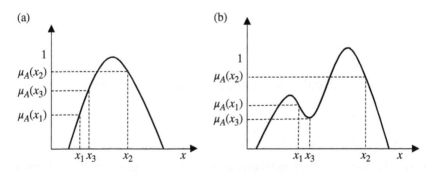

Figure 1.2 (a) Convex fuzzy set. (b) Non-convex fuzzy set.

There are few terms associated with the membership function. These are explained below.

i) Core – The core of a membership function of a fuzzy set A is the region characterized by complete and full membership in a fuzzy set A.
The core thus consists of only those elements whose membership values $\mu_A(x) = 1$.

ii) Support – The support of a membership function of a fuzzy set A is the region characterized by nonzero membership. Thus, support consists of those elements whose membership values are greater than 0, i.e. $\mu_A(x) > 0$.

iii) The boundary of membership function comprises the region where the elements possess nonzero membership but not full membership, i.e. $0 < \mu_A(x) < 1$.

iv) Alpha-level set of a fuzzy set A with $\mu_A(x)$ as the membership function is a crisp set of all elements to a degree α is given as:

$$A_\alpha = \{x \in A \,|\, \mu_A(x) \geq \alpha\}.$$

v) Strong alpha-level set, which is same as the alpha-level set, is a crisp set that consists of all elements to a degree greater than α is given as:

$$A'_\alpha = \{x \in A \,|\, \mu_A(x) > \alpha\}.$$

Example 4

$$A = \{(1,0.3),(2,0.4),(3,0.5),(4,0.7),(6,1.0),(7,0.8),(8,0.6),(10,0.4)\}$$
$$A_{0.5} = \{(3,0.5),(4,0.7),(6,1.0),(7,0.8),(8,0.6)\},$$
$$A_{0.4} = \{(2,0.4),(3,0.5),(4,0.7),(6,1.0),(7,0.8),(8,0.6),(10,0.4)\}.$$

Strong α-level set for $\alpha = 0.7$ is:

$$A'_{0.7} = \{(6,1.0),(7,0.8)\}.$$

Example 5   Consider a fuzzy set A defined on the interval $x = [0,5]$ of integers by a membership function $\mu(x) = \dfrac{x}{x+3}$. Find the α-level set corresponding to $\alpha = 0.3$.

Solution

First, we will compute the membership function in the interval $[0,5]$.

$$\mu(0) = \frac{0}{0+3} = 0, \quad \mu(1) = \frac{1}{1+3} = 0.25,$$
$$\mu(2) = 0.4, \quad \mu(3) = 0.5, \quad \mu(4) = 0.57, \quad \mu(5) = 0.625.$$

So, the α-level set corresponding to $\alpha = 0.3$ is $A_{0.3} = [2,3,4,5]$.

1) Height of A is the largest membership grade obtained by any element in A.
2) A fuzzy set is called normal if its height is 1, i.e. if there is at least one point with $\mu_A(x) = 1$. Otherwise, it is called subnormal.

1.3 Membership Function

In a fuzzy set, the degree of membership of an element signifies the extent to which the element belongs to a fuzzy set, i.e. there is a gradation of membership value of each element in a set. A membership function is a curve that defines how each point in the input space is mapped to a membership value between 0 and 1. There are different types of membership functions that may be viewed as mappings of diverse human choices to an interval [0,1]. To name a few, some membership functions are defined as follows:

i) triangular membership function (Figure 1.3)

$$\mu(x) = \begin{cases} 0, & a \leq x \\ \dfrac{x-a}{b-a}, & a \leq x \leq b \\ \dfrac{c-x}{c-b}, & b \leq x \leq c \end{cases}$$

ii) trapezoidal membership function (Figure 1.4)

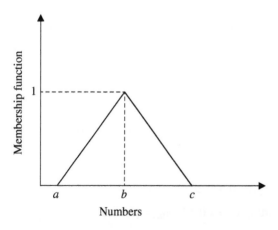

Figure 1.3 Triangular membership function.

Figure 1.4 Trapezoidal membership function.

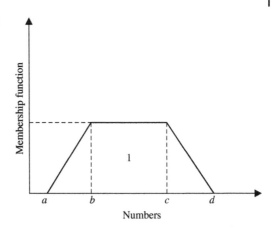

Numbers

$$\mu(x) = \begin{cases} 0, & a \leq x \\ \dfrac{x-a}{b-a}, & a \leq x \leq b \\ 1, & b \leq x \leq c \\ \dfrac{d-x}{d-c}, & c \leq x \leq d \\ 0 & x > d \end{cases}$$

a, b, c, d are the four parameters, where a, d are at the feet of the trapezium and b, c lie at the shoulders. It is to be noted that if $b = c$, then the trapezoid becomes a triangle.

iii) Gaussian membership function

$$\mu(x) = \frac{1}{\sqrt{2\pi}\sigma} \exp\left(-\frac{(x-m)^2}{2\sigma^2} \right),$$

where σ, m are the standard deviation and mean, respectively.

iv) S membership function – This membership function is given by Zadeh [3].

$$\mu(x) = \begin{cases} 0, & x < a \\ 2 \cdot \dfrac{(x-a)}{(c-a)^2}, & a < x < k\beta \\ 1 - 2 \cdot \left[\dfrac{(x-a)}{(c-a)}\right]^2, & k\beta < x < c \\ 1, & x > c \end{cases},$$

where $0 < k < 2$, β is a threshold. Generally, $\beta = 0.5x_{max}$ is chosen and x_{min} and x_{max} are the minimum and maximum values of the elements in the set. In most cases, we select $a = x_{min}$, $c = x_{max}$.

Many authors write $k\beta = \dfrac{a+c}{2}$.

v) Membership function from restricted equivalence function [4] – A function $REF : [0,1]^2 \rightarrow [0,1]$ is called restricted equivalence function if it satisfies the following conditions:

   i) $REF(x,y) = REF(y,x)$ for all $x, y \in [0,1]$,

   ii) $REF(x,y) = 1$ iff $x = y$,

   iii) $REF(x,y) = 0$, if and only if $x = 1$, $y = 0$ or $x = 0$, $y = 1$,

   iv) $REF(x,y) = REF(c(x), c(y))$ for all $x, y \in [0,1]$, c is a strong negation, where $c : [0,1] \rightarrow [0,1]$ is a negation that satisfies the following properties [4]:

     a) $c(0) = 1, c(1) = 0$,

     b) $c(x) < c(y)$, iff $x \geq y$,

     c) $c(c(x)) = x, \forall x \in [0,1]$.

   v) For all $x, y, z \in [0,1]$, if $x \leq y \leq z$, then $REF(x,y) \geq REF(x,z)$ and $REF(y,z) \geq REF(x,z)$.

Bustince et al. [4] defined a restricted equivalence function using automorphism which is as follows:

Automorphism in a unit interval $[a,b]$ is a continuously increasing function $\varphi : [a,b] \longrightarrow [a,b]$ with boundary condition $\varphi(a) = a$, $\varphi(b) = b$.

If φ_1 and φ_2 are two automorphisms in a unit interval, then

$$REF(x,y) = \varphi_1^{-1}(1 - |\varphi_2(x) - \varphi_2(y)|) \tag{1.1}$$

is a restricted equivalence function
with $c(x) = \varphi_2^{-1}(1 - \varphi_2(x))$.
If $\varphi_1(x) = \varphi_2(x) = x$,

$$\text{then } REF(x,y) = \varphi_1^{-1}(1 - |x - y|) = 1 - |x - y|. \tag{1.2}$$

Let c be a strong negation and an equilibrium point of fuzzy negation is a value such that
$c(e) = e$ which is obtained as follows:

   For $\varphi_2(x) = x$, $c(x) = 1 - x$, then $c(0) = 1$, $c(1) = 0$ and $c(e) = 1 - e$.
   If we choose $e = 0.5$, then $c(0.5) = 1 - 0.5 = 0.5$. This implies $c(e) = e$.
   For finding the membership function, consider a function $F : [0,1] \rightarrow [e,1]$ such that
   $F(x) = 1$ iff $x = 0$,
   $F(x) = e$ iff $x = 1$ and $F(x)$ is nonincreasing function.

Then the membership function is defined as [5]:

$$\mu(x) = F(c(REF(x,y))). \tag{1.3}$$

$$\text{Let the function } F(x) = 1 - (1 - e)x \tag{1.4}$$

with $F(0) = 1$, $F(1) = e$.

Substituting Eq. (1.4) in Eq. (1.3), the membership function is written as:

$$\mu(x) = 1 - (1 - e)c(REF(x,y)). \tag{1.5}$$

If $c(x) = 1 - x$ for all $x \in [0,1]$ and $REF(A,B) = 1 - |x - y|$ and $e = 0.5$, then

$$\mu(x) = 1 - 0.5 \cdot c(1 - |x - y|) = 1 - 0.5|x - y|.$$

d) Chaira [6] defined another type of membership function using restricted equivalence function.

Let $\varphi_2(x) = x$. From the definition of restricted equivalence function, we know:

$$REF(x,y) = \varphi_1^{-1}(1 - |\varphi_2(x) - \varphi_2(y)|) = \varphi_1^{-1}(1 - |x - y|),$$

so $REF(x,y) = \varphi_1^{-1}(1 - |x - y|)$.
Considering $\varphi_1(x) = \ln[x(e - 1) + 1]$, where $e = \exp(1)$ and using inverse function, we get:

$$\varphi_1^{-1}(y) = \frac{e^y - 1}{e - 1}.$$

Then, by mathematical induction we get

$$REF(x,y) = \varphi_1^{-1}(1 - |x - y|) = \frac{e^{1 - |x - y|} - 1}{e - 1}. \tag{1.6}$$

If we define membership function $\mu : [0,1]$ as:

$$\mu(x) = REF(x,y),$$

then the membership function becomes:

$$\mu(x) = \frac{e^{1 - |x - y|} - 1}{e - 1}. \tag{1.7}$$

e) Gamma membership function [7] – This membership function is derived from the probability density function of Gamma distribution. It is defined as:

$$f(x) = \frac{\left(\frac{x - \nu}{\beta}\right)^{\gamma - 1} \exp\left(-\left(\frac{x - \nu}{\beta}\right)\right)}{\Gamma(\gamma)}, x \geq \nu; \gamma, \beta > 0,$$

where ν is a location parameter, β is a scale parameter, Γ is the gamma function, and γ is a shape parameter.

When $\nu = 0$ and $\beta = 1$, then the distribution is a standard gamma distribution

$$f(x) = \frac{(x)^{\gamma - 1} \exp(-x)}{\Gamma(\gamma)};$$

if $\gamma = 1$ and $\nu \neq 0$, then $f(x) = \exp(-(x - \nu))$.
This is the Gamma membership function.

1.4 Fuzzy Relations

Fuzzy relation was initially introduced by Zadeh [3] and then by Kaufmann [8]. It represents the strength of association of the elements of fuzzy sets. Fuzzy relations are mapping elements of one universe, say X, to another universe, say Y, through a Cartesian product of two universes. Relation $\mathcal{R}$ can be "x larger than y," or "x taller that y."

Let us consider two universes, X and Y. In a crisp set, a set of ordered pairs is the product set $X \times Y$:

If $X = \{x_1, x_2, x_3, x_4\}$, $Y = \{y_1, y_2\}$, then

$$X \times Y = \{(x_1, y_1), (x_1, y_2), (x_2, y_1), (x_2, y_2), (x_3, y_1), (x_3, y_2), (x_4, y_1), (x_4, y_2)\}.$$

The notation of relation in crisp set can also extended to fuzzy set. In fuzzy set, the Cartesian product is given as follows:

Consider two fuzzy sets: $X = \{(3, 0.5), (5, 0.4), (6, 0.1)\}$, $Y = \{(4, 0.3), (8, 0.4)\}$.

The product is the set of pairs from X and Y with minimum memberships. So,

$$X \times Y = \{[(3,4), \min(0.5, 0.3)], [(3,8), \min(0.5, 0.4)], [(5,4), \min(0.4, 0.3)], [(5,8),$$
$$\min(0.4, 0.4)], [(6,4), \min(0.1, 0.3)], [(6,8), \min(0.1, 0.4)]\}$$
$$= \{[(3,4), 0.3], [(3,8), 0.4], [(5,4), 0.3], [(5,8), 0.4], [(6,4), 0.1], [(6,8), 0.1]\}.$$

Fuzzy relation is studied by a number of authors, e.g. Zadeh [3], Kaufmann [8], Klir and Yaun [9], and Zimmerman [2]. Suppose P is a product set and μ is the membership grade between x and y, then fuzzy relation $\mathcal{R}$ is a subset of the product set P taking its values of μ. Similar to ordinary sets, fuzzy relations are fuzzy subsets of $X \times Y$.

The relation $\mathcal{R}$ can be defined as:

$$\mathcal{R} = \{(x,y), \mu_R(x,y)\}, \quad \forall (x,y) \in X \times Y, \ \forall x \in X, \forall y \in Y$$

where the membership matrix of a $m \times n$ binary fuzzy relation has the form:

$$\begin{bmatrix} \mu_{\mathcal{R}}(x_1,y_1) & \mu_{\mathcal{R}}(x_1,y_2) & \cdots & \mu_{\mathcal{R}}(x_1,y_n) \\ \mu_{\mathcal{R}}(x_2,y_1) & \mu_{\mathcal{R}}(x_2,y_2) & \cdots & \mu_{\mathcal{R}}(x_2,y_n) \\ & \cdot & & \\ & \cdot & & \\ & \cdot & & \\ \mu_{\mathcal{R}}(x_m,y_1) & \mu_{\mathcal{R}}(x_m,y_2) & \cdots & \mu_{\mathcal{R}}(x_m,y_n) \end{bmatrix}$$

$\mu_{\mathcal{R}}(x_1, y_1)$ is the ordered pair in the product space $X \times Y$ and it denotes the membership grade between x_1 and y_1.

Let X and Y be two discrete fuzzy universes and the relation $\mathcal{R}$ is given as:

$$\mathcal{R}(X,Y) = \left\{ \frac{(x_1,y_1)}{0.2}, \frac{(x_1,y_2)}{0.3}, \frac{(x_1,y_3)}{0.5}, \frac{(x_2,y_1)}{0.3}, \frac{(x_2,y_2)}{0.5}, \frac{(x_2,y_3)}{0.6}, \frac{(x_3,y_1)}{0.1}, \frac{(x_3,y_2)}{0.3}, \frac{(x_3,y_3)}{0.4} \right\}.$$

The membership matrix may be represented as:

$$\begin{bmatrix} 0.2 & 0.3 & 0.5 \\ 0.3 & 0.5 & 0.6 \\ 0.1 & 0.3 & 0.4 \end{bmatrix}.$$

Just like fuzzy set, there are also few basic operations on fuzzy relation. Let us consider two fuzzy relations $\mathcal{R}_1$ and $\mathcal{R}_2$.

If $\mathcal{R}_1$ and $\mathcal{R}_2$ are two fuzzy relations in the same product space, the following operations are defined as follows:

i) The union of two fuzzy relations $\mathcal{R}_1$ and $\mathcal{R}_2$ is a new relation $\mathcal{R}_1 \cup \mathcal{R}_2$,

$$\mathcal{R}_1 \cup \mathcal{R}_2 = \int_{x \times y} \frac{\left[\mu_{\mathcal{R}_1}(x,y) \vee \mu_{\mathcal{R}_2}(x,y) \right]}{(x,y)},$$

where the membership function of $\mathcal{R}_1 \cup \mathcal{R}_2$ is given as:

$$\mu_{\mathcal{R}_1 \cup \mathcal{R}_2}(x,y) = \mu_{\mathcal{R}_1}(x,y) \vee \mu_{\mathcal{R}_2}(x,y) = \max \left[\mu_{\mathcal{R}_1}(x,y), \mu_{\mathcal{R}_2}(x,y) \right].$$

The union of the two fuzzy relations is formed by taking the maximum of the two membership grades of the corresponding elements of the two matrices.

ii) The intersection of two fuzzy relations $\mathcal{R}_1$ and $\mathcal{R}_2$ is a new relation,

$$\mathcal{R}_1 \cap \mathcal{R}_2 = \int_{x \times y} \frac{\left[\mu_{\mathcal{R}_1}(x,y) \wedge \mu_{\mathcal{R}_2}(x,y) \right]}{(x,y)}, \tag{1.8}$$

where the membership function of $\mathcal{R}_1 \cap \mathcal{R}_2$ is given as:

$$\mu_{\mathcal{R}_1 \cap \mathcal{R}_2}(x,y) = \mu_{\mathcal{R}_1}(x,y) \wedge \mu_{\mathcal{R}_2}(x,y) = \min \left[\mu_{\mathcal{R}_1}(x,y), \mu_{\mathcal{R}_2}(x,y) \right]. \tag{1.9}$$

The intersection of the two fuzzy relations is formed by taking the minimum of the two membership grades of corresponding elements of the two matrices.

Algebraic product of two fuzzy relations is a new fuzzy relation whose membership function is given as:

$$\mu_{\mathcal{R}_1 \cdot \mathcal{R}_2}(x,y) = \mu_{\mathcal{R}_1}(x,y) \cdot \mu_{\mathcal{R}_2}(x,y). \tag{1.10}$$

Algebraic sum of two relations is a new relation whose membership function is given as:

$$\mu_{\mathcal{R}_1 + \mathcal{R}_2}(x,y) = \mu_{\mathcal{R}_1}(x,y) + \mu_{\mathcal{R}_2}(x,y) - \mu_{\mathcal{R}_1}(x,y) \cdot \mu_{\mathcal{R}_2}(x,y). \tag{1.11}$$

Complement of a relation is a new relation whose membership function is given as:

$$\mu_{\bar{\mathcal{R}}_1}(x,y) = 1 - \mu_{\mathcal{R}_1}(x,y). \tag{1.12}$$

Example 6 Consider two fuzzy relations:

$$\mathcal{R}_1 = \begin{bmatrix} 0.1 & 0.3 & 0.5 \\ 0.7 & 0.2 & 0.8 \\ 0.2 & 0.6 & 0.4 \end{bmatrix}, \mathcal{R}_2 = \begin{bmatrix} 0.3 & 0.1 & 0.2 \\ 0.5 & 0.0 & 0.3 \\ 0.7 & 0.3 & 0.7 \end{bmatrix}.$$

Union: $\mu_{\mathcal{R}_1 \cup \mathcal{R}_2}(x,y) = \begin{bmatrix} 0.3 & 0.3 & 0.5 \\ 0.7 & 0.2 & 0.8 \\ 0.7 & 0.6 & 0.7 \end{bmatrix}$,

Intersection: $\mu_{\mathcal{R}_1 \cap \mathcal{R}_2}(x,y) = \begin{bmatrix} 0.1 & 0.1 & 0.2 \\ 0.5 & 0.0 & 0.3 \\ 0.2 & 0.3 & 0.4 \end{bmatrix}$,

Complement: $\mu_{\bar{\mathcal{R}}_1}(x,y) = \begin{bmatrix} 0.9 & 0.7 & 0.5 \\ 0.3 & 0.8 & 0.2 \\ 0.8 & 0.4 & 0.6 \end{bmatrix}$,

$$\mu_{\mathcal{R}_1 \cdot \mathcal{R}_2}(x,y) = \begin{bmatrix} 0.03 & 0.03 & 0.1 \\ 0.35 & 0.0 & 0.24 \\ 0.14 & 0.18 & 0.28 \end{bmatrix}.$$

Fuzzy relation can be represented in different forms. Suppose the fuzzy relation is "x is taller than y." It can be represented in (i) membership matrix, (ii) tabular form, (iii) linguistically "x is taller than y," and (iv) taking the union of fuzzy singletons.

Example 7 For the two fuzzy relations, let us define two relations as:
$\mathcal{R}_1$ = "x is larger than y" and $\mathcal{R}_2$ = "x is very close to y" where

$$\mathcal{R}_1 = \begin{bmatrix} & y_1 & y_2 & y_3 \\ x_1 & 0.9 & 0.1 & 0.1 \\ x_2 & 0.0 & 0.7 & 0.8 \\ x_3 & 1.0 & 0.1 & 0.7 \end{bmatrix}, \quad \mathcal{R}_2 = \begin{bmatrix} & y_1 & y_2 & y_3 \\ x_1 & 0.3 & 0.1 & 0.8 \\ x_2 & 0.9 & 0.2 & 0.4 \\ x_3 & 0.3 & 0.0 & 0.7 \end{bmatrix}.$$

Intersection of two fuzzy relations $\mathcal{R}_1$ and $\mathcal{R}_2$ means "x is larger than y" and "x is close to y"

$$\mu_{\mathcal{R}_1 \cap \mathcal{R}_2}(x,y) = \begin{bmatrix} 0.3 & 0.1 & 0.1 \\ 0.0 & 0.2 & 0.4 \\ 0.3 & 0.0 & 0.7 \end{bmatrix}.$$

Union of two fuzzy relations $\mathcal{R}_1$ and $\mathcal{R}_2$ means "x is larger than y" or "x is close to y."

$$\mu_{\mathcal{R}_1 \cup \mathcal{R}_2}(x,y) = \begin{bmatrix} 0.9 & 0.1 & 0.8 \\ 0.9 & 0.7 & 0.8 \\ 1.0 & 0.1 & 0.7 \end{bmatrix}.$$

1.5 Projection

Let $\mathcal{R}$ be a fuzzy relation in the Cartesian product $X \times Y$. The fuzzy relation $\mathcal{R}$ (1) defined in X is called the first projection [8, 10].

The first projection or shadow of fuzzy relation $\mathcal{R}$, $\mathcal{R}(1)$, defined in X, is a fuzzy set that results by eliminating the second set Y of $X \times Y$ by projecting the relation on X. First projection is given as:

$$\mathcal{R}(1) = \{x, \max_y[\mu_{\mathcal{R}}(x,y)] \mid (x,y) \in X \times Y\},$$

where the membership function of first projection is:

$$\mu_{\mathcal{R}_1}(x) = \vee_y[\mu_{\mathcal{R}}(x,y)].$$

First projection is the maximum over all y.

The second projection of fuzzy relation $\mathcal{R}$, $\mathcal{R}(2)$, defined in X, is a fuzzy set that results by eliminating the second set X of $X \times Y$ by projecting the relation on Y.

$$\mathcal{R}(2) = \{y, \max_x[\mu_{\mathcal{R}}(x,y)] \mid (x,y) \in X \times Y\},$$

where the membership function of second projection is:

$$\mu_{\mathcal{R}_2}(y) = \vee_x[\mu_{\mathcal{R}}(x,y)].$$

Second projection is the maximum over all x.
So, the total projection is the combined projection over X and Y.

$$\mathcal{R}(2) = \{ \max_x \max_y \mu_{\mathcal{R}}(x,y) \mid (x,y) \in X \times Y \}, \tag{1.13}$$

where the membership function of total projection is

$$\mu_{\mathcal{R}_T}(x,y) = \vee_x \vee_y [\mu_{\mathcal{R}}(x,y), (x,y)].$$

Example 8 Consider a fuzzy relation:

$$\mathcal{R} = \begin{bmatrix} 0.1 & 0.3 & 0.5 & 0.6 \\ 0.7 & 0.2 & 0.8 & 0.4 \\ 0.2 & 0.7 & 0.4 & 0.3 \end{bmatrix} x$$

$$y$$

$$\mathcal{R}(1) = \{ x, \max_y [\mu_{\mathcal{R}}(x,y)] \}$$

$$= \sum \frac{\mu_{\mathcal{R}_1}(x_i)}{x_i} = \frac{\max(0.1, 0.3, 0.5, 0.6)}{x_1} + \frac{\max(0.7, 0.2, 0.8, 0.4)}{x_2}$$

$$+ \frac{\max(0.2, 0.7, 0.4, 0.3)}{x_3} = \frac{0.6}{x_1} + \frac{0.8}{x_2} + \frac{0.7}{x_3} = \begin{bmatrix} 0.6 \\ 0.8 \\ 0.7 \end{bmatrix}.$$

$$\mathcal{R}(2) = \{ x, \max_x [\mu_{\mathcal{R}}(x,y)] \}$$

$$= \sum \frac{\mu_{\mathcal{R}_2}(y_i)}{y_i} = \frac{\max(0.1, 0.7, 0.2)}{y_1} + \frac{\max(0.3, 0.2, 0.7)}{y_2} + \frac{\max(0.5, 0.8, 0.4)}{y_3}$$

$$+ \frac{\max(0.6, 0.4, 0.3)}{y_4} = \frac{0.7}{y_1} + \frac{0.7}{y_2} + \frac{0.8}{y_3} + \frac{0.6}{y_4} = [0.7 \ 0.7 \ 0.8 \ 0.6].$$

Total projection, $\mu_{\mathcal{R}_T}(x,y) = \vee_x \vee_y [\mu_{\mathcal{R}}(x,y), (x,y)]$ (maximum of the two projections) = 0.8.

1.6 Composition of Fuzzy Relation

For two fuzzy relations, $\mathcal{R}_1 \subset X \times Y$ and $\mathcal{R}_2 \subset Y \times Z$, we cannot perform union or intersection as they are defined in different product sets. But we can perform their composition as there is a common set Y between them. Fuzzy relation in different product spaces may be combined with each other using the operator called *composition* [3, 8, 10]. Max–min composition is the most useful in applications. Let us consider any two fuzzy relations $\mathcal{R}_1$ and $\mathcal{R}_2$ such that

$$\mathcal{R}_1 \subset X \times Y$$
$$\mathcal{R}_2 \subset Y \times Z$$

There are three ways of composing a fuzzy relation.

i) Max–min composition

$$\mathcal{R}_1 \circ \mathcal{R}_2(x,z) = \max_y \left(\min \left(\mu_{\mathcal{R}_1}(x,y), \mu_{\mathcal{R}_2}(y,z) \right) \right), \quad x \in X, y \in Y, z \in Z. \quad (1.14)$$

ii) Max–product composition

$$\mathcal{R}_1 \underset{prod}{\circ} \mathcal{R}_2(x,z) = \mu_{\mathcal{R}_1}(x,y) \cdot \mu_{\mathcal{R}_2}(y,z), x \in X, y \in Y, z \in Z. \quad (1.15)$$

iii) Max–average composition

$$\mathcal{R}_1 \underset{avg}{\circ} \mathcal{R}_2(x,z) = \frac{1}{2}\max \left[\mu_{\mathcal{R}_1}(x,y) + \mu_{\mathcal{R}_2}(y,z) \right], \quad x \in X, y \in Y, z \in Z. \quad (1.16)$$

Example 9 Consider two fuzzy relations:

$$\mathcal{R}_1 = \begin{bmatrix} & y_1 & y_2 & y_3 \\ x_1 & 0.1 & 0.3 & 0.5 \\ x_2 & 0.7 & 0.2 & 0.8 \\ x_3 & 0.2 & 0.6 & 0.4 \end{bmatrix}, \mathcal{R}_2 = \begin{bmatrix} & z_1 & z_2 & z_3 \\ y_1 & 0.3 & 0.1 & 0.2 \\ y_2 & 0.5 & 0.0 & 0.3 \\ y_3 & 0.7 & 0.3 & 0.7 \end{bmatrix}$$

i) Max–min composition

$$\mathcal{R}_1 \circ \mathcal{R}_2(x,z) = \left[(x,z), \max_y \left(\min \left(\mu_{\mathcal{R}_1}(x,y), \mu_{\mathcal{R}_2}(y,z) \right) \right) \right],$$

where $\mu_{\mathcal{R}_1 \circ \mathcal{R}_2}$ is the membership function of fuzzy relation on a set. Now, we will compute max–min composition $\mathcal{R}_1 \circ \mathcal{R}_2(x,z)$.

$$\text{At } x = x_1, z = z_1, \mathcal{R}_1 \circ \mathcal{R}_2(x_1,z_1) = [0.1 \quad 0.3 \quad 0.5] \begin{bmatrix} 0.3 \\ 0.5 \\ 0.7 \end{bmatrix}$$

$$= \max[\min(0.1,0.3), \min(0.3,0.5), \min(0.5,0.7)]$$

$$= \max[0.1, 0.3, 0.5] = 0.5,$$

$$\mathcal{R}_1 \circ \mathcal{R}_2(x_1, z_2) = [0.1 \quad 0.3 \quad 0.5] \begin{bmatrix} 0.1 \\ 0.0 \\ 0.3 \end{bmatrix}$$

$$= \max[\min(0.1, 0.1), \min(0.3, 0.0), \min(0.5, 0.3)]$$

$$= \max[0.1, 0.0, 0.3] = 0.3,$$

$$\mathcal{R}_1 \circ \mathcal{R}_2(x_1, z_3) = [0.1 \quad 0.3 \quad 0.5] \begin{bmatrix} 0.2 \\ 0.3 \\ 0.7 \end{bmatrix}$$

$$= \max[\min(0.1, 0.2), \min(0.3, 0.3), \min(0.5, 0.7)]$$

$$= \max[0.1, 0.3, 0.5] = 0.5,$$

$$\mathcal{R}_1 \circ \mathcal{R}_2(x_2, z_1) = [0.7 \quad 0.2 \quad 0.8] \begin{bmatrix} 0.3 \\ 0.5 \\ 0.7 \end{bmatrix}$$

$$= \max[\min(0.7, 0.3), \min(0.2, 0.5), \min(0.8, 0.7)]$$

$$= \max[0.3, 0.2, 0.7] = 0.7.$$

Likewise, for other values of x and z, max–min composition is computed.

So, $\mathcal{R}_1 \circ \mathcal{R}_2 = \begin{bmatrix} & z_1 & z_2 & z_3 \\ x_1 & 0.5 & 0.3 & 0.5 \\ x_2 & 0.7 & 0.3 & 0.7 \\ x_3 & 0.5 & 0.3 & 0.4 \end{bmatrix}$.

ii) Max–product composition

$$\mathcal{R}_1 \underset{prod}{\circ} \mathcal{R}_2(x_1, z_1) = [0.1 \quad 0.3 \quad 0.5] \begin{bmatrix} 0.3 \\ 0.5 \\ 0.7 \end{bmatrix}$$

$$= \max[0.03, 0.15, 0.35] = 0.35,$$

$$\mathcal{R}_1 \underset{prod}{\circ} \mathcal{R}_2(x_1, z_2) = [0.1 \ \ 0.3 \ \ 0.5] \begin{bmatrix} 0.1 \\ 0.0 \\ 0.3 \end{bmatrix}$$

$$= \max[0.01, 0.0, 0.15] = 0.15.$$

Likewise, for other values of x and z, max–product composition is computed.

$$\text{So, } \mathcal{R}_1 \circ_{prod} \mathcal{R}_2(x, z) = \begin{bmatrix} 0.35 & 0.15 & 0.35 \\ 0.56 & 0.24 & 0.56 \\ 0.30 & 0.12 & 0.28 \end{bmatrix}.$$

We observe that there is difference in the elements of the matrix composition.

iii) Max–average composition

$$\mathcal{R}_1 \underset{avg}{\circ} \mathcal{R}_2(x_1, z_1) = [0.1 \ \ 0.3 \ \ 0.5] \begin{bmatrix} 0.3 \\ 0.5 \\ 0.7 \end{bmatrix}$$

$$= \left(\frac{1}{2}\right) \max[(0.1 + 0.3), (0.3 + 0.5), (0.5 + 0.7)]$$

$$= \left(\frac{1}{2}\right) \max[0.4, 0.8, 1.2] = \left(\frac{1}{2}\right) 1.2 = 0.6,$$

$$\mathcal{R}_1 \underset{avg}{\circ} \mathcal{R}_2(x_1, z_2) = [0.1 \ \ 0.3 \ \ 0.5] \begin{bmatrix} 0.1 \\ 0.0 \\ 0.3 \end{bmatrix}$$

$$= \left(\frac{1}{2}\right) \max[(0.1 + 0.1), (0.3 + 0.0), (0.5 + 0.3)]$$

$$= \left(\frac{1}{2}\right) \max[0.2, 0.3, 0.8] = \left(\frac{1}{2}\right) 0.8 = 0.4.$$

Likewise, for other combinations of x and z, max avg composition is computed.

We get,

$$\mathcal{R}_1 \underset{avg}{\circ} \mathcal{R}_2(x,z) = \begin{bmatrix} 0.60 & 0.40 & 0.60 \\ 0.75 & 0.55 & 0.75 \\ 0.55 & 0.35 & 0.55 \end{bmatrix}.$$

Example 10 Consider two fuzzy relations:
$\mathcal{R}_1$ = "x is considerably larger than y"
$\mathcal{R}_2$ = "y is very close to z"

$$\mathcal{R}_1 = \begin{bmatrix} & y_1 & y_2 & y_3 \\ x_1 & 0.8 & 0.1 & 0.2 \\ x_2 & 0.0 & 0.8 & 0.0 \\ x_3 & 0.9 & 0.2 & 0.6 \end{bmatrix}, \mathcal{R}_2 = \begin{bmatrix} & z_1 & z_2 & z_3 \\ y_1 & 0.3 & 0.9 & 0.2 \\ y_2 & 0.0 & 0.5 & 0.1 \\ y_3 & 0.9 & 0.5 & 0.7 \end{bmatrix}.$$

Using max–min composition, the composition "x is considerably larger than y and y is very close to z" is given as:

$$\mathcal{R}_1 \circ \mathcal{R}_2 = \begin{bmatrix} & z_1 & z_2 & z_3 \\ x_1 & 0.3 & 0.8 & 0.2 \\ x_2 & 0.0 & 0.5 & 0.1 \\ x_3 & 0.6 & 0.9 & 0.6 \end{bmatrix}.$$

Looking at the values, it shows that the maximum value of (x_1, z_i) is $(x_1, z_2) = 0.8$ for i = 1, 2, 3, the maximum value of (x_2, z_i) is $(x_2, z_2) = 0.5$, and the maximum value of (x_3, z_i) is $(x_3, z_2) = 0.9$. This implies that the composition suits better among x_1 and z_2, x_2 and z_2, and x_3 and z_2.

Some properties of max–min composition are as follows [2, 3]:

i) Max–min composition is reflexive – If $\mathcal{R}_1$ and $\mathcal{R}_2$ are two fuzzy relations that are reflexive, then their max–min composition, $\mathcal{R}_1 \circ \mathcal{R}_2$, is also reflexive.
ii) If $\mathcal{R}$ is reflexive, then $\mathcal{R} \subseteq \mathcal{R} \circ \mathcal{R}$.
iii) Max–min composition is symmetric – If $\mathcal{R}_1$ and $\mathcal{R}_2$ are two fuzzy relations that are symmetric, then their max–min composition, $\mathcal{R}_1 \circ \mathcal{R}_2$, is also symmetric if $\mathcal{R}_1 \circ \mathcal{R}_2 = \mathcal{R}_2 \circ \mathcal{R}_1$. In particular, if $\mathcal{R}_1$ is symmetric, then $\mathcal{R}_1 \circ \mathcal{R}_1$ is also symmetric.
iv) Max–min composition is associative – If $\mathcal{R}_1, \mathcal{R}_2$, and $\mathcal{R}_3$ are three fuzzy relations, then

$$\mathcal{R}_3 \circ (\mathcal{R}_2 \circ \mathcal{R}_1) = (\mathcal{R}_3 \circ \mathcal{R}_2) \circ \mathcal{R}_1.$$

v) Max–min composition is transitive – If $\mathcal{R}_1$ and $\mathcal{R}_2$ are two fuzzy relations that are transitive and if $\mathcal{R}_1{\circ}\mathcal{R}_2 = \mathcal{R}_2{\circ}\mathcal{R}_1$, then $\mathcal{R}_1{\circ}\mathcal{R}_1$ is transitive.

vi) If $\mathcal{R}_1$ is a fuzzy relation and $\mathcal{R}_1{\circ}\mathcal{R}_1$ is a max–min composition, then $\mathcal{R}_1{\circ}\mathcal{R}_1{\circ}\mathcal{R}_1 = \mathcal{R}_1{}^3$.

Symmetric, reflexive, and transitive relations are discussed in the next section.

1.7 Fuzzy Binary Relation

A binary relation on a set A is a set of ordered pairs of elements of A, i.e. a subset of $A \times A$. Said in another way, any relation between a set is a binary relation. In the fuzzy relation $\mathcal{R}$, $\mathcal{R} \subset X \times Y$, and if $X = Y$ then we say $\mathcal{R}$ is fuzzy binary relation on set $X \times X$. When $x, y \in X$ and the values of ordered pairs lies in the interval $[0,1]$, then a fuzzy binary relation exists in $X \times X$ [3, 8].

Let $\mathcal{R}$ be a binary fuzzy relation on X. Then, $\mathcal{R}(x,y)$ is interpreted as the degree of membership of the ordered pair (x,y) in X.

Let us take an example.

Example 11   The relation is "approximately equal" can be defined as:

$$\mathcal{R}(1,1) = 1,\ \mathcal{R}(2,2) = 1,\ \mathcal{R}(3,3) = 1,$$

$$\mathcal{R}(1,2) = \mathcal{R}(2,1) = \mathcal{R}(2,3) = \mathcal{R}(3,2) = 0.85$$

$$\mathcal{R}(1,3) = \mathcal{R}(3,1) = 0.2.$$

The membership matrix is given as:

$$\mathcal{R} = \begin{bmatrix} 1 & 0.85 & 0.2 \\ 0.85 & 1 & 0.85 \\ 0.2 & 0.85 & 1 \end{bmatrix}.$$

Properties of fuzzy binary relation:

1) Symmetric property

   Fuzzy relation $\mathcal{R}(X,X)$ is said to be symmetric if $\mu_{\mathcal{R}}(x,y) = \mu_{\mathcal{R}}(y,x)$, $\forall$, $(x,y) \in X$, where $\mu_{\mathcal{R}}(x,y)$ is the membership grade of ordered pair (x,y). An example of symmetric relation is given as:

$$\mathcal{R} = \begin{bmatrix} 1 & 0.7 & 0.5 \\ 0.7 & 0.3 & 0.4 \\ 0.5 & 0.4 & 1 \end{bmatrix}.$$

   Fuzzy relation $\mathcal{R}(X,X)$ is said to be antisymmetric if $\mu_{\mathcal{R}}(x,y) \neq \mu_{\mathcal{R}}(y,x)$, $\forall$, $(x,y) \in X$.

2) Reflexive property

Fuzzy relation $\mathcal{R}(X,X)$ is said to be reflexive if $\mu_{\mathcal{R}}(x,x) = 1$ for all $x \in X$. If $\mu_{\mathcal{R}}(x,x) \geq \varepsilon$, then the relation is called ε–reflexive. If $\max\{\mu_{\mathcal{R}}(x,y), \mu_{\mathcal{R}}(y,x)\} \leq \mu_{\mathcal{R}}(x,x)$, then the relation is locally reflexive. If $\mathcal{R}_1$ is a fuzzy reflexive relation and $\mathcal{R}_2$ is any fuzzy relation then,

$$\mu_{\mathcal{R}_2}(x,y) \leq \mu_{\mathcal{R}_1 \circ \mathcal{R}_2}(x,y), \forall x,y \in X,$$

$$\mu_{\mathcal{R}_2}(x,y) \leq \mu_{\mathcal{R}_2 \circ \mathcal{R}_1}(x,y), \forall x,y \in X.$$

An example of a fuzzy reflexive relation is given. If $\mathcal{R}$ is a fuzzy reflexive relation,

$$\mathcal{R} = \begin{bmatrix} 1 & 0.7 & 0.5 \\ 0.7 & 1 & 0.4 \\ 0.5 & 0.4 & 1 \end{bmatrix},$$

then their max–min composition $\mathcal{R} \subseteq \mathcal{R} \circ \mathcal{R}$ is also reflexive.

$$\mathcal{R} \circ \mathcal{R} = \begin{bmatrix} 1 & 0.7 & 0.5 \\ 0.7 & 1 & 0.4 \\ 0.5 & 0.4 & 1 \end{bmatrix} \circ \begin{bmatrix} 1 & 0.7 & 0.5 \\ 0.7 & 1 & 0.4 \\ 0.5 & 0.4 & 1 \end{bmatrix} = \begin{bmatrix} 1 & 0.7 & 0.5 \\ 0.7 & 1 & 0.4 \\ 0.5 & 0.4 & 1 \end{bmatrix}.$$

3) Transitive property

Fuzzy relation is said to be transitive if $\forall (x,y), (y,z), (x,z) \in X$, the max–min transitivity is defined as:

$$\mu_{\mathcal{R}}(x,z) \geq \max_{y}[\min(\mu_{\mathcal{R}}(x,y), \mu_{\mathcal{R}}(y,z))].$$

Fuzzy relation is also a min-transitivity

$$\mu_{\mathcal{R}}(x,z) \geq \min(\mu_{\mathcal{R}}(x,y), \mu_{\mathcal{R}}(y,z)). \tag{1.17}$$

From application point of view, the min-transitivity property is used. A few methods for constructing fuzzy min-transitive relations are there in the literature, which have important applications in image understanding and fuzzy decision making [11].

Fuzzy relation $\mathcal{R}$ is called transitive if:

$$\mathcal{R} \circ \mathcal{R} \subseteq \mathcal{R} \text{ or } \mathcal{R}^2 \subseteq \mathcal{R}.$$

Generally, it is defined as $\mathcal{R}^{n+1} = \mathcal{R}^n \circ \mathcal{R}$.

Transitivity can also be defined in another way as max–product transitivity. It is defined as:

$$\mu_{\mathcal{R}}(x,z) \geq \max_{y}[(\mu_{\mathcal{R}}(x,y) \cdot \mu_{\mathcal{R}}(y,z))]. \tag{1.18}$$

A binary relation that is reflexive, symmetric, and transitive is called an equivalence relation.

1.8 Transitive Closure of Fuzzy Binary Relation

In mathematics, transitive closure is a binary relation $\mathcal{R}$ on X is the smallest relation on X that is transitive. Let us take an example to define transitive closure. Suppose X is a set of places and $x\mathcal{R}y$ means that there is a direct bus from one place x to another place y, $\forall x, y \in X$. Then, transitive closure of a relation $\mathcal{R}$ is a relation $\acute{\mathcal{R}}$ on X such that $x\acute{\mathcal{R}}y$ means it is possible to go from one place to another with one or more buses. This means that the transitive closure gives us a set of all places we can go from any starting place.

A relation $\mathcal{R}$ is said to be transitive if for every $(x,y) \in \mathcal{R}$ and $(y,z) \in \mathcal{R}$, there is $(x,z) \in \mathcal{R}$. Transitive closure is the transitive relation that contains $\mathcal{R}(X,X)$ with smallest possible membership. For any fuzzy relation, transitive closure exists. The elements of the transitive closure have the smallest possible membership grades. If $\mathcal{R}$ is a fuzzy relation defined on a set X and that $\mathcal{R}$ is not transitive, then the transitive closure of $\mathcal{R}$ is a connectivity relation $\acute{\mathcal{R}}$.

If the set has a cardinality n, then *transitive closure* of a fuzzy binary relation is a relation $\acute{\mathcal{R}}$ [8]:

$$\acute{\mathcal{R}} = \mathcal{R} \cup \mathcal{R}^2 \cup \mathcal{R}^3 \cup \ldots \cup \mathcal{R}^n.$$

For $i > 0$, $\mathcal{R}^{i+1} = \mathcal{R} \circ \mathcal{R}^i$, where $\mathcal{R}^i$ is the ith power of $\mathcal{R}$ and $\mathcal{R}^1 = \mathcal{R}$.

Transitive closure of any fuzzy binary relation is a transitive binary relation, i.e. $\acute{\mathcal{R}}^2 \subseteq \acute{\mathcal{R}}$.

This denotes that $\acute{\mathcal{R}}$ is transitive and it helps in constructing a transitive relation from any relation.

Also, another characteristic of transitive closure is $\mathcal{R} \subseteq \acute{\mathcal{R}}$.

A fuzzy relation $\mathcal{R}$ is called tolerance relation $\mathcal{R}_1$ if it is both reflexive and symmetric [12].

Any fuzzy tolerance relation may be reformed to a fuzzy equivalence $\mathcal{R}$ relation using $(n-1)$ fuzzy compositions, i.e.

$$\mathcal{R}_1{}^{n-1} = \mathcal{R}_1 \circ \mathcal{R}_1 \circ \mathcal{R}_1 \ldots \circ \mathcal{R}_{1n-1} = \mathcal{R}.$$

Example 12 Example for transitivity. Let us consider a fuzzy relation $\mathcal{R}$,

$$\mathcal{R} = \begin{array}{c} \\ x_1 \\ x_2 \\ x_3 \end{array} \begin{array}{ccc} x_1 & x_2 & x_3 \\ \left[\begin{array}{ccc} 0.2 & 0.1 & 0.3 \\ 0.1 & 1.0 & 0.2 \\ 0.3 & 0.0 & 0.7 \end{array}\right] \end{array}.$$

For transitivity, we see $\mu_{\mathcal{R}}(x_1,x_2) = 0.1$ and $\mu_{\mathcal{R}}(x_2,x_3) = 0.2$ and we compute $\min(\mu_{\mathcal{R}}(x_1,x_2),\mu_{\mathcal{R}}(x_2,x_3)) = 0.1$ (min-transitivity).

From the data, $\mu_{\mathcal{R}}(x_1,x_3) = 0.3 \geq 0.1$. On checking different combinations, the min-transitive definition follows. So, the relation follows transitivity.

Transitivity can also be shown using the max–min transitive property.

$$\mathcal{R}\circ\mathcal{R} = \begin{bmatrix} 0.2 & 0.1 & 0.3 \\ 0.1 & 1.0 & 0.2 \\ 0.3 & 0.0 & 0.7 \end{bmatrix} \circ \begin{bmatrix} 0.2 & 0.1 & 0.3 \\ 0.1 & 1.0 & 0.2 \\ 0.3 & 0.0 & 0.7 \end{bmatrix} = \begin{bmatrix} 0.3 & 0.1 & 0.3 \\ 0.2 & 1.0 & 0.2 \\ 0.3 & 0.1 & 0.7 \end{bmatrix}.$$

It is observed that $\mathcal{R}\circ\mathcal{R} \subseteq \mathcal{R}$.

Example 13 Transitive closure

Let $\mathcal{R}$ be a fuzzy relation, $\mathcal{R} = \begin{bmatrix} 0.2 & 0.1 & 0.4 \\ 0.1 & 1.0 & 0.1 \\ 0.3 & 0.0 & 0.7 \end{bmatrix}$.

$$\mathcal{R}^2 = \mathcal{R}\circ\mathcal{R} = \begin{bmatrix} 0.2 & 0.1 & 0.4 \\ 0.1 & 1.0 & 0.1 \\ 0.3 & 0.0 & 0.7 \end{bmatrix} \circ \begin{bmatrix} 0.2 & 0.1 & 0.4 \\ 0.1 & 1.0 & 0.1 \\ 0.3 & 0.0 & 0.7 \end{bmatrix} = \begin{bmatrix} 0.3 & 0.1 & 0.4 \\ 0.1 & 1.0 & 0.1 \\ 0.3 & 0.1 & 0.7 \end{bmatrix},$$

as $\mathcal{R}^2 \neq \mathcal{R}$, we will compute $\mathcal{R}^3$

$$\mathcal{R}^3 = \mathcal{R}^2\circ\mathcal{R} = \begin{bmatrix} 0.3 & 0.1 & 0.4 \\ 0.1 & 1.0 & 0.1 \\ 0.3 & 0.1 & 0.7 \end{bmatrix} \circ \begin{bmatrix} 0.2 & 0.1 & 0.4 \\ 0.1 & 1.0 & 0.1 \\ 0.3 & 0.1 & 0.7 \end{bmatrix} = \begin{bmatrix} 0.3 & 0.1 & 0.4 \\ 0.1 & 1.0 & 0.1 \\ 0.3 & 0.1 & 0.7 \end{bmatrix},$$

as $\mathcal{R}^2 = \mathcal{R}^3$. So, the transitive closure

$$\acute{\mathcal{R}} = \mathcal{R}\cup\mathcal{R}^2 = \begin{bmatrix} 0.2 & 0.1 & 0.4 \\ 0.1 & 1.0 & 0.1 \\ 0.3 & 0.0 & 0.7 \end{bmatrix} \cup \begin{bmatrix} 0.3 & 0.1 & 0.4 \\ 0.1 & 1.0 & 0.1 \\ 0.3 & 0.1 & 0.7 \end{bmatrix} = \begin{bmatrix} 0.3 & 0.1 & 0.4 \\ 0.1 & 1.0 & 0.1 \\ 0.3 & 0.1 & 0.7 \end{bmatrix}.$$

So, $\acute{\mathcal{R}} = \begin{bmatrix} & x_1 & x_2 & x_3 \\ x_1 & 0.3 & 0.1 & 0.4 \\ x_2 & 0.1 & 1.0 & 0.1 \\ x_3 & 0.3 & 0.1 & 0.7 \end{bmatrix}.$

For transitivity, we see $\mu_{\acute{\mathcal{R}}}(x_1,x_2) = 0.1$ and $\mu_{\acute{\mathcal{R}}}(x_2,x_3) = 0.1$.

Then, we compute $\min\left(\mu_{\acute{\mathcal{R}}}(x_1,x_2),\mu_{\acute{\mathcal{R}}}(x_2,x_3)\right) = 0.1$.

From the data, we see $\mu_{\acute{\mathcal{R}}}(x_1,x_3) = 0.4 \geq 0.1$. This is done for all the data and it is observed that $\mu_{\acute{\mathcal{R}}}(x,z) \geq \min\left(\mu_{\acute{\mathcal{R}}}(x,y),\mu_{\acute{\mathcal{R}}}(y,z)\right)$.

So, the fuzzy relation $\acute{\mathcal{R}}$ is transitive.

Transitivity can also be verified using max–min transitivity,

$$\begin{bmatrix} 0.3 & 0.1 & 0.4 \\ 0.1 & 1.0 & 0.1 \\ 0.3 & 0.1 & 0.7 \end{bmatrix} \circ \begin{bmatrix} 0.3 & 0.1 & 0.4 \\ 0.1 & 1.0 & 0.1 \\ 0.3 & 0.1 & 0.7 \end{bmatrix} = \begin{bmatrix} 0.3 & 0.1 & 0.4 \\ 0.1 & 1.0 & 0.1 \\ 0.3 & 0.1 & 0.7 \end{bmatrix}.$$

It is observed that $\acute{\mathcal{R}}^2 \subseteq \acute{\mathcal{R}}$ implies $\acute{\mathcal{R}}$ is transitive.

1.9 Fuzzy Equivalence Relation

A fuzzy relation $\mathcal{R}$ in $X \times X$ is an equivalence relation if it satisfies the following conditions:

i) Reflexive property – $\mu_{\mathcal{R}}(x_i,x_i) = 1$
ii) Symmetric property – $\mu_{\mathcal{R}}(x_i,x_j) = \mu_{\mathcal{R}}(x_j,x_i)$
iii) Transitive property.

Example 14 Consider a fuzzy relation $\mathcal{R}$

$$\mathcal{R} = \begin{bmatrix} & x_1 & x_2 & x_3 & x_4 \\ x_1 & 1.0 & 0.7 & 0.2 & 0.3 \\ x_2 & 0.7 & 1.0 & 0.4 & 0.0 \\ x_3 & 0.2 & 0.4 & 1.0 & 0.0 \\ x_4 & 0.3 & 0.0 & 0.0 & 1.0 \end{bmatrix}.$$

It is observed that the relation is reflexive as $\mu_{\mathcal{R}}(x,x) = 1$.

The relation is symmetric – $\mu_{\mathcal{R}}(x,y) = \mu_{\mathcal{R}}(y,x)$.

For transitive property, we see $\mu_{\mathcal{R}}(x_1,x_2) = 0.7$ and $\mu_{\mathcal{R}}(x_2,x_3) = 0.4$ and we compute $\min(\mu_{\mathcal{R}}(x_1,x_2),\mu_{\mathcal{R}}(x_2,x_3)) = 0.4$.

From the data we see, $\mu_{\mathcal{R}}(x_1,x_3) = 0.2 \leq 0.4$.

So, the fuzzy relation does not follow transitivity. That means the relation is not an equivalence relation. We can make the relation equivalence relation using composition.

First composition – $\mathcal{R}^2 = \mathcal{R} \circ \mathcal{R} =$

$$
\begin{bmatrix} 1.0 & 0.7 & 0.2 & 0.3 \\ 0.7 & 1.0 & 0.4 & 0.0 \\ 0.2 & 0.4 & 1.0 & 0.0 \\ 0.3 & 0.0 & 0.0 & 1.0 \end{bmatrix} \circ \begin{bmatrix} 1.0 & 0.7 & 0.2 & 0.3 \\ 0.7 & 1.0 & 0.4 & 0.0 \\ 0.2 & 0.4 & 1.0 & 0.0 \\ 0.3 & 0.0 & 0.0 & 1.0 \end{bmatrix} = \begin{array}{c|cccc} & x_1 & x_2 & x_3 & x_4 \\ \hline x_1 & 1.0 & 0.7 & 0.4 & 0.3 \\ x_2 & 0.7 & 1.0 & 0.4 & 0.3 \\ x_3 & 0.4 & 0.4 & 1.0 & 0.2 \\ x_4 & 0.3 & 0.3 & 0.2 & 1.0 \end{array}.
$$

It is seen that the relation $\mathcal{R}^2$ is reflexive and symmetric. For transitivity, we see $\mu_{\mathcal{R}^2}(x_3, x_1) = 0.4$ and $\mu_{\mathcal{R}^2}(x_1, x_4) = 0.3$ and we compute $\min(\mu_{\mathcal{R}^2}(x_2, x_4), \mu_{\mathcal{R}^2}(x_4, x_1)) = 0.3$.

From the data, $\mu_{\mathcal{R}^2}(x_3, x_4) = 0.2 \leq 0.3$.

This implies that the relation $\mathcal{R}^2$ is not transitive.

Now, we compute $\mathcal{R}^3$, $\mathcal{R}^3 = \mathcal{R}^2 \circ \mathcal{R}$ (as the cardinality of the set is 4, i.e. $n = 4$, so we will compute till $n - 1 = 3$) as,

$$
\mathcal{R}^3 = \begin{bmatrix} 1.0 & 0.7 & 0.4 & 0.3 \\ 0.7 & 1.0 & 0.4 & 0.3 \\ 0.4 & 0.4 & 1.0 & 0.2 \\ 0.3 & 0.3 & 0.2 & 1.0 \end{bmatrix} \circ \begin{bmatrix} 1.0 & 0.7 & 0.2 & 0.3 \\ 0.7 & 1.0 & 0.4 & 0.0 \\ 0.2 & 0.4 & 1.0 & 0.0 \\ 0.3 & 0.0 & 0.0 & 1.0 \end{bmatrix} = \begin{bmatrix} 1.0 & 0.7 & 0.4 & 0.3 \\ 0.7 & 1.0 & 0.4 & 0.3 \\ 0.4 & 0.4 & 1.0 & 0.3 \\ 0.3 & 0.3 & 0.3 & 1.0 \end{bmatrix}.
$$

For transitivity, we will look for min transitivity and it is observed that $\mathcal{R}^3$ is transitive. We can also check using max–min transitivity. It is observed that

$$
\mathcal{R}^3 \circ \mathcal{R}^3 = \begin{bmatrix} 1.0 & 0.7 & 0.4 & 0.3 \\ 0.7 & 1.0 & 0.4 & 0.3 \\ 0.4 & 0.4 & 1.0 & 0.3 \\ 0.3 & 0.3 & 0.3 & 1.0 \end{bmatrix} \circ \begin{bmatrix} 1.0 & 0.7 & 0.4 & 0.3 \\ 0.7 & 1.0 & 0.4 & 0.3 \\ 0.4 & 0.4 & 1.0 & 0.3 \\ 0.3 & 0.3 & 0.3 & 1.0 \end{bmatrix} = \begin{bmatrix} 1.0 & 0.7 & 0.4 & 0.3 \\ 0.7 & 1.0 & 0.4 & 0.3 \\ 0.4 & 0.4 & 1.0 & 0.3 \\ 0.3 & 0.3 & 0.3 & 1.0 \end{bmatrix},
$$

implies $\mathcal{R}^3 \circ \mathcal{R}^3 \subseteq \mathcal{R}^3$. Thus, the fuzzy relation $\mathcal{R}^3$ is a fuzzy equivalence relation. It follows symmetric, reflexive, and transitive properties.

1.10 Intuitionistic Fuzzy Set

Fuzzy set theory, as described earlier, takes into account membership degree and the nonmembership degree is the complement of the membership degree. However, in real life, this linguistic negation does not satisfy the logical negation. While selecting the membership degree, there may be some kind of hesitation while defining the membership function. Membership function may be Gaussian, triangular, exponential, or any other membership function. So, due to the hesitation, nonmembership degree is less than or equal to the complement of the membership degree. This is the reason why different results are obtained with different membership functions. Due to this reason, Atanassov in 1985 [13] suggested an intuitionistic fuzzy set (IFS) where the nonmembership degree is not equal to the complement of the membership degree due to the fact that some kind of hesitation or lack of knowledge is present while defining the membership function.

So, compared to fuzzy set theory, IFS considers more (two) uncertainties – membership and nonmembership degrees.

An IFS A in a finite set X may be mathematically represented as:

$$A = \{x, \mu_A(x), \nu_A(x) | x \in X\},$$

where $\mu_A(x)$, $\nu_A(x) : X \rightarrow [0,1]$ are, respectively, the membership and the non-membership functions of an element x in a finite set X with the condition

$$0 \le \mu_A(x) + \nu_A(x) \le 1.$$

The hesitation degree, $\pi_A(x)$, as mentioned above, is also known as the intuitionistic fuzzy index that arises due to the lack of knowledge or "personal error" while assigning the membership degree.

The condition for the hesitation degree is:

$$\mu_A(x) + \nu_A(x) + \pi_A(x) = 1$$

and $0 \le \pi_A(x) \le 1$, for each $x \in X$.

Like fuzzy set, there also hold few definitions for IFSs. For two IFSs A and B, with $\mu_A(x)$ and $\nu_A(x)$ are the membership and nonmembership degrees of the elements in set A, and $\mu_B(x)$ and $\nu_B(x)$ are the membership and nonmembership degrees of the elements in set B, then the following conditions hold:

i) $A \cup B = \{\max(\mu_A, \mu_B), \min(\nu_A, \nu_B)\}$,
ii) $A \cap B = \{\min(\mu_A, \mu_B), \max(\nu_A, \nu_B)\}$,
iii) $\bar{A} = \{x, \mu_A(x), \nu_A(x)\}$, where $\bar{A}$ is the complement of A,
iv) $A \le B = \{A, \mu_A(x) \le \mu_B(x), \nu_A(x) \ge \nu_B(x)\}$,
v) $A \cdot B = \{x, \mu_A(x) \cdot \mu_B(x), \nu_A(x) + \nu_B(x) - \nu_A(x) \cdot \nu_B(x)\}$,
   We see *that*

$$A \cdot A = \{x, \mu_A(x) \cdot \mu_A(x), \nu_A(x) + \nu_A(x) - \nu_A(x) \cdot \nu_A(x)\}$$

$$= \{x, [\mu_A(x)]^2, 2\nu_A(x) - [\nu_A(x)]^2\}$$

$$= \{x, [\mu_A(x)]^2, 1 - [1 - \nu_A(x)]^2\}.$$

So, in general, for any positive integer n,

$$A^n = \left\{ x, [\mu_A(x)]^n, 1 - (1 - \nu_A(x))^n \right\}.$$

Just like fuzzy set, there are terms to describe linguistic variables known linguistic hedges.

$$A^2 = A \cdot A = \left\{ x, [\mu_A(x)]^2, 1 - (1 - \nu_A(x))^2 \right\}.$$

Here, for $n = 2$ and A^2 is called concentration, $CON(A) = very(A)$.

$$A^{1/2} = \left\{ x, [\mu_A(x)]^{\frac{1}{2}}, 1 - (1 - \nu_A(x))^{\frac{1}{2}} \right\}.$$

Here, $n = 1/2$ and $A^{1/2}$ is called the dilation of set A, i.e. $DIL(A)$.

$$\text{plus}(A) = A^{1.25},$$

$$\text{minus}(A) = A^{0.75}.$$

Plus and minus are also called artificial hedges [2].

vi) A and $B = A \wedge B = \{x, \min(\mu_A(x), \mu_B(x)), \max(\nu_A(x), \nu_B(x))\}$,

vii) A or $B = A \vee B = \{x, \max(\mu_A(x), \mu_B(x)), \min(\nu_A(x), \nu_B(x))\}$.

1.11 Construction of Intuitionistic Fuzzy Set

For constructing an IFS, fuzzy generators are used. Fuzzy generators are a type of fuzzy complements with some conditions. Before discussing intuitionistic fuzzy generator, fuzzy complements are reviewed. Fuzzy complement is a fuzzy negation. If $\mu(x)$ is the membership degree of an element of set A, then $c(\mu(x))$ denotes a fuzzy complement of the element and it signifies the degree to which x belongs to the fuzzy complement set $c(A)$. Fuzzy complement has the properties:

Boundary condition – $c(0) = 1$, $c(1) = 0$,
Involutive property – $c(a) = a, \forall a \in [0,1]$,
Monotonicity – For all $a, b \in [0,1]$, if $a \leq b$, then $c(a) \geq c(b)$
and $c(a)$ is continuous.

Equilibrium of a fuzzy complement c is any value "a" for which $c(a) = a$.
At a certain value, a, $0 < a < 1$, its complement is $c(a) = 1 - a$ (using the standard fuzzy complement).
Now, if $a = 0.5$, then $c(0.5) = 1 - 0.5 = 0.5$.
So, 0.5 is the equilibrium of the fuzzy complement.
There are many fuzzy complements. The most common and simplest fuzzy complement is the Zadeh's negation:

$$c(\mu(x)) = 1 - \mu(x). \tag{1.19}$$

Fuzzy complement is computed from fuzzy complement functional which is defined as:

$$N(\mu(x)) = g^{-1}(g(1) - g(\mu(x))), \tag{1.20}$$

where $g(.)$ is an increasing function with $g(0) = 0$.

Sugeno-type fuzzy complement [14] is generated using an increasing function, which is written as:

$$g(\mu(x)) = \frac{1}{\lambda}\log(1 + \lambda \cdot \mu(x)). \tag{1.21}$$

Sugeno complement is computed using Eq. (1.20) as follows:

$$N(\mu(x)) = g^{-1}\left(\frac{1}{\lambda}\log\left(\frac{1 + \lambda}{1 + \lambda \cdot \mu(x)}\right)\right) \text{ with } g(1) = \frac{1}{\lambda}\log(1 + \lambda).$$

From Eq. (1.21), we compute $g^{-1}(\mu(y)) = \frac{e^{\lambda \cdot \mu(y)} - 1}{\lambda}$, and finally we get,

$$g^{-1}\left(\frac{1}{\lambda}\log\left(\frac{1 + \lambda}{1 + \lambda \cdot \mu(x)}\right)\right) = \frac{1 - \mu(x)}{1 + \lambda\mu(x)}.$$

So,

$$N(\mu(x)) = \frac{1 - \mu(x)}{1 + \lambda\mu(x)}, \qquad \lambda \in (-1, \infty). \tag{1.22}$$

Yager [15] suggested a generator where the increasing function is given as: $g(\mu(x)) = \mu(x)^{\alpha}$ and the fuzzy complement is:

$$N(\mu(x)) = (1 - \mu(x)^{\alpha})^{1/\alpha}, \alpha \in (0, \infty). \tag{1.23}$$

Another form of increasing function is given by an author as:
$g(\mu(x)) = \frac{\mu(x)}{\gamma + (1-\gamma)\mu(x)}$ and its complement function is

$$N(\mu(x)) = \frac{\gamma^2(1 - \mu(x))}{\gamma + (1 - \mu(x)) + \mu(x)}, \gamma > 0. \tag{1.24}$$

When $\alpha = 1$, $\lambda = 0$, and $\gamma = 1$, Eqs. (1.22)–(1.23) reduce to standard fuzzy complement, i.e. $(1 - \mu(x))$.

Klir and Yuan [16] suggested a dual generator:

$$N(\mu(x)) = f^{-1}(f(0)) - f(\mu(x))),$$

where $f(.)$ is a decreasing function.

Not all the fuzzy complements are intuitionistic fuzzy generator. From the definition of intuitionistic fuzzy generator given by Bustince et al. [17]:

A function $\varphi : [0,1]$ is called an intuitionistic fuzzy generator if

$$\varphi(x) \leq 1 - x, \forall x \in [0,1].$$

So, $\varphi(0) \leq 1$ and $\varphi(1) = 0$.

An intuitionistic fuzzy generator is called continuous, decreasing, and increasing if φ is continuous, decreasing, and increasing, respectively.

Nonmembership values for the IFSs are computed using intuitionistic fuzzy generator. Let us take an example of Sugeno's fuzzy complement:

$$N(\mu(x)) = \varphi_\lambda(\mu(x)) = \frac{1 - \mu(x)}{1 + \lambda\mu(x)}.$$

At $-1 < \lambda < 0$, it is observed that the condition for intuitionistic fuzzy generator does not hold, i.e.

$$\varphi_\lambda(\mu(x)) \geq 1 - \mu(x).$$

Also, for Yager's fuzzy complement,

$$N(\mu(x)) = \varphi_\alpha(\mu(x)) = \left(1 - \mu(x)^\alpha\right)^{1/\alpha}, \alpha \in (0, \infty).$$

At $\alpha > 1$, condition for intuitionistic fuzzy generator does not hold.

With these conditions mentioned above, i.e. $-1 < \lambda < 0$ and $\alpha > 1$, it follows $\mu(x) + N(\mu(x)) > 1$, which is not true.

To satisfy the condition for intuitionistic fuzzy generator, λ, α values are changed.

So, Yager-type intuitionistic fuzzy generator is written as:

$$\varphi_\alpha(\mu(x)) = \left(1 - \mu(x)^\alpha\right)^{1/\alpha}, 0 < \alpha \leq 1,$$

and Sugeno-type intuitionistic fuzzy generator is written as:

$$\varphi_\lambda(\mu(x)) = \frac{1 - \mu(x)}{1 + \lambda\mu(x)}, \lambda \geq 0.$$

Thus, with the help of Sugeno-type intuitionistic fuzzy complement, IFS becomes:

$$A^{IFS} = \left\{ x, \mu_A(x), \frac{1 - \mu_A(x)}{1 + \lambda\mu_A(x)} \mid x \in X \right\}, \lambda \geq 0.$$

Likewise, with Yager intuitionistic fuzzy generator, IFS becomes:

$$A^{IFS} = \left\{ x, \mu_A(x), \left(1 - \mu_A(x)^\alpha\right)^{1/\alpha} \mid x \in X \right\}, 0 < \alpha \leq 1.$$

Chaira [18] also suggested an intuitionistic fuzzy generator which is as follows:

$$N(\mu(x)) = \frac{1-\mu(x)}{1-(1-e^{\lambda})\mu(x)} = \frac{1-\mu(x)}{1+(e^{\lambda}-1)\mu(x)}, \lambda \geq 0,$$ (1.25)

with an increasing generating function:

$$f(\mu(x)) = \frac{1}{\lambda}\ln\left[1-\mu(x)(1-e^{\lambda})\right].$$

With Chaira's intuitionistic fuzzy generator, IFS may be written as:

$$A^{IFS} = \left\{x, \mu_A(x), \frac{1-\mu_A(x)}{1-(1-e^{\lambda})\mu_A(x)} \mid x \in X\right\}.$$

1.12 Intuitionistic Fuzzy Relations

After the introduction of fuzzy set theory, researchers modeled fuzzy relation in different fields. In 1983, when Atanassov [13] introduced IFS, many researchers extended fuzzy relation to intuitionistic fuzzy relation (IFR) using IFS. Burillo and Bustince [19–21] had given the definition of IFRs and their properties. Lei et al. [22] further explored IFRs and their compositional operations. They used T-norm and T-conorm for IFR.

Let X and Y be two universes of discourse and $IFS(X \times Y)$ represents a family of all IFSs in $(X \times Y)$. Let $R \in (X \times Y)$ be the IFR which is a subset in $(X \times Y)$, which is given as:

$$R = \{(x,y), \mu_R(x,y), \nu_R(x,y) \mid x \in X, y \in Y\},$$

where $\mu_R : X \times Y \to [0,1]$, $\nu_R : X \times Y \to [0,1]$ denotes the membership and non-membership function of R, respectively, with the condition $0 \leq \mu_R(x,y) + \nu_R(x,y) \leq 1$ and

$\pi_R(x,y) = 1 - \mu_R(x,y) - \nu_R(x,y)$ is the hesitation index.

The complementary relation of R is:

$$R = \{(x,y), \nu_R(x,y), \mu_R(x,y) \mid x \in X, y \in Y\}.$$

The following operations on IFRs hold [19]:

i) If R is an IFR on $X \times Y$, then R^{-1} is an IFR on $Y \times X$. In terms of membership function, it may be written as:

$$\mu_{R^{-1}}(y,x) = \mu_R(x,y) \text{ and } \nu_{R^{-1}}(y,x) = \nu_R(x,y), \forall(y,x) \in Y \times X.$$

This is also called inverse relation on R.

If R and S are two IFRs between X and Y, then

i) If $R \leq S$, then $R^{-1} \leq S^{-1}$.

In terms of membership function, $R \leq S$ implies $\mu_R(x,y) \leq \mu_S(x,y)$.

Then, $\mu_{R^{-1}}(y,x) = \mu_R(x,y) \leq \mu_S(x,y) = \mu_{S^{-1}}(y,x)$.

Hence, $\mu_{R^{-1}}(y,x) \leq \mu_{S^{-1}}(y,x)$.

Likewise, for the nonmembership function, $R \leq S$ implies $\nu_R(x,y) \geq \nu_S(x,y)$,

so $\nu_{R^{-1}}(y,x) = \nu_R(x,y) \geq \nu_S(x,y) = \nu_{S^{-1}}(y,x)$.

Hence, $\nu_{R^{-1}}(y,x) \geq \nu_{S^{-1}}(y,x)$

ii) Union

$$\mu_{R\vee S} = \max(\mu_R(x,y),\mu_S(x,y)),$$

$$\nu_{R\vee S} = \min(\nu_R(x,y),\nu_S(x,y)).$$

iii) Intersection

$$\mu_{R\wedge S} = \min(\mu_R(x,y),\mu_S(x,y)),$$

$$\nu_{R\wedge S} = \max(\nu_R(x,y),\nu_S(x,y)).$$

iv) For any three relations, R, S, T on relation $IFR \in (X \times Y)$

a) $\left(R^{-1}\right)^{-1} = R$,

b) $(R \wedge S)^{-1} = R^{-1} \wedge S^{-1}$

In terms of membership function, we can write

$$\mu_{(R\wedge S)^{-1}}(x,y) = \mu_{R\wedge S}(y,x) = \mu_R(y,x) \wedge \mu_S(y,x) = \mu_{R^{-1}}(x,y)$$
$$\wedge \mu_{S^{-1}}(x,y) = \mu_{R^{-1}\wedge S^{-1}}(x,y).$$

c) $(R \vee S)^{-1} = R^{-1} \vee S^{-1}$,

In terms of membership function, we can write

$$\mu_{(R\vee S)^{-1}}(x,y) = \mu_{R\vee S}(y,x) = \mu_R(y,x) \vee \mu_S(y,x) = \mu_{R^{-1}}(x,y)$$
$$\vee \mu_{S^{-1}}(x,y) = \mu_{R^{-1}\vee S^{-1}}(x,y).$$

d) $R \vee (S \wedge T) = (R \vee S) \wedge (R \vee T)$ and $R \wedge (S \vee T) = (R \wedge S) \vee (R \wedge T)$. $\qquad$ (1.26)

It follows distributive property. Considering the membership function

$$\mu_{R\vee(S\wedge T)}(x,y) = \mu_R(x,y) \vee [\mu_S(x,y) \wedge \mu_T(x,y)]$$

$$= [\mu_R(x,y) \vee \mu_S(x,y)] \wedge [\mu_R(x,y) \vee \mu_T(x,y)]$$

$$= \mu_{R\vee S}(x,y) \wedge \mu_{R\vee T}(x,y) = \mu_{(R\vee S)\wedge(R\vee T)}(x,y).$$

Likewise, a similar procedure is followed for the nonmembership function.

e) $S \vee R \geq R$, $\;\; S \vee R \geq S$, $\;\; S \wedge R \leq R$, $\;\; S \wedge R \leq S$.

Example 15   If R and S are two IFRs:

$$R = \begin{bmatrix} (0.1,0.5) & (0.3,0.4) & (0.5,0.3) \\ (0.7,0.1) & (0.2,0.5) & (0.4,0.2) \\ (0.2,0.3) & (0.1,0.3) & (0.2,0.4) \end{bmatrix}, \quad S = \begin{bmatrix} (0.2,0.3) & (0.5,0.3) & (0.4,0.5) \\ (0.3,0.4) & (0.5,0.3) & (0.3,0.4) \\ (0.4,0.4) & (0.3,0.6) & (0.6,0.2) \end{bmatrix},$$

$$R \cup S = \begin{bmatrix} (0.2,0.3) & (0.5,0.3) & (0.5,0.3) \\ (0.7,0.1) & (0.5,0.3) & (0.4,0.2) \\ (0.4,0.3) & (0.3,0.3) & (0.6,0.2) \end{bmatrix},$$

$$R \cap S = \begin{bmatrix} (0.1,0.5) & (0.3,0.4) & (0.4,0.5) \\ (0.3,0.4) & (0.2,0.5) & (0.3,0.4) \\ (0.2,0.4) & (0.1,0.6) & (0.2,0.4) \end{bmatrix}.$$

1.13   Composition of Intuitionistic Fuzzy Relation

Like fuzzy relation, an IFR R from X to Y, $R \in X \times Y$ is characterized by membership function μ_R and nonmembership function ν_R. Let $R \in X \times Y$, $S \in Y \times Z$ be two IFRs. Composition of IFR is given as:

$$\mu_{S \circ R}(x,z) = \max_y\{\min[\mu_R(x,y),\mu_S(y,z)]\},$$

$$\nu_{S \circ R}(x,z) = \min_y\{\max[\nu_R(x,y),\nu_S(y,z)]\}.$$

Example 16   Consider two IFRs R and S

$$R = \begin{bmatrix} (0.7,0.2) & (0.3,0.3) & (0.2,0.8) \\ (0.6,0.1) & (0.3,0.1) & (0.2,0.8) \\ (0.1,0.9) & (0.1,0.9) & (1.0,0.0) \end{bmatrix},$$

$$R \circ R = \begin{bmatrix} (0.7,0.2) & (0.3,0.3) & (0.2,0.8) \\ (0.6,0.1) & (0.3,0.1) & (0.2,0.8) \\ (0.1,0.9) & (0.1,0.9) & (1.0,0.0) \end{bmatrix} \circ \begin{bmatrix} (0.7,0.2) & (0.3,0.3) & (0.2,0.8) \\ (0.6,0.1) & (0.3,0.1) & (0.2,0.8) \\ (0.1,0.9) & (0.1,0.9) & (1.0,0.0) \end{bmatrix}$$

$$= \begin{bmatrix} (0.7,0.2) & (0.3,0.3) & (0.2,0.8) \\ (0.6,0.1) & (0.3,0.1) & (0.2,0.8) \\ (0.1,0.9) & (0.1,0.9) & (1.0,0.0) \end{bmatrix}.$$

It is observed that $R \circ R \subseteq R$. From the definition of transitivity, the relation R is transitive.

1.13.1 Composition of IFR Using T-norms and T-conorms [19–21]

T-norms and T-conorms are described in Chapter 3 and they represent intersection and union in fuzzy set theory, respectively.

Let α, β, ρ, δ be T-norms or T-conorms and the relations be $R \in X \times Y$ and $S \in Y \times Z$. The composed IFR $S \overset{\alpha,\beta}{\underset{\rho,\delta}{\circ}} R \in IFR(X \times Z)$ is written as [19]:

$$S \overset{\alpha,\beta}{\underset{\rho,\delta}{\circ}} R = \left\{ (x,z), \mu_{\underset{\rho,\delta}{S \circ R}}^{\alpha,\beta}(x,z), \nu_{\underset{\rho,\delta}{S \circ R}}^{\alpha,\beta}(x,z) \,\middle|\, x \in X, z \in Z \right\}, \tag{1.27}$$

where

$$\mu_{\underset{\rho,\delta}{S \circ R}}^{\alpha,\beta}(x,z) = \alpha_y\{\beta[\mu_R(x,y),\mu_S(y,z)]\} \text{ and} \tag{1.28}$$

$$\nu_{\underset{\rho,\delta}{S \circ R}}^{\alpha,\beta}(x,z) = \rho_y\{\delta[\nu_R(x,y),\nu_S(y,z)]\} \tag{1.29}$$

with the condition,

$$0 \le \mu_{\underset{\rho,\delta}{S \circ R}}^{\alpha,\beta}(x,z) + \nu_{\underset{\rho,\delta}{S \circ R}}^{\alpha,\beta}(x,z) \le 1, \forall (x,z) \in X \times Z.$$

α, β are applied to membership functions and ρ, δ are applied to nonmembership functions. As per the authors [19–21], composition of IFR satisfies most of the properties for $\alpha = \vee, \beta$ T-norm, $\rho = \wedge, \delta$ T-conorm. Few properties/theorems [19–21] are written as follows:

i) If $\alpha = \vee, \beta = \wedge$ and $\rho = \wedge, \delta = \vee$, then Eqs. (1.28) and (1.29) become

$$\mu_{\underset{\rho,\delta}{S \circ R}}^{\alpha,\beta}(x,z) = \vee_y\{\mu_R(x,y) \wedge \mu_S(y,z)\},$$

$$\nu_{\underset{\rho,\delta}{S \circ R}}^{\alpha,\beta}(x,z) = \wedge_y\{\nu_R(x,y) \vee \nu_S(y,z)\}.$$

ii) For each $S \in IFR(Y \times Z)$ and $R \in IFR(X \times Y)$ and α, β, ρ, δ are T-norms and T-conorms,

$$\left(S \overset{\alpha,\beta}{\underset{\rho,\delta}{\circ}} R \right)^{-1} = R^{-1} \overset{\alpha,\beta}{\underset{\rho,\delta}{\circ}} S^{-1}.$$

In terms of membership function, condition (ii) is written as:

$$\mu_{\left(\underset{\rho,\delta}{\overset{\alpha,\beta}{S\circ R}}\right)^{-1}}(z,x) = \mu_{\left(\underset{\rho,\delta}{\overset{\alpha,\beta}{S\circ R}}\right)}(x,z) = \alpha_y\{\beta[\mu_R(x,y),\mu_S(y,z)]\}$$

$$= \alpha_y\{\beta[\mu_{R^{-1}}(y,x),\mu_{S^{-1}}(z,y)]\}$$

$$= \alpha_y\{\beta[\mu_{S^{-1}}(z,y),\mu_{R^{-1}}(y,x)]\}$$

$$= \mu_{\underset{\rho,\delta}{\overset{\alpha,\beta}{R^{-1}\circ S^{-1}}}}(z,x).$$

A similar procedure is followed for the nonmembership function.

$$\nu_{\left(\underset{\rho,\delta}{\overset{\alpha,\beta}{S\circ R}}\right)^{-1}}(z,x) = \nu_{\left(\underset{\rho,\delta}{\overset{\alpha,\beta}{S\circ R}}\right)}(x,z) = \rho_y\{\delta[\nu_R(x,y),\nu_S(y,z)]\}$$

$$= \rho_y\{\delta[\nu_{R^{-1}}(y,x),\nu_{S^{-1}}(z,y)]\}$$

$$= \rho_y\{\delta[\nu_{S^{-1}}(z,y),\nu_{R^{-1}}(y,x)]\}$$

$$= \nu_{\underset{\rho,\delta}{\overset{\alpha,\beta}{R^{-1}\circ S^{-1}}}}(z,x).$$

iii) If $R_1 \le R_2$, then $R_1 \underset{\rho,\delta}{\overset{\alpha,\beta}{\circ}} S \le R_2 \underset{\rho,\delta}{\overset{\alpha,\beta}{\circ}} S$, for every $R \in IFR(Y \times Z)$, $S \in IFR(X \times Y)$

$R_1 \le R_2$ implies $\mu_{R_1}(y,z) \le \mu_{R_2}(y,z)$ and $\nu_{R_1}(y,z) \ge \nu_{R_2}(y,z)$

$$\mu_{\underset{\rho,\delta}{R_1 \overset{\alpha,\beta}{\circ} S}}(x,z) = \alpha_y\{\beta[\mu_S(x,y),\mu_{R_1}(y,z)]\}$$

$$\le \alpha_y\{\beta[\mu_S(x,y),\mu_{R_2}(y,z)]\} = \mu_{\underset{\rho,\delta}{R_2 \overset{\alpha,\beta}{\circ} S}}(x,z).$$

$$\nu_{\underset{\rho,\delta}{R_1 \overset{\alpha,\beta}{\circ} S}}(x,z) = \rho_y\{\delta[\nu_S(x,y),\nu_{R_1}(y,z)]\}$$

$$\ge \rho_y\{\delta[\nu_S(x,y),\nu_{R_2}(y,z)]\} = \nu_{\underset{\rho,\delta}{R_2 \overset{\alpha,\beta}{\circ} S}}(x,z).$$

iv) Likewise, if $S_1 \le S_2$, then $R \underset{\rho,\delta}{\overset{\alpha,\beta}{\circ}} S_1 \le R \underset{\rho,\delta}{\overset{\alpha,\beta}{\circ}} S_2$, for every $R \in IFR$.

v) If $R, S \in IFR(X \times X)$ and $S \le R$, then $S \underset{\rho,\delta}{\overset{\alpha,\beta}{\circ}} S \le R \underset{\rho,\delta}{\overset{\alpha,\beta}{\circ}} R$.

From (iii), we find if $S \le R$ then $S \underset{\rho,\delta}{\overset{\alpha,\beta}{\circ}} S \le R \underset{\rho,\delta}{\overset{\alpha,\beta}{\circ}} S$.

Again from (iv) we can write if $S \le R$ then $R \underset{\rho,\delta}{\overset{\alpha,\beta}{\circ}} S \le R \underset{\rho,\delta}{\overset{\alpha,\beta}{\circ}} R$.

Hence, $S \underset{\rho,\delta}{\overset{\alpha,\beta}{\circ}} S \le R \underset{\rho,\delta}{\overset{\alpha,\beta}{\circ}} S \le R \underset{\rho,\delta}{\overset{\alpha,\beta}{\circ}} R \Longrightarrow S \underset{\rho,\delta}{\overset{\alpha,\beta}{\circ}} S \le R \underset{\rho,\delta}{\overset{\alpha,\beta}{\circ}} R.$

vi) If $S, R \in IFR(Y \times Z)$ and $T \in IFR(X \times Y)$ and $\alpha, \beta, \rho, \delta$ are the T-norms or T-conorms, then

$$(S \vee R) \overset{\alpha,\beta}{\underset{\rho,\delta}{\circ}} T \geq \left(S \overset{\alpha,\beta}{\underset{\rho,\delta}{\circ}} T \right) \vee \left(R \overset{\alpha,\beta}{\underset{\rho,\delta}{\circ}} T \right)$$

$$(S \wedge R) \overset{\alpha,\beta}{\underset{\rho,\delta}{\circ}} T \geq \left(S \overset{\alpha,\beta}{\underset{\rho,\delta}{\circ}} T \right) \wedge \left(R \overset{\alpha,\beta}{\underset{\rho,\delta}{\circ}} T \right).$$

The above conditions show the distributive property. This can be verified as follows:

From condition (v) in Section 1.11, we have $S \vee R \geq R$, $S \vee R \geq S$.

This implies, $(S \vee R) \overset{\alpha,\beta}{\underset{\rho,\delta}{\circ}} T \geq R \overset{\alpha,\beta}{\underset{\rho,\delta}{\circ}} T$ and $(S \vee R) \overset{\alpha,\beta}{\underset{\rho,\delta}{\circ}} T \geq S \overset{\alpha,\beta}{\underset{\rho,\delta}{\circ}} T$.

So, $(S \vee R) \overset{\alpha,\beta}{\underset{\rho,\delta}{\circ}} T \geq \left(R \overset{\alpha,\beta}{\underset{\rho,\delta}{\circ}} T \right) \vee \left(S \overset{\alpha,\beta}{\underset{\rho,\delta}{\circ}} T \right)$.

vii) If $S, R \in IFR(Y \times Z)$ and $T \in IFR(X \times Y)$ and α, ρ, are the T-norms and T-conorms, then

$$(S \vee R) \overset{\alpha,\beta}{\underset{\rho,\delta}{\circ}} T = \left(S \overset{\alpha,\beta}{\underset{\rho,\delta}{\circ}} T \right) \vee \left(R \overset{\alpha,\beta}{\underset{\rho,\delta}{\circ}} T \right) \text{ iff } \alpha = \vee \text{ and } \rho = \wedge.$$

In terms of membership function, $(S \vee R) \overset{\alpha,\beta}{\underset{\rho,\delta}{\circ}} T$ is written as:

$$\mu_{(S \vee R) \overset{\alpha,\beta}{\underset{\rho,\delta}{\circ}} T}(x,z) = \alpha_y \{\beta \left[\mu_T(x,y), \mu_S(y,z) \vee \mu_R(y,z) \right]\}$$

$$= \alpha_y \{\beta \left[\mu_T(x,y), \mu_S(y,z) \right] \vee \beta \left[\mu_T(x,y), \mu_R(y,z) \right]\}$$

$$= \alpha_y \{\beta \left[\mu_T(x,y), \mu_S(y,z) \right]\} \vee \alpha_y \{\beta \left[\mu_T(x,y), \mu_R(y,z) \right]\}$$

$$= \left(S \overset{\alpha,\beta}{\underset{\rho,\delta}{\circ}} T \right) \vee \left(R \overset{\alpha,\beta}{\underset{\rho,\delta}{\circ}} T \right).$$

1.14 Intuitionistic Fuzzy Binary Relation

1.14.1 Reflexive Property

IFR R is said to be reflexive : iff $\forall x \in X, \mu_R(x,x) = 1$ and $\nu_R(x,x) = 0$.

IFR R is said to be antireflexive : iff $\forall x \in X, \mu_R(x,x) = 0$ and $\nu_R(x,x) = 1$.

$$(1.30)$$

Theorem 1 [19] For α T-conorm, β T-conorm and ρ T-norm, δ T-norm

i) If relation $R \in IFR(X \times X)$ is reflexive, then $R \leq R \overset{\alpha,\beta}{\underset{\rho,\delta}{\circ}} R$.

For reflexive property we see that (in terms of membership function)

$$\mu_{\left(R \overset{\alpha,\beta}{\underset{\rho,\delta}{\circ}} R \right)} (x,z) = \alpha_y \{\beta \left[\mu_R(x,y), \mu_R(y,z) \right] \}$$

$$= \alpha_{y \neq x} \{ \beta \left[\mu_R(x,x), \mu_R(x,z) \right], \beta \left[\mu_R(x,y), \mu_R(y,z) \right] \}$$

$$= \alpha_{y \neq x} \{ \beta \left[1, \mu_R(x,z) \right], \beta \left[\mu_R(x,y), \mu_R(y,z) \right] \}, \text{as } \mu_R(x,x) = 1$$

$$= \alpha_{y \neq x} \{ 1, \beta \left[\mu_R(x,y), \mu_R(y,z) \right] \}, \text{as } \beta \text{ is a T-conorm}$$

$$= 1 \geq \mu_R(x,z), \text{as } \alpha \text{ is a T-conorm.}$$

A similar procedure is followed for the nonmembership value.

$$\nu_{\left(R \overset{\alpha,\beta}{\underset{\rho,\delta}{\circ}} R \right)} (x,z) = \rho_y \{ \delta \left[\nu_R(x,y), \nu_R(y,z) \right] \}$$

$$= \rho_{y \neq x} \{ \delta \left[\nu_R(x,x), \nu_R(x,z) \right], \delta \left[\nu_R(x,y), \nu_R(y,z) \right] \}$$

$$= \rho_{y \neq x} \{ \delta \left[0, \nu_R(x,z) \right], \delta \left[\nu_R(x,y), \nu_R(y,z) \right] \}, \text{as } \nu_R(x,x) = 0$$

$$= \rho_{y \neq x} \{ 0, \delta \left[\nu_R(x,y), \nu_R(y,z) \right] \}, \text{as } \delta \text{ is a T-norm}$$

$$= 0 \leq \nu_R(x,z), \text{as } \rho \text{ is a T-norm.}$$

ii) If relation R is antireflexive, then $R \geq R \overset{\rho,\delta}{\underset{\alpha,\beta}{\circ}} R$ is also antireflexive. This can be proved similar to (i).

Theorem 2 [19] For α T-conorm and ρ T-norm:

i) If relation $R \in IFR(X \times Y)$ is reflexive, then $R \overset{\alpha,\beta}{\underset{\rho,\delta}{\circ}} R$ is also reflexive.

In terms of membership function,

$$\mu_{\left(\underset{\rho,\delta}{\overset{\alpha,\beta}{R \circ R}}\right)}(x,x) = \alpha_y\{\beta[\mu_R(x,y),\mu_R(y,x)]\}$$

$$= \alpha_{y \neq x}\{\beta[\mu_R(x,x),\mu_R(x,x)],\beta[\mu_R(x,y),\mu_R(y,x)]\}$$

$$= \alpha_{y \neq x}\{\beta[1,1],\beta[\mu_R(x,y),\mu_R(y,x)]\}, \text{as } \mu_R(x,x) = 1$$

$$= \alpha_{y \neq x}\{1,\beta[\mu_R(x,y),\mu_R(y,x)]\}$$

$$= 1, \text{as } \alpha \text{ is a T-conorm}$$

In terms of nonmembership function,

$$\nu_{\left(\underset{\rho,\delta}{\overset{\alpha,\beta}{R \circ R}}\right)}(x,x) = \rho_y\{\delta[\nu_R(x,y),\nu_R(y,x)]\}$$

$$= \rho_{y \neq x}\{\delta[\nu_R(x,x),\nu_R(x,x)],\delta[\nu_R(x,y),\nu_R(y,x)]\}$$

$$= \rho_{y \neq x}\{\delta[0,0],\delta[\nu_R(x,y),\nu_R(y,x)]\}, \text{as } \nu_R(x,x) = 0$$

$$= \rho_{y \neq x}\{0,\delta[\nu_R(x,y),\nu_R(y,x)]\}$$

$$= 0, \text{ as } \rho \text{ is a T-norm}$$

ii) If relation R is antireflexive, then $R \overset{\rho,\delta}{\underset{\alpha,\beta}{\circ}} R$ is also antireflexive.

The proof is similar to that of reflexivity.

iii) If relation R is reflexive IFR, then R^{-1} is also reflexive.

iv) For every $R_2 \in IFR(X \times X)$, if R_1 is reflexive IFR, then $R_1 \vee R_2$ is also reflexive. We know for R_1 to be reflexive, $\mu_{R_1}(x,x) = 1$ and $\nu_{R_1}(x,x) = 0$.

In terms of membership function : $\mu_{R_1 \vee R_2}(x,x) = \mu_{R_1}(x,x) \vee \mu_{R_2}(x,x)$

$$= 1 \vee \mu_{R_2}(x,x) = 1$$

and using nonmembership function: $\nu_{R_1 \vee R_2}(x,x) = \nu_{R_1}(x,x) \wedge \nu_{R_2}(x,x)$

$$= 0 \wedge \nu_{R_2}(x,x) = 0.$$

v) $R_1 \wedge R_2$ is also reflexive if and only if $R_2 \in IFR(X \times X)$ is reflexive.

Using membership function, $\mu_{R_1 \wedge R_2}(x,x) = \mu_{R_1}(x,x) \wedge \mu_{R_2}(x,x) = 1 \wedge \mu_{R_2}(x,x)$

$$= \mu_{R_2}(x,x), (\text{as } R_1 \text{ is reflexive})$$

and using nonmembership function, $\nu_{R_1 \wedge R_2}(x,x) = \nu_{R_1}(x,x) \vee \nu_{R_2}(x,x)$

$$= 0 \vee \nu_{R_2}(x,x) = \nu_{R_2}(x,x).$$

This implies $R_2 \in IFR(X \times X)$ is reflexive.

Example 17 The following example shows that the relation R, $R \in IFR(X \times X)$ is not reflexive but follows reflexive property.

The membership and nonmembership matrices are:

$$\mu_R = \begin{bmatrix} 0.3 & 0.7 & 0.3 \\ 0.5 & 0.8 & 0.5 \\ 0.1 & 0.4 & 0.2 \end{bmatrix}, \nu_R = \begin{bmatrix} 0.6 & 0.2 & 0.6 \\ 0.2 & 0.0 & 0.4 \\ 0.6 & 0.3 & 0.7 \end{bmatrix}.$$

For $\alpha = \vee$, $\beta = \wedge$, $\rho = \wedge$, $\delta = \vee$,

$$\mu_{R \underset{\wedge,\vee}{\overset{\vee,\wedge}{\circ}} R}(x,z) = \begin{bmatrix} 0.3 & 0.7 & 0.3 \\ 0.5 & 0.8 & 0.5 \\ 0.1 & 0.4 & 0.2 \end{bmatrix} \circ \begin{bmatrix} 0.3 & 0.7 & 0.3 \\ 0.5 & 0.8 & 0.5 \\ 0.1 & 0.4 & 0.2 \end{bmatrix} = \begin{bmatrix} 0.5 & 0.7 & 0.5 \\ 0.5 & 0.8 & 0.5 \\ 0.4 & 0.4 & 0.4 \end{bmatrix},$$

$$\nu_{R \underset{\vee,\wedge}{\overset{\wedge,\vee}{\circ}} R}(x,z) = \begin{bmatrix} 0.6 & 0.2 & 0.6 \\ 0.2 & 0.0 & 0.4 \\ 0.6 & 0.3 & 0.7 \end{bmatrix} \circ \begin{bmatrix} 0.6 & 0.2 & 0.6 \\ 0.2 & 0.0 & 0.4 \\ 0.6 & 0.3 & 0.7 \end{bmatrix} = \begin{bmatrix} 0.2 & 0.2 & 0.4 \\ 0.2 & 0.0 & 0.4 \\ 0.3 & 0.3 & 0.4 \end{bmatrix}.$$

It is observed that the reflexive property is satisfied:

$\mu_{R \underset{\wedge,\vee}{\overset{\vee,\wedge}{\circ}} R}(x,z) \geq \mu_R(x,z)$ and $\nu_{R \underset{\vee,\wedge}{\overset{\wedge,\vee}{\circ}} R}(x,z) \leq \nu_R(x,z)$. But the relation R is not reflexive as $\mu_R(x,x) \neq 1$ and $\nu_R(x,x) \neq 0$.

1.14.2 Symmetric Property

An IFR R is said to be symmetric if R^{-1} is symmetric [19].

Consider R^{-1} to be symmetric, so $\mu_{R^{-1}}(y,x) = \mu_{R^{-1}}(x,y)$

$$\implies \mu_R(x,y) = \mu_R(y,x) \tag{1.31}$$

The same rule follows for the nonmembership function $\nu_R(x,y)$.
This implies that R is symmetric.
Again, $\mu_{R^{-1}}(x,y) = \mu_R(y,x) = \mu_R(x,y)$, and $\nu_{R^{-1}}(x,y) = \nu_R(y,x) = \nu_R(x,y)$.
This implies, $R = R^{-1}$.
So, from the above discussion, an IFR, R is said to be symmetric iff $R = R^{-1}$,
that is

$$\forall x,y \in X, \mu_R(x,y) = \mu_R(y,x),$$

$$\nu_R(x,y) = \nu_R(y,x).$$

Relation R is π-symmetric or antisymmetric if
$\forall (x, y) \in X \times X, x \neq y$, then $\mu_R(x,y) \neq \mu_R(y,x)$

$$\nu_R(x,y) \neq \nu_R(y,x),$$
$$\pi_R(x,y) = \pi_R(y,x).$$

If $\alpha, \beta, \rho, \delta$ are any T-norms or T-conorms, and $R, S \in IFR(X \times X)$ is symmetrical, then

$$S \underset{\rho,\delta}{\overset{\alpha,\beta}{\circ}} R = \left(R \underset{\rho,\delta}{\overset{\alpha,\beta}{\circ}} S \right)^{-1}. \tag{1.32}$$

Proof: As R, S are symmetrical, so $R = R^{-1}$, $S = S^{-1}$, then

$$S \underset{\rho,\delta}{\overset{\alpha,\beta}{\circ}} R = S^{-1} \underset{\rho,\delta}{\overset{\alpha,\beta}{\circ}} R^{-1} = \left(R \underset{\rho,\delta}{\overset{\alpha,\beta}{\circ}} S \right)^{-1}.$$

Also, if R is symmetrical, $R \underset{\rho,\delta}{\overset{\alpha,\beta}{\circ}} R$ is also symmetrical. But composition of two symmetrical relations will not always be symmetrical.

1.14.3 Transitive Property

For α T-conorm, β T-norm, ρ T-norm, and δ T-conorm [19], relation $R \in IFR$ $(X \times X)$ is transitive, if $R \geq R \underset{\rho,\delta}{\overset{\alpha,\beta}{\circ}} R$ or $R \geq R \underset{\wedge,\delta}{\overset{\vee,\beta}{\circ}} R$.

Relation R is c-transitive, if $R \leq R \underset{\rho,\delta}{\overset{\alpha,\beta}{\circ}} R$ or $R \leq R \underset{\wedge,\delta}{\overset{\vee,\beta}{\circ}} R$.

Transitive closure of relation R is the minimum IFR $\hat{R}$ on $X \times X$ which contains R, i.e. $R \leq \hat{R}$ and it is transitive, i.e. $\hat{R} \underset{\rho,\delta}{\overset{\alpha,\beta}{\circ}} \hat{R} \leq \hat{R}$.

c-Transitive closure of the relation R is the biggest c-transitive relation $\check{R}$ on $X \times X$, which contains R and is transitive.

Transitive closure $\hat{R}$ of IFR R is defined as:

$$\hat{R} = R \cup R^2 \cup R^3 \cup \ldots \tag{1.33}$$

For every $R \in IFR(X \times X)$, if $\alpha = \vee$, $\beta = \wedge$, $\rho = \wedge$, $\delta = \vee$, then

For transitive closure, $\hat{R} = R \vee R \overset{\vee,\wedge}{\circ} R \overset{\vee,\wedge}{\circ} R \overset{\vee,\wedge}{\circ} R \vee ...R^n$.

For c-transitive closure, $\check{R} = R \wedge R \overset{\wedge,\vee}{\circ} R \wedge R \overset{\wedge,\vee}{\circ} R \overset{\wedge,\vee}{\circ} R \wedge ...R^n$,

where $R^2 = R \overset{\vee,\wedge}{\circ} R$ and $R^3 = R \overset{\vee,\wedge}{\circ} R \overset{\vee,\wedge}{\circ} R$.

1.15 Summary

This chapter describes the definition and operations of fuzzy and IFS theory with examples. Different types of membership functions are described. Fuzzy projections, composition of two fuzzy relations are explained with examples. Fuzzy relations that include reflexivity, symmetricity, transitivity, and equivalence relations are discussed along with examples. Intuitionistic fuzzy operations, relations, compositions, and intuitionistic fuzzy binary relations are explained with examples.

References

1 Zadeh, L.A. (1965). Fuzzy sets. *Information and Control* 8: 338–353.
2 Zimmerman, H.J. (1996). *Fuzzy Set Theory and Its Application*. Netherlands: Kluwer Academic Publishers.
3 Zadeh, L.A. (1971). Similarity relations and fuzzy orderings. *Information Sciences* 3: 177–200.
4 Bustince, H., Barrenechea, E., and Pagola, M. (2006). Restricted equivalence function. *Fuzzy Sets and Systems* 157 (17): 2333–2346.
5 Bustince, H., Mohedano, V., Barrenechea, E., and Pagola, M. (2006). An algorithm for calculating the threshold of an image representing uncertainty through A-IFSs, IPMU'2006, pp. 2383–2390
6 Chaira, T. (2010). Intuitionistic fuzzy segmentation of medical images. *IEEE Transaction of Biomedical Engineering* 57 (6): 1430–1436.
7 Chaira, T. and Ray, A.K. (2003). Segmentation using fuzzy divergence. *Pattern Recognition Letters* 24 (12): 1837–1844.
8 Kaufmann, A. (1980). *Introduction to the Theory of Fuzzy Subsets-vol-1: Fundamental Theoretical Elements*. New York: Academic Press.
9 Klir, G. and Yaun, B. (1977). *Fuzzy Set and Fuzzy logic: Theory and Application*. Upper Saddle River, NJ: Prentice Hall.
10 Zadeh, L.A. (1975a). Calculus of fuzzy restrictions. In: *Fuzzy Sets and Their Applications to Cognitive and Decision Processes* (ed. L.A. Zadeh, K.S. Fu, K. Tanaka and M. Shimura), 1–39. New York: Academic Press.

11 Kundu, S. (2000). A representation theorem for min-transitive fuzzy relations. *Fuzzy Sets and Systems* 109: 453–457.

12 Das, M., Chakraborty, M.K., and Ghosal, T.K. (1998). Fuzzy tolerance relation, fuzzy tolerance space and basis. *Fuzzy Sets and Systems* 97: 361–369.

13 Atanassov, K.T. (1986). Intuitionistic fuzzy set. *Fuzzy Sets and Systems* 20: 87–97.

14 Sugeno, M. (1977). Fuzzy measures and fuzzy integral: a survey. In: *Fuzzy Automata and Decision Processes* (ed. M.M. Gupta, G.S. Sergiadis and B.R. Gaines), 89–102. North Holland, Amsterdam: Elsevier.

15 Yager, R.R. (1980). On the measures of fuzziness and negation. Part II: lattices. *Information and Control* 44: 236–260.

16 Klir, G.J. and Yuan, B. (1995). *Fuzzy Sets and Fuzzy Logic*. Upper Saddle River, NJ: Prentice-Hall.

17 Bustince, H., Kacpryzk, J., and Mohedano, V. (2000). Intuitionistic fuzzy generators: application to intuitionistic fuzzy complementation. *Fuzzy Sets and Systems* 114: 485–504.

18 Chaira, T. (2013). Enhancement of medical images using Atanassov's intuitionistic fuzzy domain using an alternative intuitionistic fuzzy generator with application to medical image segmentation. *International Journal of Intelligent and Fuzzy systems* 27 (3): 1347–1359.

19 Burillo, P. and Bustince, H. (1995). Intuitionistic fuzzy relations. *Mathware and Soft Computing* 3 (Part – I): 5–38.

20 Bustince, H. (2000). Construction of intuitionistic fuzzy relations with predetermined properties. *Fuzzy Sets and Systems* 109 (3): 379–403.

21 Bustince, H. and Burillo, P. (1996). Structures of intuitionistic fuzzy relations. *Fuzzy sets and Systems* 78: 293–303.

22 Lei, Y.J., Wang, B.S., and Miao, Q.G. (2005). On the intuitionistic fuzzy relations with compositional operations. *Systems Engineering —Theory and Practice* 25 (2): 113–118.

2

Playing with Fuzzy/Intuitionistic Fuzzy Numbers

2.1 Introduction

This chapter provides an approach to model unknown data by means of fuzzy set theory. As precise data are out of reach, fuzzy numbers can address the problem of deriving uncertainty on a sum of variables whose values lie within fuzzy intervals. Such data may be modeled by a fuzzy set which acts as "more or less" on non-fuzzy values of the data. The term "more or less" is used here to emphasize the fact that not only there are two usual degrees: "complete possibility" and "impossibility" but it also contains intermediate values.

Fuzzy numbers are used widely in decision-making problems, where values of parameters or decision variables are not precisely fixed or assessed. To name a few, applications could be multi-criteria optimization and decision making under uncertainty, where fuzzy-expected utilities can be obtained out of incomplete assessed probabilities. Sometimes, it may happen that the data are dependent and in that case, the variables are interactive. To deal with such type of interaction in real-life problems, fuzzy arithmetic is an important tool in aggregating the values. It is used in sensitivity analysis in systems modeling, computer-aided design, and operations research.

2.2 Fuzzy Numbers

Uncertainty is frequently encountered in day-to-day life. Concrete problems often involve many quantities that are idealizations of inaccurate information involving numerical values. For instance, when we measure the height or weight of a person, many people will get a numerical value. But the values may not be equal due to imprecision in measuring instrument or the person who took the

Fuzzy Set and Its Extension: The Intuitionistic Fuzzy Set, First Edition. Tamalika Chaira.
© 2019 John Wiley & Sons, Inc. Published 2019 by John Wiley & Sons, Inc.

measurement or many other reasons. This is the reason we use the word "around" something.

Fuzzy numbers are the basis for fuzzy arithmetic. A fuzzy number is a fuzzy subset of the universe of numerical numbers. The concept of fuzzy number was introduced by Zadeh. Since then several authors investigated the properties of fuzzy numbers.

A fuzzy number A is a fuzzy set on real number $\mathcal{R}$, such that [1]

a) A is piecewise continuous,
b) Fuzzy number is convex.
c) Fuzzy number A is normal, i.e. if m is a mean value of the fuzzy number A, then $\mu(m) = 1$. It has exactly one $m \in \mathcal{R}$, where $\mu(m) = 1$.
d) A is monotone ascending in the interval $(-\infty, m)$ and monotone descending in the interval (m, ∞).
e) α-cut of the fuzzy number A_α is a closed interval for $\alpha \in (0,1]$.

Fuzzy number A is positive denoted by $A > 0$ and fuzzy number A is negative denoted by $A < 0$ if its membership function $\mu_A(x)$ satisfies $\mu_A(x) = 0$ for each $x < 0$ and $x > 0$, respectively.

2.3 Fuzzy Intervals

Interval computation has various real-life applications starting from robotics, automatic control, image processing, astrophysics, traffic control, and expert systems. Also, in many practical situations, in addition to the intervals that are guaranteed by the manufacturer, it is wise to consider subintervals, which the manufacturer cannot guarantee, but the designers or producers or experts of the manufacturing instrument claim to be true.

Suppose μ_R is a convex fuzzy set and has a unimodal shape. When there is only one real number, such that $\mu_R(m) = 1$, then μ_R is called a fuzzy number. But if it models a number whose value is "approximately m" as shown in Figure 2.1, then μ_R is represented as a fuzzy interval. Then, fuzzy number is expressed as a fuzzy

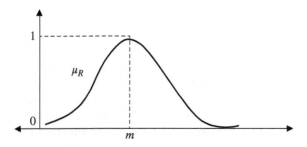

Figure 2.1 Fuzzy number.

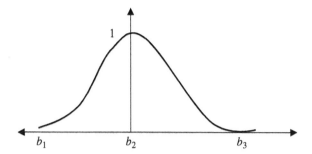

Figure 2.2 Fuzzy number with fuzzy intervals.

set defining a fuzzy interval. Since the boundary of this interval is ambiguous, the interval is also a fuzzy set.

Generally, a fuzzy interval is represented by two end points b_1 and b_3 and a peak point b_2 and is written as $[b_1, b_2, b_3]$ as shown in Figure 2.2.

2.4 Zadeh's Extension Principle

There is a need to extend the concepts from classical set theory to fuzzy set theory. Extension principle is the most basic concept of fuzzy set theory that can be used to generalize crisp mathematical concepts to fuzzy sets. This extension was proposed by Zadeh and is known as Extension principle [1]. The Extension principle for a function $f: X \rightarrow Y$ indicates how the image of a fuzzy subset A of X should be computed when the function f is applied. It is expected that this image will be a fuzzy subset of Y.

Definition 1 Following the definition of extension principle by Zadeh [1], Zimmerman [2], and Dubois and Prade [3], which is defined as:

Let X be a Cartesian product of universes, $X = X_1 \times X_2 \times \cdots \times X_n$ and let A_1, $A_2, \ldots, A_n$ be n fuzzy subsets in $X_1, X_2, \ldots, X_n$. Let f be a function such that $f: X \rightarrow Y$, i.e. a mapping from X to Y such that $y = f(x_1, x_2, \ldots, x_n)$. The extension principle allows a mapping from n fuzzy subsets A to a fuzzy subset B on Y through a function f such that

$$B = \{y, \mu_B(y) \mid y = f(x_1, x_2, \ldots, x_n) \text{ and } x_1, x_2, \ldots, x_n \in X\},$$

whose membership function $\mu_B(y)$ is given as:

$$\mu_B(y) = \begin{cases} \sup_{\substack{x_1, x_2 \ldots, x_n \\ y = f(x_1, x_2 \ldots, x_n)}} \min\left(\mu_{A_1}(x_1), \mu_{A_2}(x_2), \ldots, \mu_{A_n}(x_n)\right), & \text{if } f^{-1}(y) \neq \emptyset \\ 0 & \text{otherwise} \end{cases},$$

$$(2.1)$$

where f^{-1} is inverse of f.

$\mu_B(y)$ is the greatest membership value in $\mu_A(x_1, x_2, x_3, \ldots, x_n)$. When $n = 1$, then $B = \{y, \mu_B(y) \mid y = f(x)\}$, where

$$
\mu_B(y) = \begin{cases} \sup\limits_{x \in f^{-1}(y)} \mu_A(x), & \text{if } f^{-1}(y) \neq \emptyset \\ \\ 0 & \text{otherwise} \end{cases}
$$

It is to be noted that the image of A is denoted by $B = f(A)$. Zadeh [1] writes Eq. (2.1) as

$$
B = f(A_1, A_2, \ldots, A_n)
$$
$$
= \int_{X = X_1 \times X_2 \times \cdots \times X_n} \min\left(\mu_{A_1}(x_1), \mu_{A_2}(x_2), \mu_{A_3}(x_3), \ldots, \mu_{A_n}(x_n)\right) / f(x_1, x_2, \ldots, x_n)
$$

So, if $f : X \longrightarrow Y$ and A is a fuzzy subset of X, then A is defined as:

$$
A = \frac{\mu_A(x_1)}{x_1} + \frac{\mu_A(x_2)}{x_2} + \frac{\mu_A(x_3)}{x_3} + \cdots + \frac{\mu_A(x_4)}{x_4} = \sum_{i=1}^{\infty} \frac{\mu_A(x_i)}{x_i}.
$$

Then, according to the extension principle, the image of fuzzy subset A under the mapping f is given as:

$$
B = f \sum_{i=1}^{\infty} \frac{\mu_A(x_i)}{x_i} = \sum_{i=1}^{\infty} \frac{\mu_A(x_i)}{f(x_i)}.
$$

Therefore, the image of A, which is denoted by B can be derived from the knowledge of the images of x_i in X using the function f.

2.4.1 Extension Principle for Two Variables

L.C. de Barros et al. [4] suggested an extension principle for two variables. Let f be a mapping function such that $f : X \times Y \to Z$ and let A, B be two fuzzy subsets of X and Y. Then, the extension principle allows a mapping from fuzzy subsets A and B to a fuzzy subset C of Z through a function f such that the membership function $\mu_C(z)$ is given as:

$$
\mu_c(z) = \sup_{f^{-1}(z)} \min(\mu_A(x), \mu_B(y)), \tag{2.2}
$$

where $z = f(x,y)$ and f^{-1} is inverse of f. C may be written as $f(A,B)$.

Let us take few examples on single and two variables:

Example 1 Let f is a function from X to Y and A be a fuzzy set on X where

$$A = \frac{0.4}{-2} + \frac{0.6}{-1} + \frac{1}{0} + \frac{0.8}{1} + \frac{0.5}{2},$$

find $f(A)$, the image of A using the mapping function is $f(x) = x + 2$.

Solution

Using the extension principle, fuzzy subset B may be written as:

$$B = f(A) = \frac{0.4}{0} + \frac{0.6}{1} + \frac{1}{2} + \frac{0.8}{3} + \frac{0.5}{4}$$

So, $B = \dfrac{0.4}{0} + \dfrac{0.6}{1} + \dfrac{1}{2} + \dfrac{0.8}{3} + \dfrac{0.5}{4}$.

Figure 2.3 shows a mapping diagram of image of A using the function f.

Example 2 Let

$$A = \sum_{i=1}^{\infty} \frac{\mu_A(x_i)}{x_i} = \left\{ \frac{0.2}{-2} + \frac{0.5}{-1} + \frac{0.7}{0} + \frac{1}{1} + \frac{0.6}{2} \right\},$$

find $f(A)$, the image of A using the mapping function is $f(x) = x^2$.

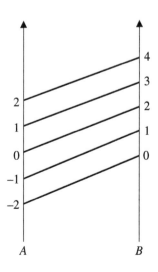

Figure 2.3 Image of a fuzzy subset using the extension principle for a function f.

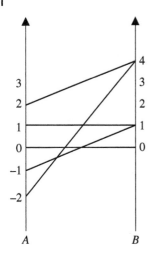

Figure 2.4 Image of a fuzzy subset from the extension principle for a function *f*.

Solution

Using the extension principle,

$$B = f(A) = \sum_{i=1}^{\infty} \frac{\mu_A(x_i)}{f(x_i)} = \sum_{i=1}^{5} \frac{\mu_A(x_i)}{x^2}$$

$$= \left\{ \frac{0.2}{4} + \frac{0.5}{1} + \frac{0.7}{0} + \frac{1}{1} + \frac{0.6}{4} \right\}$$

$$= \left\{ \frac{\max(0.2, 0.6)}{4} + \frac{\max(0.5, 1)}{1} + \frac{0.7}{0} \right\}$$

$$= \left\{ \frac{0.7}{0} + \frac{1}{1} + \frac{0.6}{4} \right\}.$$

Figure 2.4 shows a mapping diagram of image of *A* using the function *f*.

Example 3   Let *f* be a function from $f(x,y) = x + y$. Two fuzzy sets

$$A = \frac{0.4}{2} + \frac{0.6}{3} + \frac{1}{4} + \frac{0.7}{5},$$

$$B = \frac{0.5}{3} + \frac{0.7}{4} + \frac{1}{5} + \frac{0.8}{6},$$

find the membership degree of fuzzy subset $f(A,B)$ through a function $f(x,y) = x + y$.

Solution

Using Eq. (2.2), $\mu_c(z) = \sup_{f^{-1}(z)} \min(\mu_A(x), \mu_B(y))$

The membership function of image or fuzzy subset $f(A, B)$ through a function $f(x, y) = x + y$ is given as:

$$\mu_c(z) = \frac{\min(0.4, 0.5)}{5} + \frac{\min(0.4, 0.7)}{6} + \frac{\min(0.4, 1)}{7} + \frac{\min(0.4, 0.8)}{8}$$
$$+ \frac{\min(0.6, 0.5)}{6} + \frac{\min(0.6, 0.7)}{7} + \frac{\min(0.6, 1)}{8} + \frac{\min(0.6, 0.8)}{9}$$
$$+ \frac{\min(1, 0.5)}{7} + \frac{\min(1, 0.7)}{8} + \frac{\min(1, 1)}{9} + \frac{\min(1, 0.8)}{10}$$
$$+ \frac{\min(0.7, 0.5)}{8} + \frac{\min(0.7, 0.7)}{9} + \frac{\min(0.7, 1)}{10} + \frac{\min(0.7, 0.8)}{11}$$

$$= \frac{0.4}{5} + \frac{0.4}{6} + \frac{0.4}{7} + \frac{0.4}{8} + \frac{0.5}{6} + \frac{0.6}{7} + \frac{0.6}{8} + \frac{0.6}{9} + \frac{0.5}{7} + \frac{0.7}{8}$$
$$+ \frac{1}{9} + \frac{0.8}{10} + \frac{0.5}{8} + \frac{0.7}{9} + \frac{0.7}{10} + \frac{0.7}{11}$$

$$\Rightarrow \mu_c(z) = \frac{0.4}{5} + \frac{\max(0.4, 0.5)}{6} + \frac{\max(0.4, 0.6, 0.5)}{7}$$
$$+ \frac{\max(0.4, 0.6, 0.7, 0.5)}{8} + \frac{\max(0.6, 1, 0.7)}{9} + \frac{\max(0.8, 0.7)}{10} + \frac{0.7}{11}$$

$$\Rightarrow \mu_c(z) = \frac{0.4}{5} + \frac{0.5}{6} + \frac{0.6}{7} + \frac{0.7}{8} + \frac{1}{9} + \frac{0.8}{10} + \frac{0.7}{11}.$$

The membership degree $\mu_c(z)$ of $z = 10$ in fuzzy subset C is 0.8.
It can be done directly as: $\mu_c(10) = \sup_{x+y=10} \min(\mu_A(x), \mu_B(y))$

$$= \sup \min(\mu_A(4), \mu_A(6)), \min(\mu_A(5), \mu_A(5))$$
$$= \sup \min(1, 0.8), \min(0.7, 1.0)$$
$$= \sup(0.8, 0.7) = 0.8.$$

Example 4 The above example may be done with different functions. If $f(x, y) = x + 2y$, then compute the membership degree at $z = 16$ in C.

Solution

Consider two finite fuzzy sets

$$A = \frac{0.4}{4} + \frac{0.7}{5} + \frac{1}{6} + \frac{0.6}{7} + \frac{0.2}{8},$$

$$B = \frac{0.2}{2} + \frac{0.6}{3} + \frac{1}{4} + \frac{0.6}{5} + \frac{0.2}{6}.$$

Compute the membership degree of $z = 16$ in C.

Solution: Using the function $f(x,y) = x + 2y$, we have

$$\mu_C(z) = \frac{\min(0.4,0.2)}{8} + \frac{\min(0.4,0.6)}{10} + \frac{\min(0.4,1)}{12} + \frac{\min(0.4,0.6)}{14}$$
$$+ \frac{\min(0.4,0.2)}{16} + \frac{\min(0.7,0.2)}{9} + \frac{\min(0.7,0.6)}{11} + \frac{\min(0.7,1)}{13}$$
$$+ \frac{\min(0.7,0.6)}{15} + \frac{\min(0.7,0.2)}{17} + \frac{\min(1,0.2)}{10} + \frac{\min(1,0.6)}{12}$$
$$+ \frac{\min(1,1)}{14} + \frac{\min(1,0.6)}{16} + \frac{\min(1,0.2)}{18} + \frac{\min(0.6,0.2)}{11}$$
$$+ \frac{\min(0.6,0.6)}{13} + \frac{\min(0.6,1)}{15} + \frac{\min(0.6,0.6)}{17} + \frac{\min(0.6,0.2)}{19}$$
$$+ \frac{\min(0.2,0.2)}{12} + \frac{\min(0.2,0.6)}{14} + \frac{\min(0.2,1)}{16} + \frac{\min(0.2,0.6)}{18}$$
$$+ \frac{\min(0.2,0.2)}{20}.$$

So, $\mu_{\tilde{f}(A,B)}(16) = \sup_{16 = x + 2y} \min(\mu_A(x), \mu_B(y))$

$$= \sup[\min(0.4,0.2), \min(1,0.6), \min(0.2,1)]$$

$$= \sup[0.2, 0.6, 0.2] = 0.6$$

2.5 Fuzzy Numbers with α-Levels

A fuzzy α-cut can be visualized as a set of elements whose membership values are greater than "approximately α," i.e. belong to a fuzzy interval $(\alpha,1]$ and it is a continuous increasing function [1]. We get a crisp interval using α-cut operation. Suppose A is a fuzzy set and A_α be its α-cut. Then, $A_\alpha = \mu_A^{-1}[(\alpha,1]]$.

A fuzzy number A is a fuzzy set in a set of real numbers $\mathcal{R}$ and $\alpha \in [0,1]$, then α-cut of set A, denoted by A_α is a crisp set which is written as:

$$A_\alpha = \{x \in \mathcal{R} \mid A(x) \geq \alpha\}.$$

The strong a cut, denoted by A_{α^+} is the crisp set, which is written as:

$$A_{\alpha^+} = \{x \in \mathcal{R} \mid A(x) > \alpha\}$$

For all $\alpha \in [0,1]$, crisp interval $[a_1^{\alpha}, b_1^{\alpha}]$ is determined by:

$$a_1^{\alpha} = \inf\{x \in \mathcal{R} \,|\, A(x) \geq \alpha\}$$

$$b_1^{\alpha} = \sup\{x \in \mathcal{R} \,|\, A(x) \geq \alpha\}$$

Let us represent α levels of a fuzzy number A as:

$$A^{\alpha} = [a_1^{\alpha}, b_1^{\alpha}] \text{ with } a_1^{\alpha} = \mu^{-1}(\alpha) \text{ and } b_1^{\alpha} = \mu^{-1}(\alpha)$$

Following this, Nguyen [5] suggested an extension principle on α levels. If $f: X \longrightarrow Z$ is a continuous function and let the images of $A_1, A_2, \ldots, A_n$ be denoted by $(A_1, A_2, \ldots, A_n)$, then $[f(A_1, A_2, \ldots, A_n)]^{\alpha} = f([A_1^{\alpha}, A_2^{\alpha}, A_3^{\alpha}, \ldots, A_n^{\alpha}])$, $\alpha \in [0,1]$. This implies $[B]^{\alpha} = f([A^{\alpha}])$, where B is the image $f(A)$. It means that α–levels of the fuzzy set obtained by the Zadeh's Extension Principle coincides with the images of the α-levels by the crisp function. $[f(A_1, A_2, \ldots, A_n)]^{\alpha}$ is α level set of $f(A_1, A_2, \ldots, A_n)$.

i) Fuzzy number A is said to be triangular (defined on a triplet $[a,b,c]$) if its membership function is given as (Figure 2.5):

$$\mu_A(x) = \begin{cases} 0 & \text{if } x \leq a \\ \dfrac{x-a}{b-a} & \text{if } a < x \leq b \\ \dfrac{c-x}{c-b} & \text{if } b < x \leq c \\ 0 & \text{if } x > c, \end{cases} \tag{2.3}$$

where (a,c) is the base of the triangle and b is the vertex.

Figure 2.5 Triangular fuzzy number.

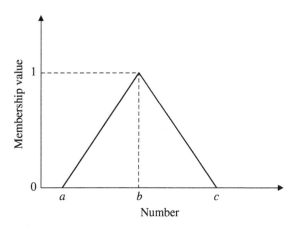

α-cut of a fuzzy number is computed by setting $\alpha \in [0,1]$ to both left and right reference functions of A. We know $A^\alpha = [a_1{}^\alpha, b_1{}^\alpha]$ with $a_1{}^\alpha = \mu^{-1}(\alpha)$ and $b_1{}^\alpha = \mu^{-1}(\alpha)$.

So, for the region $a < x \le b$, $\alpha = \mu(a_1{}^\alpha) = \dfrac{a_1{}^\alpha - a}{b - a}$, substituting $x = a_1{}^\alpha$

and for the region $b < x \le c$, $\alpha = \mu(b_1{}^\alpha) = \dfrac{c - b_1{}^\alpha}{c - b}$, substituting $x = b_1{}^\alpha$.

Expressing $a_1{}^\alpha$, $b_1{}^\alpha$ in terms of α, we get two α levels (lower and upper levels)

$$a_1{}^\alpha = \alpha(b - a) + a$$

$$b_1{}^\alpha = c - \alpha(c - b).$$

So, the α-cut of A is written as:

$$A^\alpha = [a_1{}^\alpha, b_1{}^\alpha] = [\alpha(b - a) + a, c - \alpha(c - b)]. \tag{2.4}$$

Example 5   We wish to express a number around 6 using a triangular fuzzy number A. Its membership function is given by (Figure 2.6):

$$\mu_A(x) = \begin{cases} 1 - \dfrac{|x - 6|}{0.4}, & 5.6 < x \le 6.4 \\[2mm] 0 & \text{otherwise.} \end{cases}$$

Here, $a = 5.6$, $b = 6$, $c = 6.4$.

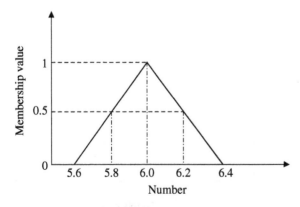

Figure 2.6 Triangular fuzzy number with α-levels.

It can also be written as:

$$\mu_A(x) = \begin{cases} \dfrac{x-5.6}{6-5.6} = \dfrac{x-5.6}{0.4}, & 5.6 < x \le 6 \\[2mm] \dfrac{6.4-x}{6.4-6} = \dfrac{6.4-x}{0.4}, & 0.6 < x \le 6.4. \end{cases}$$

The α-levels of a fuzzy number is:

$$A^\alpha = [a_1{}^\alpha, b_1{}^\alpha] = [\alpha(b-a)+a, c-\alpha(c-b)] = [5.6+0.4\alpha, \ 6.4-0.4\alpha].$$

$$a_1{}^\alpha = 5.6 + 0.4\alpha, \ b_1{}^\alpha = 6.4 - 0.4\alpha, \alpha \in [0,1].$$

At $\alpha = 0$, $A^0 = [5.6, 6.4]$, at $\alpha = 1$, $A^1 = [6]$ and at $\alpha = 0.5$, $A^{0.5} = [5.8, 6.2]$. These are shown in Figure 2.6.

It is to be noted that triangular fuzzy number may not be symmetric, i.e. $(b - a)$ and $(c - b)$ may not be equal.

ii) Fuzzy number A may be trapezoidal if the membership function has the form of trapezoid. The membership function is given by:

$$\mu_A(x) = \begin{cases} \dfrac{x-a}{b-a} & \text{if } a \le x < b \\[2mm] 1 & \text{if } b \le x \le c \\[2mm] \dfrac{d-x}{d-c} & \text{if } c < x \le d \\[2mm] 0 & \text{otherwise,} \end{cases} \tag{2.5}$$

where a, b, c, and d are the numbers.

Following a similar procedure as that of triangular membership function, expressing $a_1{}^\alpha$, $b_1{}^\alpha$ in terms of α, we get

$$a_1{}^\alpha = a + \alpha(b-a),$$

$$b_1{}^\alpha = d - \alpha(d-c)$$

for all $\alpha \in [0,1]$.

So the α-cut of A is written as:

$$A^\alpha = [a_1{}^\alpha, b_1{}^\alpha] = [a + \alpha(b-a), d - \alpha(d-c)] \tag{2.6}$$

It is to be noted that trapezoidal fuzzy number may not be symmetric, i.e. $(b - a)$ and $(d - c)$ may not be equal.

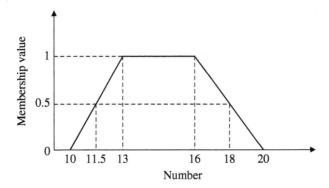

Figure 2.7 Trapezoidal fuzzy number.

Example 6 Consider a fuzzy set of some students that are represented by trapezoidal fuzzy number.

$$\mu_A(x) = \begin{vmatrix} \dfrac{x-10}{3} & \text{if } 10 \le x < 13 \\ 1 & \text{if } 13 \le x \le 16 \\ \dfrac{20-x}{4} & \text{if } 16 < x \le 20 \\ 0 & \text{otherwise.} \end{vmatrix}$$

Its α-levels using Eq. (2.6) are:

$$A^\alpha = [a_1{}^\alpha, b_1{}^\alpha] = [10 + 3\alpha, 20 - 4\alpha].$$

It is observed that at $\alpha = 1$, $A^1 = [13,16]$ which are the shoulder of the trapezium.

At $\alpha = 0$, $A^0 = [10,20]$ which is the feet of the trapezium.

At $\alpha = 0.5$, $A^{0.5} = [11.5,18]$.

So, we can find different fuzzy intervals at different α cuts.

It is graphically shown in Figure 2.7.

2.6 Operations on Fuzzy Numbers with Intervals

In this section, arithmetic operations on fuzzy numbers with intervals are given [2].

Let A and B be two fuzzy numbers expressed in intervals $[a_1,a_3]$ and $[b_1,b_3]$, respectively, and these are defined in real numbers, i.e.

$A = [a_1, a_3]$ and $B = [b_1, b_3]$, then

i) Addition of two fuzzy numbers $A(+)B$

$$[a_1, a_3] + [b_1, b_3] = [a_1 + b_1, a_3 + b_3]. \tag{2.7}$$

ii) Subtraction:

$$A(-)B: \quad [a_1, a_3] - [b_1, b_3] = [a_1 - b_3, a_3 - b_1]. \tag{2.8}$$

iii) Symmetry: $-A = (-a_3, -a_1)$.

iv) Multiplication of A by a scalar $= [\gamma a_1, \gamma a_3]$, if $\gamma \geq 0$.

v) Multiplication: $A(\cdot)B$

$$[a_1, a_3] \cdot [b_1, b_3] = [a_1 \cdot b_1 \wedge a_1 \cdot b_3 \wedge a_3 \cdot b_1 \wedge a_3 \cdot b_3, a_1 \cdot b_1 \vee a_1 \cdot b_3 \vee a_3 \cdot b_1 \vee a_3 \cdot b_3] \tag{2.9}$$

vi) Division: $A(/)B$

$$\frac{[a_1, a_3]}{[b_1, b_3]} = \left[\frac{a_1}{b_1} \wedge \frac{a_1}{b_3} \wedge \frac{a_3}{b_1} \wedge \frac{a_3}{b_3}, \frac{a_1}{b_1} \vee \frac{a_1}{b_3} \vee \frac{a_3}{b_1} \vee \frac{a_3}{b_3} \right]. \tag{2.10}$$

vii) Inverse of the interval:

$$[a_1, a_3]^{-1} = \left[\frac{1}{a_1} \wedge \frac{1}{a_3}, \frac{1}{a_1} \vee \frac{1}{a_3} \right]. \tag{2.11}$$

If the operations are performed on positive real numbers, i.e. on R^+, then

Multiplication: $A(\cdot)B = [a_1, a_3] \cdot [b_1, b_3] = [a_1 \cdot b_1, a_3 \cdot b_3]$.

Inverse of the interval: $[a_1, a_3]^{-1} = \left[\dfrac{1}{a_3}, \dfrac{1}{a_1} \right]$.

Division: $A(/)B = \dfrac{[a_1, a_3]}{[b_1, b_3]} = [a_1, a_3] \cdot \left[\dfrac{1}{b_3}, \dfrac{1}{b_1} \right] = \left[\dfrac{a_1}{b_3}, \dfrac{a_3}{b_1} \right]$.

Maximum of two sets: $[a_1, a_3] \vee [b_1, b_3] = [a_1 \vee b_1, a_3 \vee b_3]$.

Minimum of two sets: $[a_1, a_3] \wedge [b_1, b_3] = [a_1 \wedge b_1, a_3 \wedge b_3]$.

Example 7 Let $A = [3, 4]$ and $B = [-3, 5]$, we compute

$$A(+)B = [3 + (-3), 4 + 5] = [0, 9].$$

$$A(-)B = [3 - 5, 4 - (-3)] = [-2, 7].$$

$$A(\cdot)B = [3 \cdot (-3) \wedge 3 \cdot 5 \wedge 4 \cdot (-3) \wedge 4 \cdot 5, \ 3 \cdot (-3) \vee 3 \cdot 5 \vee 4 \cdot (-3) \vee 4 \cdot 5]$$

$$= [(-9) \wedge 15 \wedge (-12) \wedge 20, (-9) \vee 15 \vee (-12) \vee (20) = [-12, 20].$$

$$A(/)B = \left[\frac{3}{(-3)} \wedge \frac{3}{5} \wedge \frac{4}{(-3)} \wedge \frac{4}{5}, \frac{3}{(-3)} \vee \frac{3}{5} \vee \frac{4}{(-3)} \vee \frac{4}{5} \right]$$

$$= \left[\frac{4}{(-3)}, \frac{4}{5} \right] = [-1.33, 0.8].$$

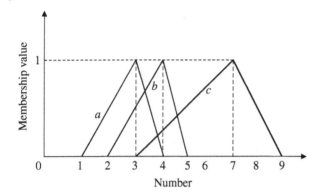

Figure 2.8 Addition of two triangular fuzzy numbers.

$$[A]^{-1} = [3,4]^{-1} = \left[\frac{1}{3} \wedge \frac{1}{4}, \frac{1}{3} \vee \frac{1}{4}\right] = \left[\frac{1}{4}, \frac{1}{3}\right].$$

$$[B]^{-1} = [-3,5]^{-1} = \left[\frac{1}{-3} \wedge \frac{1}{5}, \frac{1}{-3} \vee \frac{1}{5}\right] = \left[\frac{1}{-3}, \frac{1}{5}\right].$$

Example 8 Consider two triangular fuzzy numbers:

$A = [1,3,4]$ and $B = [2,4,5]$. Find $A(+)B$, $A(-)B$. Plot $A(+)B$.

Solution

$$A(+)B = [3,7,9].$$
$$A(-)B = [-4,-1,2].$$

We see that on adding two fuzzy triangular numbers, marked as (a) and (b) in Figure 2.8, it is observed that the final fuzzy number is enlarged, marked as (c).

2.7 Operations with Fuzzy Numbers based on α-Levels

Let A and B be fuzzy numbers with α levels, $A^{\alpha} = [a_1{}^{\alpha}, a_2{}^{\alpha}]$ and $B^{\alpha} = [b_1{}^{\alpha}, b_2{}^{\alpha}]$, then the following operations hold [4].

Addition of two fuzzy numbers A and B: $[A + B]^{\alpha} = [A^{\alpha}] + [B^{\alpha}] = [a_1{}^{\alpha} + b_1{}^{\alpha}, a_2{}^{\alpha} + b_2{}^{\alpha}]$.

Subtraction: $[A - B]^{\alpha} = [A^{\alpha}] - [B^{\alpha}] = [a_1{}^{\alpha} - b_2{}^{\alpha}, a_2{}^{\alpha} - b_1{}^{\alpha}]$.

Multiplication:

$[A.B]^\alpha =$

$[A^\alpha]\cdot[B^\alpha] = [\min(a_1{}^\alpha b_1{}^\alpha, a_1{}^\alpha b_2{}^\alpha, a_2{}^\alpha b_1{}^\alpha, a_2{}^\alpha b_2{}^\alpha), \max(a_1{}^\alpha b_1{}^\alpha, a_1{}^\alpha b_2{}^\alpha, a_2{}^\alpha b_1{}^\alpha, a_2{}^\alpha b_2{}^\alpha)].$

Division: $\left(\dfrac{A}{B}\right)^\alpha = \dfrac{[A^\alpha]}{[B^\alpha]} = [a_1{}^\alpha, a_2{}^\alpha]\cdot\left[\dfrac{1}{b_2{}^\alpha}, \dfrac{1}{b_1{}^\alpha}\right].$

Consider two triangular fuzzy numbers $A = [a,b,c]$ and $B = [l,m,n]$ whose membership function is given by:

$$\mu_A(x) = \begin{cases} 0 & \text{if } x \le a \\ \dfrac{x-a}{b-a} & \text{if } a < x \le b \\ \dfrac{c-x}{c-b} & \text{if } b < x \le c \\ 0 & \text{if } x > c, \end{cases}$$

$$\mu_B(x) = \begin{cases} 0 & \text{if } x \le l \\ \dfrac{x-l}{m-l} & \text{if } l < x \le m \\ \dfrac{n-x}{n-m} & \text{if } m < x \le n \\ 0 & \text{if } x > n. \end{cases}$$

Then, α-cut of A is written as:

$A^\alpha = [a_1{}^\alpha, a_2{}^\alpha] = [a + \alpha(b-a), c - \alpha(c-b)] = [a + \alpha(b-a), c + \alpha(b-c)]$ and

$B^\alpha = [b_1{}^\alpha, b_2{}^\alpha] = [l + \alpha(m-l), n - \alpha(n-m)].$

Addition of two fuzzy numbers:

$$[A+B]^\alpha = A^\alpha + B^\alpha = [a + \alpha(b-a), c - \alpha(c-b)] + [l + \alpha(m-l), n - \alpha(n-m)]$$

$$= [a + \alpha(b-a) + l + \alpha(m-l), c - \alpha(c-b) + n - \alpha(n-m)]$$

$$= [(a+l) + \alpha\{(b+m) - (a+l)\}, (c+n) - \alpha\{(c+n) - (b+m)\}].$$

It is observed that these intervals are the α-levels of the following triangular fuzzy number:

$[a+l, b+m, c+n].$

Example 9   Two triangular fuzzy numbers $A = [2,3,4]$ and $B = [4,6,8]$. Compute addition, subtraction, multiplication, and division.

Solution

$$A^\alpha = [\alpha(3-2) + 2, 4 - \alpha(4-3)] = [\alpha + 2, 4 - \alpha]$$
$$B^\alpha = [\alpha(6-4) + 4, 8 - \alpha(8-6)] = [2\alpha + 4, 8 - 2\alpha]$$

We know $\alpha \in [0,1]$, so at $\alpha = 0$, $A^0 = [2,4]$, and at $\alpha = 1$, $A^1 = [3]$. Likewise, at $\alpha = 0$, $B^0 = [4,8]$ and at $\alpha = 1$, $B^1 = [6]$.

i) $[A + B]^\alpha = [A^\alpha] + [B^\alpha] = [\alpha + 2, 4 - \alpha] + [2\alpha + 4, 8 - 2\alpha] = [3\alpha + 6, 12 - 3\alpha]$
   We see that at $\alpha = 0$, $[A + B]^0 = [6,12]$ and at $\alpha = 1$, $[A + B]^1 = [9]$.
   Thus, $A(+)B = (6,9,12)$.

ii) $[A - B]^\alpha = [A^\alpha] - [B^\alpha] = [\alpha + 2, 4 - \alpha] - [2\alpha + 4, 8 - 2\alpha] = [3\alpha - 6, -3\alpha]$.
   We see that at $\alpha = 0$, $[A - B]^0 = [-6,0]$, at $\alpha = 1$, $[A - B]^1 = [-3]$.
   Thus, $A - B = (-6, -3, 0)$.
   As the fuzzy numbers, $A = [2,3,4]$ and $B = [4,6,8]$ are real positive numbers, multiplication and division for positive real numbers are performed accordingly as mentioned earlier in Section 2.6:

i) $[A \cdot B]^\alpha = [A^\alpha] \cdot [B^\alpha] = [\alpha + 2, 4 - \alpha] \cdot [2\alpha + 4, 8 - 2\alpha] = [(\alpha + 2)(2\alpha + 4), (4 - \alpha)(8 - 2\alpha)]$.

ii) $\left(\dfrac{A}{B}\right)^\alpha = \dfrac{[A^\alpha]}{[B^\alpha]} = [\alpha + 2, 4 - \alpha] \cdot \left[\dfrac{1}{8 - 2\alpha}, \dfrac{1}{2\alpha + 4}\right] = \left[\dfrac{(\alpha + 2)}{(8 - 2\alpha)}, \dfrac{(4 - \alpha)}{(2\alpha + 4)}\right]$.

It is seen that the results of two triangular fuzzy numbers after addition and subtraction will give triangular fuzzy numbers, but the results of two triangular fuzzy numbers after multiplication or division will not produce triangular fuzzy numbers. An example will explain this clearly.

Example 10   Let A and B be triangular fuzzy numbers, $A = (1,3,5)$, $B = (2,4,7)$. Compute addition, multiplication, and division using the membership function.

Solution

The membership functions of fuzzy numbers A and B are:

$$\text{Then, } \mu_A(x) = \begin{cases} 0 & \text{if } x \leq 1 \\ \dfrac{x-1}{2} & \text{if } 1 < x \leq 3 \\ \dfrac{5-x}{2} & \text{if } 3 < x \leq 5 \\ 0 & \text{if } x > 5, \end{cases}$$

$$\mu_B(x) = \begin{vmatrix} 0 & \text{if } x \leq 2 \\ \dfrac{x-2}{2} & \text{if } 2 < x \leq 4 \\ \dfrac{7-x}{3} & \text{if } 4 < x \leq 7 \\ 0 & \text{if } x > 7. \end{vmatrix}$$

Let $A^\alpha = [a_1{}^\alpha, a_2{}^\alpha]$, $B^\alpha = [b_1{}^\alpha, b_2{}^\alpha]$.
Then, α-cut of the fuzzy number A is computed as:

$\alpha = \dfrac{a_1{}^\alpha - 1}{2} \Longrightarrow a_1{}^\alpha = 2\alpha + 1$, and

$\alpha = \dfrac{5 - a_2{}^\alpha}{2} \Longrightarrow a_2{}^\alpha = 5 - 2\alpha$. Substituting $x = a_1{}^\alpha$, $a_2{}^\alpha$ for α levels.

So, $A^\alpha = [a_1{}^\alpha, a_2{}^\alpha] = [2\alpha + 1, 5 - 2\alpha]$.
Likewise, α-cut of the fuzzy number B is:

$\alpha = \dfrac{b_1{}^\alpha - 2}{2} \Longrightarrow b_1{}^\alpha = 2\alpha + 2$,

$\alpha = \dfrac{7 - b_2{}^\alpha}{3} \Longrightarrow b_2{}^\alpha = 7 - 3\alpha$.

$B^\alpha = [b_1{}^\alpha, b_2{}^\alpha] = [2\alpha + 2, 7 - 3\alpha]$.

i) Addition: $[A + B]^\alpha = A^\alpha + B^\alpha = [a_1{}^\alpha + b_1{}^\alpha, a_2{}^\alpha + b_2{}^\alpha] = [4\alpha + 3, 12 - 5\alpha]$.
Substituting the value of $\alpha = 0$ and 1, we get, $A(+)B = [3,7,12]$.
Using mathematical calculation, we can compute the membership function of $A(+)B$ as:

$$\mu_{A(+)B}(x) = \begin{vmatrix} 0 & \text{if } x \leq 3 \\ \dfrac{x-3}{4} & \text{if } 3 < x \leq 7 \\ \dfrac{12-x}{5} & \text{if } 7 < x \leq 12 \\ 0 & \text{if } x \geq 12. \end{vmatrix}$$

At $x = 7$, $\mu_{A(+)B}(x) = 1$.
As the membership function is linear, $A(+)B$ is a triangular fuzzy number.
From Figure 2.9, we can see that $A(+)B$ is a triangular fuzzy number.

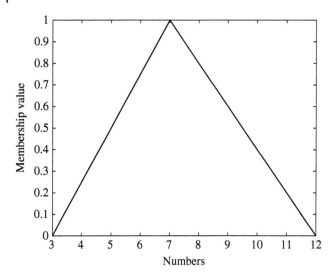

Figure 2.9 Addition of two triangular fuzzy numbers using alpha level.

ii) Multiplication:

$$[A \cdot B]^\alpha = A^\alpha \cdot B^\alpha = [2\alpha + 1, 5 - 2\alpha] \cdot [2\alpha + 2, 7 - 3\alpha]$$

$$= [(2\alpha + 1)(2\alpha + 2), (5 - 2\alpha)(7 - 3\alpha)]$$

$$= [4\alpha^2 + 6\alpha + 2, 6\alpha^2 - 29\alpha + 35] = [c_1{}^\alpha, c_2{}^\alpha] \text{ (say)}.$$

We see that at $\alpha = 0$, $[A \cdot B]^0 = [2,35]$ and at $\alpha = 1$, $[A \cdot B]^1 = [12]$.
Using mathematical calculation we compute the membership function as:

$$\mu_{A(\cdot)B}(x) = \begin{cases} 0 & \text{if } x \le 2 \\ \dfrac{-3 + \sqrt{4x + 1}}{4} & \text{if } 2 < x \le 12 \\ \dfrac{29 - \sqrt{24x + 1}}{12} & \text{if } 12 < x \le 35 \\ 0 & \text{if } x \ge 35. \end{cases}$$

The calculation is done by solving: $c_1{}^\alpha = x = 4\alpha^2 + 6\alpha + 2$

$$\Longrightarrow 4\alpha^2 + 6\alpha + 2 - x = 0 \Longrightarrow \alpha = \frac{-3 + \sqrt{4x + 1}}{4}.$$

Likewise, $c_2{}^\alpha = x = 6\alpha^2 - 29\alpha + 35 \Longrightarrow \alpha = \dfrac{29 - \sqrt{24x + 1}}{12}$.

It is observed that at $x = 12$, $\mu_{A(\cdot)B}(x) = 1$.

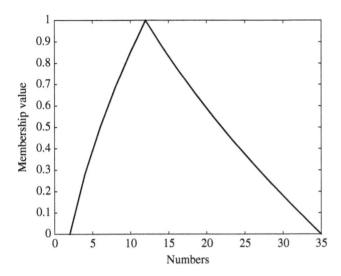

Figure 2.10 Multiplication of two triangular fuzzy numbers.

As the membership function is not linear, $A(\cdot)B$ is not a triangular fuzzy number.

It is seen from Figure 2.10 that $A(\cdot)B$ is not a triangular fuzzy number.

iii) Division: $\left[\dfrac{A}{B}\right]^{\alpha} = \left[\dfrac{A^{\alpha}}{B^{\alpha}}\right] = \dfrac{[2\alpha + 1, 5 - 2\alpha]}{[2\alpha + 2, 7 - 3\alpha]}$

$$= \left[\dfrac{(2\alpha + 1)}{(7 - 3\alpha)}, \dfrac{(5 - 2\alpha)}{(2\alpha + 2)}\right].$$

We see that $[A/B]^{0} = [1/7, 5/2]$, $[A.B]^{1} = [3/4]$.

Using mathematical calculation we compute the membership function as:

$$\text{Then, } \mu_{A(/)B}(x) = \begin{vmatrix} 0 & \text{if } x \le \dfrac{1}{7} \\[2mm] \dfrac{7x-1}{2+3x} & \text{if } \dfrac{1}{7} < x \le \dfrac{3}{4} \\[2mm] \dfrac{5-2x}{2x+2} & \text{if } \dfrac{3}{4} \le x \le \dfrac{5}{2} \\[2mm] 0 & \text{if } x \ge \dfrac{5}{2}. \end{vmatrix}$$

At $x = 3/4$, $\mu_{A(/)B}(x) = 1$.

As the membership function is not linear, $A(/)B$ is not a triangular fuzzy number.

Also from Figure 2.11, it is observed that $A(/)B$ is not a triangular fuzzy number.

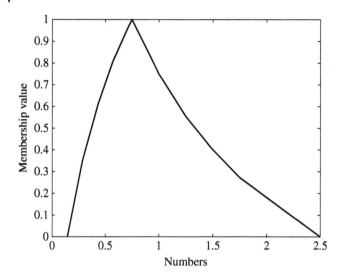

Figure 2.11 Division of two triangular fuzzy numbers.

Example 11 (a): Compute the image of a triangular fuzzy number by a known function, $f(x) = \exp(x/2)$, using an extension principle for a triangular fuzzy number $P = (0, 1, 2)$.

Solution

The α-level of P are the intervals

$$A^\alpha = [a + \alpha(b-a), c - \alpha(c-b)] = [\alpha, 2-\alpha].$$

α-levels of image $\tilde{f}(A)$ are obtained from

$$\left[\tilde{f}(A)\right]^\alpha = f([A]^\alpha) = f([\alpha, 2-\alpha])$$

$$= \left[e^{\alpha/2}, e^{(2-\alpha)/2}\right]$$

where $\tilde{f}(A)$ is the image of triangular fuzzy number with $\alpha \in [0,1]$.
Cases:

i) When $\alpha = 0$, then $\left[\tilde{f}(A)\right]^0 = [e^0, e^1] = [1, 2.7]$.

ii) When $\alpha = 1/2$, then $\left[\tilde{f}(A)\right]^{1/2} = \left[e^{1/4}, e^{3/4}\right]$

$$= [1.28, 2.11].$$

iii) When $\alpha = 1$, then $\left[\tilde{f}(A)\right]^1 = \left[e^{1/2}, e^{1/2}\right] = [1.65]$.

A graph is plotted for α = 0, 0.5, 1, which is shown in Figure 2.12a. It is observed that the three points $(1,0), \left(1.28, \frac{1}{2}\right), (1.65,1)$ are not aligned, which means they do not lie in a straight line. This means that $\tilde{f}(A)$ is not a triangular fuzzy number.

(b) The above example can also be done using a logarithmic function, $f(x)$ = ln(x), for a positive real value numbers $(x > 0)$ using an extension principle

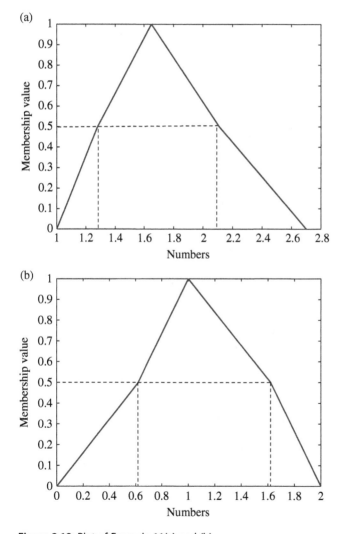

Figure 2.12 Plot of Example 11(a) and (b).

for a triangular fuzzy number $P = (1, e^1, e^2)$. It is to be noted that $\ln : \mathcal{R}^+ \longrightarrow \mathcal{R}, \mathcal{R}$ is a real number.

Solution

The α-level of P are the intervals

$$A^\alpha = [a + \alpha(b-a), c - \alpha(c-b)] = \left[1 + \alpha(e-1), e^2 - \alpha(e^2 - e)\right]$$

$$= [1 + \alpha(e-1), e(e - \alpha(e-1))].$$

α-levels of image $\tilde{f}(A)$ are obtained from

$$\left[\tilde{f}(A)\right]^\alpha = f([A]^\alpha) = f([1 + \alpha(e-1), e(e - \alpha(e-1))])$$

$$= [\ln(1 + \alpha(e-1)), \ln(e(e - \alpha(e-1)))]$$

$$= [\ln(1 + \alpha(e-1)), 1 + \ln(e - \alpha(e-1))],$$

where $\tilde{f}(A)$ is the image of triangular fuzzy number with $\alpha \in [0,1]$.
Cases:

i) When $\alpha = 0$, then $\left[\tilde{f}(A)\right]^0 = [\ln(1), 1 + \ln(e)] = [0, 2]$.

ii) When $\alpha = 1/2$, then $\left[\tilde{f}(A)\right]^{1/2} = \left[\ln\left(\dfrac{1 + (e-1)}{2}\right), 1 + \ln\left(\dfrac{2e - (e-1)}{2}\right)\right]$

$$= \left[\ln\left(\dfrac{e+1}{2}\right), 1 + \ln\left(\dfrac{e+1}{2}\right)\right] = [0.62, 1.62].$$

iii) When $\alpha = 1$, then $\left[\tilde{f}(A)\right]^1 = [\ln(e), 1 + \ln(1)] = [1]$.

A graph is plotted for $\alpha = 0, 0.5, 1$, which is shown in Figure 2.12b. It is observed that the three points $(0,0)$, $\left(\ln\left(\dfrac{e+1}{2}\right), \dfrac{1}{2}\right)$, $(1,1) = (0,0)$, $(0.62, 0.5)$, $(1,1)$ are not aligned, which means they do not lie in a straight line. This means that $\tilde{f}(A)$ is not a triangular fuzzy number.

2.8 Operations on Fuzzy Numbers Using Extension Principle

The arithmetic operations for fuzzy numbers may be defined from the extension principle for fuzzy sets in an analogous way.

A binary operation $*$ in R is said to be increasing iff [2, 3]:
If $a_1 > a_2$ and $b_1 > b_2$, then $a_1 * b_1 > a_2 * b_2$.
Likewise, a binary operation $*$ in R is said to be decreasing iff

If $a_1 > a_2$ and $b_1 > b_2$, then $a_1 * b_1 < a_2 * b_2$.

In normal algebraic operation, $f(a_1, a_2) = a_1 + a_2$ is an increasing operation on real number R.

$f(a_1, a_2) = a_1 . a_2$ is an increasing operation on positive real number R.

$f(a_1, a_2) = -(a_1 + a_2)$ is a decreasing operation.

The operations such as "+," "_," ".," "/" can be extended to operations on fuzzy numbers and these are denoted as: $\oplus, \ominus, \odot, \oslash$.

Zimmerman [2]: Let A and B be two continuous fuzzy numbers whose membership functions are continuous and surjective (onto) from $\Re \longrightarrow [0,1]$ and the operation $*$ be a continuous increasing (decreasing) binary operation. Then, the extended operation, $A \circledast B$, is a fuzzy number whose membership function is also continuous and surjective from $\Re \longrightarrow [0,1]$.

Dubois and Prade [3] suggested a method to determine the membership functions $\mu_{A \circledast B}(z)$ using Zadeh's extension principle. Using the principle, operation $*$ can be extended to $\circledast$ to combine two fuzzy numbers A and B with membership functions $\mu_A(x)$ and $\mu_B(y)$, respectively.

$$\mu_{A \circledast B}(z) = \sup_{z = x \circledast y} \min(\mu_A(x), \mu_B(y)) \qquad (2.12)$$

If the operation $*$ is commutative, then the extended operation $\circledast$ is also commutative and similarly if the operation $*$ is associative, then $\circledast$ is also associative [3].

2.8.1 Operations

For any three fuzzy numbers x, y, $z \in R$, the following operations are as follows [1]:

Extended addition,

$$A \oplus B : \mu_{(A \oplus B)}(z) = \sup_{z = x + y} \min(\mu_A(x), \mu_B(y)). \qquad (2.13)$$

Extended subtraction,

$$A \ominus B : \mu_{A \ominus B}(z) = \sup_{z = x - y} \min(\mu_A(x), \mu_B(y)). \qquad (2.14)$$

Extended multiplication,

$$A \odot B : \mu_{A \odot B}(z) = \sup_{z = x \cdot y} \min(\mu_A(x), \mu_B(y)). \qquad (2.15)$$

Extended division,

$$A \oslash B : \mu_{A \oslash B}(z) = \sup_{z = x / y} \min(\mu_A(x), \mu_B(y)). \qquad (2.16)$$

2.8.2 Examples on Operations of Fuzzy Numbers Using Extension Principle

i) Extended addition: It is an increasing operation so we can get extended operation. Addition of two fuzzy numbers whose membership function is given as:

$$\mu_{A\oplus B} = \sup_{z=x+y} \min(\mu_A(x), \mu_B(y)).$$

Example 12 Consider two finite fuzzy sets A and B

$$A = \frac{0.4}{2} + \frac{1}{3} + \frac{0.6}{4}, B = \frac{0.6}{1} + \frac{1}{2} + \frac{0.5}{3}, \text{ compute extended addition.}$$

Solution

$$A\oplus B = \left[\frac{\min(0.4, 0.6)}{3} + \frac{\min(0.4, 1)}{4} + \frac{\min(0.4, 0.5)}{5} + \frac{\min(1, 0.6)}{4}\right.$$

$$\left. + \frac{\min(1, 1)}{5} + \frac{\min(1, 0.5)}{6} + \frac{\min(0.6, 0.6)}{5} + \frac{\min(0.6, 1)}{6} + \frac{\min(0.6, 0.5)}{7}\right]$$

$$= \frac{0.4}{3} + \frac{0.4}{4} + \frac{0.4}{5} + \frac{0.6}{4} + \frac{1}{5} + \frac{0.5}{6} + \frac{0.6}{5} + \frac{0.6}{6} + \frac{0.5}{7}$$

$$= \frac{0.4}{3} + \frac{\max(0.4, 0.6)}{4} + \frac{\max(0.4, 1, 0.6)}{5} + \frac{\max(0.5, 0.6)}{6} + \frac{0.5}{7}$$

$$= \frac{0.4}{3} + \frac{0.6}{4} + \frac{1}{5} + \frac{0.6}{6} + \frac{0.5}{7}$$

It is observed that the fuzzy number after addition is also a fuzzy number as it follows convexity.

ii) Extended subtraction: It is neither an increasing nor decreasing operation. We can write the operation $A \ominus B = A \oplus (\ominus B)$ [3], so $A \ominus B$ is a fuzzy number. Difference of two fuzzy numbers whose membership function is given as:

$$\mu_{A\ominus B} = \sup_{z=x-y} \min(\mu_A(x), \mu_B(y))$$

Example 13 Consider two finite fuzzy sets A and B,

$$A = \frac{0.4}{4} + \frac{1}{5} + \frac{0.6}{6},$$

$B = \dfrac{0.6}{2} + \dfrac{1}{3} + \dfrac{0.5}{4}$, find extended subtraction.

Solution

$$A \ominus B = \left[\dfrac{\min(0.4, 0.6)}{2} + \dfrac{\min(0.4, 1)}{1} + \dfrac{\min(0.4, 0.5)}{0} + \dfrac{\min(1, 0.6)}{3} \right.$$

$$\left. + \dfrac{\min(1, 1)}{2} + \dfrac{\min(1, 0.5)}{1} + \dfrac{\min(0.6, 0.6)}{4} + \dfrac{\min(0.6, 1)}{3} + \dfrac{\min(0.6, 0.5)}{2} \right]$$

$$= \dfrac{0.4}{2} + \dfrac{0.4}{1} + \dfrac{0.4}{0} + \dfrac{0.6}{3} + \dfrac{1}{2} + \dfrac{0.5}{1} + \dfrac{0.6}{4} + \dfrac{0.6}{3} + \dfrac{0.5}{2}$$

$$= \dfrac{0.4}{0} + \dfrac{\max(0.4, 0.5)}{1} + \dfrac{\max(0.4, 1, 0.5)}{2} + \dfrac{\max(0.6, 0.6)}{3} + \dfrac{0.6}{4}$$

$$= \dfrac{0.4}{0} + \dfrac{0.4}{1} + \dfrac{1}{2} + \dfrac{0.6}{3} + \dfrac{0.6}{4}.$$

The fuzzy number after subtraction is a fuzzy number as the convexity follows.

iii) Extended multiplication: Multiplication is an increasing operation on positive real numbers and a decreasing operation on negative real numbers. Thus, we get positive fuzzy numbers when the fuzzy numbers are all either positive or negative. If A is a positive fuzzy number and B is a negative fuzzy number, then we get a negative fuzzy number [3]. Multiplication of two fuzzy numbers $A \odot B$ whose membership function is given by:

$$\mu_{A \odot B} = \sup_{z = xy} \min(\mu_A(x) \cdot \mu_B(y))$$

Example 14 Consider A and B are fuzzy numbers:

$A = \left\{ \dfrac{0.4}{1}, \dfrac{1}{2}, \dfrac{0.5}{3} \right\}$ and $B = \left\{ \dfrac{0.5}{3}, \dfrac{1}{4}, \dfrac{0.6}{5} \right\}$. Compute extended multiplication.

It is to be noted that operations for multiplication and division require extensive computation.
An example is given where the resulting fuzzy number after multiplication is no longer a fuzzy number.

Solution

$$A \odot B = \sup_{z = x.y} \left\{ \frac{\min(0.4, 0.5)}{3}, \frac{\min(0.4, 1)}{4}, \frac{\min(0.4, 0.6)}{5}, \frac{\min(1, 0.5)}{6}, \frac{\min(1, 1)}{8}, \frac{\min(1, 0.6)}{10}, \right. $$
$$\left. \frac{\min(0.5, 0.5)}{9}, \frac{\min(0.5, 1)}{12}, \frac{\min(0.5, 0.6)}{15} \right\}$$

$$= \sup_{z = x.y} \left\{ \frac{0.4}{3}, \frac{0.4}{4}, \frac{0.4}{5}, \frac{0.5}{6}, \frac{1}{8}, \frac{0.6}{10}, \frac{0.5}{9}, \frac{0.5}{12}, \frac{0.5}{15} \right\}$$

$$= \left\{ \frac{0.4}{3}, \frac{0.4}{4}, \frac{0.4}{5}, \frac{0.5}{6}, \frac{1}{8}, \frac{0.5}{9}, \frac{0.6}{10}, \frac{0.5}{12}, \frac{0.5}{15} \right\}.$$

It is seen that the fuzzy number after multiplication is not a fuzzy number as the fuzzy set does not follow convexity. The fuzzy number $\frac{0.5}{9}$ does not follow the convexity.

2.9   L–R Representation of Fuzzy Numbers

LR fuzzy number was first introduced by Dubois and Prade [3] and is the commonly used type of fuzzy numbers. It is represented by its mean value, left and right spreads, and shape functions. Extended operation on fuzzy numbers requires extensive computation unless some restriction is put on the membership function. They proposed some algorithms for extended operation. For practical purposes, it will be more appropriate if some specific kind of fuzzy numbers such as extended operations to fuzzy numbers in LR representation or triangular fuzzy numbers are used. Computational efficiency is very increased on using these specific types of fuzzy numbers, which is required while dealing with real-time applications.

Dubois and Prade [3] and Zimmerman [2] suggested that L or R is a fuzzy number which is a reference function for fuzzy number that maps $\mathbb{R}^+ \rightarrow [0,1]$ iff

a) $L(x) = L(-x)$,
b) L is nonincreasing on $[0, +\infty)$.
c) $L(0) = 1, L(x) < 1 \ \forall x > 0$,
d) $L(x) > 0 \ \forall x < 1$,
e) $L(1) = 0$ or $L(x) > 0 \ \forall x$ and $L(+\infty) = 0$.

$L(x)$ or $R(x)$ may take different functions, which are as follows [3, 6]:
$L(x) = \max(0, 1 - |x|^p)$ with $p \geq 0$,

$$L(x) = e^{-|x|^p}, p > 0 \text{ and } L(x) = \frac{1}{1 + |x|^p}, p \geq 0.$$

A convenient way, where a fuzzy number can be represented, is using a reference function L, which is a nonincreasing mapping in the interval $(0, +\infty)$.

Let $P = (p_1, p_2, p_3, p_4)$ be a generalized LR fuzzy number. Using two reference functions L and R, membership function of a generalized fuzzy number P may be given as [3, 7]:

$$\mu_P(x) = \begin{cases} L\left(\dfrac{p_2 - x}{p_2 - p_1}\right), & p_1 \leq x < p_2 \\[2mm] 1, & p_2 \leq x \leq p_3 \\[2mm] R\left(\dfrac{x - p_3}{p_4 - p_3}\right), & p_3 < x \leq p_4, \end{cases} \tag{2.17}$$

where L and R are strictly decreasing functions defined on $[0, 1]$ such that

$L(t) = R(t) = 1$, if $t \leq 0$

$L(t) = R(t) = 0$, if $t \geq 1$.

When $p_2 = p_3$, and it becomes a classical L–R fuzzy number whose membership function is written as:

$$\mu_P(x) = \begin{cases} L\left(\dfrac{p_2 - x}{p_2 - p_1}\right), & p_1 \leq x < p_2 \\[2mm] R\left(\dfrac{x - p_2}{p_3 - p_2}\right), & p_2 < x \leq p_3, \end{cases}$$

or it may be written as:

$$\mu_P(x) = \begin{cases} L\left(\dfrac{p - x}{\alpha}\right), & p - \alpha \leq x \leq p, \alpha > 0 \text{ or } x \leq p, \alpha > 0 \\[2mm] R\left(\dfrac{x - p}{\beta}\right), & p \leq x \leq p + \beta, \beta > 0 \text{ or } x \geq p, \beta > 0, \end{cases} \tag{2.18}$$

where α, β are the left and right spread of μ_P, respectively, and are nonnegative real numbers. These are shown in Figure 2.13. Generally, L and R are called left and right reference functions of μ_P. p is the mean value of the fuzzy number P. A fuzzy number P can be denoted as: $P = (p, \alpha, \beta)_{LR}$, which is shown in Figure 2.13. When $\alpha = \beta$, it is called a symmetric fuzzy number. If the spreads are zero, then the fuzzy number P is a non-fuzzy number. As the spreads increase, P becomes more and more fuzzy.

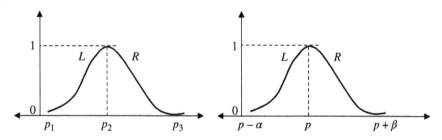

Figure 2.13 Fuzzy number with fuzzy intervals.

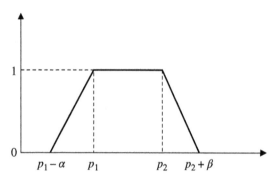

Figure 2.14 L–R type trapezoidal fuzzy number.

The support of P is $(p - \alpha, p + \beta)$ and p is the core of P.

A fuzzy number P is called L–R type trapezoidal fuzzy number if [7]:

$$
\mu_P(x) = \begin{cases} L\left(\dfrac{p_1 - x}{\alpha}\right), & p_1 - \alpha \le x < p_1 \\[2mm] 1, & p_1 \le x \le p_2 \\[2mm] R\left(\dfrac{x - p_2}{\beta}\right), & p_2 < x \le p_2 + \beta \end{cases} \tag{2.19}
$$

p_1, p_2 are the shoulders of P. Symbolically P is written as $P = (p_1, p_2, \alpha, \beta)_{LR}$. The support of P is $(p_1 - \alpha, p_2 + \beta)$ shown in Figure 2.14.

Most shape functions L (or R) are strictly decreasing in practical applications and these may be $L(x) = R(x) = \max(0, 1 - x)$ for triangular fuzzy number or $L(x) = R(x) = e^{-x^2}$ for Gaussian fuzzy number.

Another point is that with $L(x) = R(x) = 1 - x$, the generalized LR fuzzy number in Eq. (2.17) becomes a trapezoidal fuzzy number in Eq. (2.5) [7, 8].

$$
L\left(\frac{p_2 - x}{p_2 - p_1}\right) = 1 - \frac{p_2 - x}{p_2 - p_1} = \frac{p_2 - p_1 - p_2 + x}{p_2 - p_1} = \frac{x - p_1}{p_2 - p_1}.
$$

$$R\left(\frac{x-p_3}{p_4-p_3}\right) = 1 - \frac{x-p_3}{p_4-p_3} = \frac{p_4-p_3-x+p_3}{p_4-p_3} = \frac{p_4-x}{p_4-p_3}.$$

Similarly, with $L(x) = R(x) = 1 - x$ and $p_2 = p_3$, generalized LR fuzzy number in Eq. (2.17) becomes triangular fuzzy number.

Example 15   Let $L(x) = \dfrac{1}{1+3|x|}$ and $R(x) = \dfrac{1}{1+x^2}$ and the fuzzy number P, where $p = 5$, $\alpha = 2$, $\beta = 3$. Compute the membership function of LR fuzzy number.

Solution

Using the equations in (2.17), the membership function of the fuzzy number P is written as:

$$\mu_P(x) = \begin{cases} L\left(\dfrac{5-x}{2}\right), & x \le 5 \\[3mm] R\left(\dfrac{x-5}{3}\right), & x \ge 5 \end{cases}$$

It is given as $L(x) = \dfrac{1}{1+3|x|}$, $L\left(\dfrac{5-x}{2}\right) = \dfrac{1}{1+3\left|\dfrac{5-x}{2}\right|}$.

Likewise, for $R(x)$, $R(x) = \dfrac{1}{1+x^2}$, so $R\left(\dfrac{x-5}{3}\right) = \dfrac{1}{1+\left(\dfrac{x-5}{3}\right)^2}$.

So, the membership function of LR fuzzy number $(5,2,3)_{LR}$ is given as:

$$\mu_P(x) = \begin{cases} \dfrac{1}{1+3\left|\dfrac{5-x}{2}\right|}, & x \le 5 \\[5mm] \dfrac{1}{1+\left(\dfrac{x-5}{3}\right)^2}, & x \ge 5. \end{cases}$$

Example 16   Let $L(x) = \dfrac{1}{1+2|x|}$ and $R(x) = e^{-|x|}$. The fuzzy number $P = (p,\alpha,\beta)_{LR} = (5,2,3)_{LR}$ where $p = 5$, $\alpha = 2$, $\beta = 3$. Compute the membership function of LR fuzzy number.

Solution

Using the equations in (2.18), the membership function of LR fuzzy number is given as:

$$\mu_P(x) = \begin{cases} L\left(\dfrac{5-x}{2}\right), & x \le 5 \\[2mm] R\left(\dfrac{x-5}{3}\right), & x \ge 5 \end{cases}$$

Given $L(x) = \dfrac{1}{1+2|x|}$, so $L\left(\dfrac{5-x}{2}\right) = \dfrac{1}{1+2\left|\dfrac{5-x}{2}\right|}$

Likewise, for $R(x) = e^{-|x|}$, so $R\left(\dfrac{x-5}{3}\right) = e^{-\left|\frac{x-5}{3}\right|}$.

So the membership function of LR fuzzy number $(5,2,3)_{LR}$ is

$$\mu_P(x) = \begin{cases} \dfrac{1}{1+2\left|\dfrac{5-x}{2}\right|}, & x \le 5 \\[4mm] e^{-\left|\frac{x-5}{3}\right|}, & x \ge 5. \end{cases}$$

The membership function plot is shown in Figure 2.15.

Example 17   Consider a fuzzy number $(p,\alpha,\beta)_{LR} = (p,q,q)_{LR}$ where $\alpha = \beta = q$ and the shape function is:

$$L(x) = R(x) = e^{-x^2}.$$

Compute the membership function of LR fuzzy number.

Solution

The membership function of LR fuzzy number is given as:

$$\mu_P(x) = \begin{cases} L\left(\dfrac{p-x}{q}\right), & x \le p \\[2mm] R\left(\dfrac{x-p}{q}\right), & x \ge p \end{cases}$$

Then, $L\left(\dfrac{p-x}{q}\right) = e^{-\left(\frac{p-x}{q}\right)^2}$ and $R\left(\dfrac{x-p}{q}\right) = e^{-\left(\frac{x-p}{q}\right)^2}$.

This is called a normal fuzzy number.

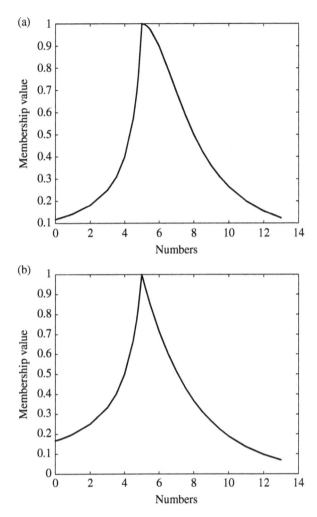

Figure 2.15 Membership function plot for LR fuzzy number: (a) Example 15 and (b) Example 16.

Next, we will proceed with addition, subtraction, and multiplication of L–R type fuzzy numbers Dubois and Prade [3].

Let us consider two fuzzy numbers $P = (p,\alpha,\beta)_{LR}$ and $Q = (q,\gamma,\delta)_{LR}$ where p, q are the means of the fuzzy numbers P and Q, respectively. Let x and y be real numbers and considering the increasing parts of P and Q, we write

$$L\left(\frac{p-x}{\alpha}\right) = \rho = L\left(\frac{q-y}{\gamma}\right), \qquad (2.20)$$

where ρ is a value in the interval $[0,1]$.

On solving Eq. (2.20), we get

$$x = p - \alpha L^{-1}(\rho) \quad \text{and} \quad y = q - \gamma L^{-1}(\rho).$$

Adding x and y, we get $s = (p + q) - (\alpha + \gamma)L^{-1}(\rho)$,

which implies $L\left(\dfrac{p + q - s}{\alpha + \gamma}\right) = \rho$.

The same rule follows for the decreasing part of the fuzzy numbers P and Q.

$$R\left(\frac{x - p}{\beta}\right) = \rho = R\left(\frac{y - q}{\delta}\right). \tag{2.21}$$

On solving, we get

$$x = p + \beta R^{-1}(\rho) \quad \text{and} \quad y = q + \delta R^{-1}(\rho)$$

Adding, we get $s = (p + q) + (\beta + \delta)R^{-1}(\rho)$

implies $R\left(\dfrac{s - (p + q)}{\beta + \delta}\right) = \rho$.

i) *Addition* of two LR fuzzy numbers:

$$(p, \alpha, \beta)_{LR} \oplus (q, \gamma, \delta)_{LR} = (p + q, \alpha + \gamma, \beta + \delta)_{LR}. \tag{2.22}$$

ii) *Opposite* of a fuzzy number:

$$-(p, \alpha, \beta)_{LR} = (-p, \beta, \alpha)_{LR}. \tag{2.23}$$

iii) *Subtraction* of two LR fuzzy numbers:

$$\begin{aligned}(p, \alpha, \beta)_{LR} \ominus (q, \gamma, \delta)_{LR} &= (p, \alpha, \beta)_{LR} + (-q, \delta, \gamma)_{LR} \\ &= (p - q, \alpha + \delta, \beta + \gamma)_{LR}.\end{aligned} \tag{2.24}$$

iv) *Multiplication of two fuzzy numbers* Dubois and Prade [3]: $s = x. y$

Considering the increasing part of the fuzzy number

$$\begin{aligned}s &= \left[p - \alpha L^{-1}(\rho)\right] \cdot \left[q - \gamma L^{-1}(\rho)\right] \\ &= p \cdot q - (p\gamma + q\alpha)L^{-1}(\rho) + \alpha\gamma\left(L^{-1}(\rho)\right)^2.\end{aligned}$$

Neglecting the term $\alpha\gamma(L^{-1}(\rho))^2$ provided that α and γ are very small compared to p and q, we get

$$s = p \cdot q - (p\gamma + q\alpha)L^{-1}(\rho). \tag{2.25}$$

Likewise, for the decreasing part of the fuzzy numbers we get

$$s = x \cdot y = \left[p + \beta R^{-1}(\rho) \right] \cdot \left[q + \delta L R^{-1}(\rho) \right]$$

$$= p \cdot q + (p\delta + q\beta) R^{-1}(\rho) + \beta\delta \left(L^{-1}(\rho) \right)^2.$$

Neglecting the term $\beta\delta(L^{-1}(\rho))^2$ provided that α and γ are very small compared with p and q, we get

$$s = p \cdot q + (p\delta + q\beta) R^{-1}(\rho). \tag{2.26}$$

Multiplication:

$$(p,\alpha,\beta)_{LR} \otimes (q,\gamma,\delta)_{LR} = (p \cdot q, p\gamma + q\alpha, p\delta + q\beta)_{LR}, P > 0, Q > 0 \tag{2.27}$$

When $P < 0$, $Q > 0$, Eq. (2.27) becomes

$$(p,\alpha,\beta)_{LR} \otimes (q,\gamma,\delta)_{LR} = (p \cdot q, q\alpha - p\delta, q\beta - p\gamma)_{LR}.$$

When $P < 0$, $Q < 0$, Eq. (2.27) becomes

$$(p,\alpha,\beta)_{LR} \otimes (q,\gamma,\delta)_{LR} = (p \cdot q, -p\delta - q\beta, q\alpha - p\gamma)_{LR}.$$

LR number $P = (p,\alpha,\beta)_{LR}$ is positive if $p - \alpha > 0$.

Two LR numbers $P = (p,\alpha,\beta)_{LR}$ and $Q = (q,\gamma,\delta)_{LR}$ are said to be equal if $p = q$, $\alpha = \gamma$, $\beta = \delta$.

LR numbers $P = (p,\alpha,\beta)_{LR}$ is said to be a subset of another LR fuzzy number, $Q = (q,\gamma,\delta)_{LR}$ iff $p - \alpha \geq q - \gamma$, $p + \beta \leq q + \delta$.

Example 18 Consider two fuzzy LR fuzzy numbers as

$$P = (2,0.4,0.7)_{LR},$$

$$Q = (4,0.6,0.2)_{LR}.$$

Then, $P \oplus Q = (6,1,0.9)_{LR}$.

$$\ominus Q = (-4,0.2,0.6)_{LR},$$

Then, $P \ominus Q = (-2,0.6,1.3)_{LR}$.

2.10 Intuitionistic Fuzzy Numbers

After the introduction of fuzzy number that handles imprecise numerical quantities, triangular fuzzy numbers and trapezoidal fuzzy numbers have been widely studied, developed, and applied to various fields. As fuzzy number is not capable of dealing with any application where there is a lack of knowledge about membership degree, Mohedano and Burillo [9] later generalized it to intuitionistic fuzzy number. It is similar to fuzzy number, the only exception is that both membership and nonmembership functions are considered. In this section both triangular and trapezoidal fuzzy numbers are discussed. As an extension of fuzzy set, i.e. intuitionistic fuzzy to triangular and trapezoidal fuzzy numbers, the triangular intuitionistic fuzzy numbers (TIFNs) and trapezoidal intuitionistic fuzzy numbers have not received adequate attention till today. Research studies are carried out using arithmetic and logical operation of triangular/trapezoidal intuitionistic fuzzy number. An intuitionistic fuzzy set, $A = (\mu_A, \nu_A)$, is characterized by membership and nonmembership functions and is said to be an intuitionistic fuzzy number where the membership and nonmembership values are fuzzy numbers.

2.11 Triangular Intuitionistic Fuzzy Number

Let a TIFN, $A = [(a,b,c); w_A], [(d,b,f); u_A])$, where $0 \leq w_A \leq 1$, $0 \leq u_A \leq 1$ and $0 \leq w_A + u_A \leq 1$. The membership function is defined as follows [10, 11]:

$$\mu_A(x) = \begin{cases} \dfrac{x-a}{b-a} w_A, & a \leq x \leq b \\[2mm] w_A, & x = b \\[2mm] \dfrac{c-x}{c-b} w_A, & b \leq x \leq c \\[2mm] 0, & x < a \text{ or } x > c \end{cases} \tag{2.28}$$

and the nonmembership function is written as:

$$\nu_A(x) = \begin{cases} \dfrac{b-x+u_A(x-d)}{b-d}, & d \leq x \leq b \\[2mm] u_A, & x = b \\[2mm] \dfrac{x-b+u_A(f-x)}{f-b}, & b \leq x \leq f \\[2mm] 1 & x < d \text{ or } x > f, \end{cases} \tag{2.29}$$

where the values w_A and u_A represent the maximum membership degree and the minimum nonmembership degree such that they satisfy the conditions: $0 \leq w_A \leq 1$, $0 \leq u_A \leq 1$ and $0 \leq w_A + u_A \leq 1$.

It may be noted that if $a = d$, $c = f$, then TIFN is written as:
$A = [(a,b,c; w_A, u_A)]$. This is shown in Figure 2.16.

If $w_A = 1$, $u_A = 0$, then TIFN, A, becomes a triangular fuzzy number [3], i.e. $= [(a,b,c); 1, 0]$.

Score function and accuracy function [5] of a TIFN, $A = ([(a,b,c; w_A)], [(d,b,f; u_A)])$, is given as:

Figure 2.16 Representation of an intuitionistic fuzzy triangular number.

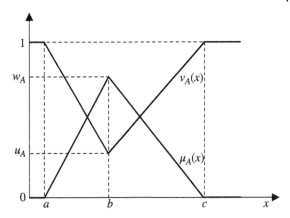

Score function: $s(A) = \dfrac{a+2b+c}{4} w_A - \dfrac{d+2b+f}{4} u_A.$

Accuracy function: $h(A) = \dfrac{a+2b+c}{4} w_A + \dfrac{d+2b+f}{4} u_A,$

where $\dfrac{a+2b+c}{4}$ is the graded mean of triangular fuzzy number (explained in Chapter 3).

2.12 Operations Using Triangular Intuitionistic Fuzzy Numbers

There are few work by authors [10–14] on operations on intuitionistic fuzzy numbers.

Consider two TIFNs, $A = ([(a,b,c; w_A)], [(d,b,f; u_A)])$, $B = ([(a',b',c'; w_B)], [(d', b',f'; u_B)])$, the following operations are:

i) Addition:

$$A + B = ([(a+a',b+b',c+c');w_A \wedge w_B],[(d+d',b+b',f+f');u_A \vee u_B]).$$

$$(2.30)$$

ii) Subtraction:

$$A - B = ([(a-c',b-b',c-a');w_A \wedge w_B],[(d-f',b-b',f-d');u_A \vee u_B]).$$

$$(2.31)$$

iii) Multiplication:

$$A.B = ([(aa',bb',cc');w_A \wedge w_B],[(dd',bb',ff');u_A \vee u_B]).$$ (2.32)

iv) Division:

$$\frac{A}{B} = \left(\left[\frac{a}{c'}, \frac{b}{b'}, \frac{c}{a'}; w_A \wedge w_B \right], \left[\frac{d}{f'}, \frac{b}{b'}, \frac{f}{d'}; u_A \vee u_B \right] \right).$$ (2.33)

If we consider two intuitionistic triangular fuzzy numbers: $A = [(a,b,c);$ $w_A,u_A]$, $B = [(a',b',c'); w_B,u_B]$ (considering $a = d$, $c = f$ and $a' = d'$, $c' = f'$ in the above-mentioned intuitionistic fuzzy number for simplicity), then the above operations are as follows [13]:

i) Addition: $A + B = [(a + a',b + b',c + c'); w_A \wedge w_B, u_A \vee u_B]$.
ii) Subtraction: $A - B = [(a - c',b - b',c - a'); w_A \wedge w_B, u_A \vee u_B]$.
iii) Multiplication: $= [(aa',bb',cc'); w_A \wedge w_B, u_A \vee u_B]$.
iv) Division: $\dfrac{A}{B} = \left[\left(\dfrac{a}{c'}, \dfrac{b}{b'}, \dfrac{c}{a'} \right); w_A \wedge w_B, u_A \vee u_B \right]$.
v) $A^{-1} = \left[\dfrac{1}{c}, \dfrac{1}{b}, \dfrac{1}{a}; w_A, u_A \right]$.

But there are some limitations on the operations (2.29–2.32) [10]:
Consider two intuitionistic triangular fuzzy numbers: $A = [(a,b,c); 1,0]$, $B = [(a,b,c); 0,1]$, Then, the addition and subtraction operations become:

i) Addition:

$$A + B = ([(2a,2b,2c);0,1]).$$ (2.34)

ii) Subtraction:

$$A - B = ([(a - c,0,c - a);0,1]).$$ (2.35)

It is observed that the complementary effect on maximum to minimum membership values is ignored. Again from the operations $A \times B$, A/B in Eqs. (2.31) and (2.32), it is seen that they do not consider any membership or nonmembership values but only the minimum and maximum membership values and that might lead to biased results and loss of information. So, it would be reasonable if the membership and nonmembership values lie within the interval range of $(0,1)$.

So, Wang et al. [11] suggested an improvement on the definitions of operations.

The new definitions of arithmetic operation of triangular fuzzy number are as follows:

Given two TIFNs, $A = ([(a,b,c; w_A)], [(d,b,f; u_A)]), B = ([(a',b',c'; w_B)], [(d',b',f'; u_B)])$, the following operations are computed:

$$\left\| \tilde{A}' \right\| = \frac{|a| + 2|b| + |c|}{4}, \left\| \tilde{A} \right\| = \frac{|d| + 2|b| + |f|}{4},$$

$$\left\| \tilde{B}' \right\| = \frac{|a'| + 2|b'| + |c'|}{4}, \left\| \tilde{B} \right\| = \frac{|d'| + 2|b'| + |f'|}{4}, \text{ then}$$

i) $A + B = \left(\left[(a+a',b+b',c+c'); \dfrac{\left\| \tilde{A}' \right\| w_A + \left\| \tilde{B}' \right\| w_B}{\left\| \tilde{A}' \right\| + \left\| \tilde{B}' \right\|} \right], \right.$

$\left. \left[(d+d',b+b',f+f'); \dfrac{\left\| \tilde{A} \right\| u_A + \left\| \tilde{B} \right\| u_B}{\left\| \tilde{A} \right\| + \left\| \tilde{B} \right\|} \right] \right).$

ii) $A.B = ([(aa',bb',cc'); w_A w_B], [(dd',bb',ff'); u_A + u_B - u_A u_B]).$

iii) $A - B = \left(\left[(a-c',b-b',c-a'); \dfrac{\left\| \tilde{A}' \right\| w_A + \left\| \tilde{B}' \right\| w_B}{\left\| \tilde{A}' \right\| + \left\| \tilde{B}' \right\|} \right], \right.$

$\left. \left[(d-f',b-b',f-d'); \dfrac{\left\| \tilde{A} \right\| u_A + \left\| \tilde{B} \right\| u_B}{\left\| \tilde{A} \right\| + \left\| \tilde{B} \right\|} \right] \right).$

iv) $\dfrac{A}{B} = \left(\left[\dfrac{a}{c'}, \dfrac{b}{b'}, \dfrac{c}{a'}; w_A w_B \right], \left[\dfrac{d}{f'}, \dfrac{b}{b'}, \dfrac{f}{d'}; u_A + u_B - u_A u_B \right] \right).$

v) Scalar multiplication, $\lambda A = ([(\lambda a, \lambda b, \lambda c); w_A], [(\lambda d, \lambda b, \lambda f); u_A]), \lambda \geq 0.$

vi) $A^\lambda = ([(a^\lambda, b^\lambda, c^\lambda); w_A], [(d^\lambda, b^\lambda, f^\lambda); 1 - (1 - u_A)^\lambda]).$

The above operations become easier if $a = d, c = f$ and $a' = d', c' = f'$.

2.13 Trapezoidal Intuitionistic Fuzzy Numbers

Like triangular fuzzy numbers, few work on trapezoidal fuzzy numbers are there in the literature [15–17]. Trapezoidal has a better capability to process ill-defined quantities. Let a trapezoidal intuitionistic fuzzy number: $A = ([(a,b,c,d); w_A], [(e,b,c,h); u_A])$ where $0 \leq w_A \leq 1, 0 \leq u_A \leq 1$ and $0 \leq w_A + u_A \leq 1$.

Then a trapezoidal intuitionistic fuzzy number whose membership function is written as follows [11, 14]:

$$\mu_A(x) = \begin{cases} \dfrac{x-a}{b-a} w_A, & a \le x < b \\[2mm] w_A, & b \le x \le c \\[2mm] \dfrac{d-x}{d-c} w_A, & c < x \le d \\[4mm] 0, & \text{otherwise} \end{cases} \tag{2.36}$$

and the nonmembership function is defined as:

$$\nu_A(x) = \begin{cases} \dfrac{b-x+u_A(x-e)}{b-e}, & e \le x < b \\[2mm] u_A, & b \le x \le c \\[2mm] \dfrac{x-c+u_A(h-x)}{h-c}, & c < x \le h \\[4mm] 1 & \text{otherwise.} \end{cases} \tag{2.37}$$

Let $[a,b,c,d] = [e,b,c,h]$, then trapezoidal intuitionistic fuzzy number, A, will be written as

$$A = [(a,b,c,d); w_A, u_A].$$

Few operations on trapezoidal fuzzy numbers are given as follows:
Consider two trapezoidal fuzzy numbers $A = [(a,b,c,d); w_A, u_A]$, $B = [(a',b',c', d'); w_B, u_B]$.

i) Addition: $A + B = [(a + a', b + b', c + c', d + d'); w_A \wedge w_B, u_A \vee u_B]$.
ii) Multiplication: $= [(aa', bb', cc', dd'); w_A \wedge w_B, u_A \vee u_B]$.
iii) Scalar multiplication, $\lambda A = [(\lambda a, \lambda b, \lambda c, \lambda d); w_A, u_A]$.
iv) $A^\lambda = [(a^\lambda, b^\lambda, c^\lambda, d^\lambda); w_A, u_A]$.

2.14 Cut Set of Intuitionistic Fuzzy Number

In this section, we will show cut set of both triangular and trapezoidal intuitionistic fuzzy numbers. The computation is similar to that of computation of fuzzy α-cut.

a) Triangular fuzzy number: A (α,β) cut set of a TIFN, $A = [(a,b,c); w_A, u_A]$ is a crisp subset, which is defined as [13, 14]:

$$A_{\alpha, \beta} = \{x \mid \mu_A(x) \ge \alpha, \nu_A(x) \le \beta\} \text{ where } A_\alpha = \{x \mid \mu_A(x) \ge \alpha\} \text{ and } A_\beta = \{x \mid \mu_A(x) \le \beta\}.$$

$$0 \le \alpha \le w_A, u_A \le \beta \le 1 \text{ and } 0 \le \alpha + \beta \le 1.$$

A_α is computed using Eq. (2.28) and the computation of α-cut is similar to that fuzzy α-cut, i.e.

$\alpha = \dfrac{x-a}{b-a}w_A$ and $\alpha = \dfrac{c-x}{c-b}w_A$. Substituting $x = a_1^\alpha$ and $x = a_2^\alpha$ in the two equations, we get $a_1^\alpha = a + \dfrac{(b-a)\alpha}{w_A}$ and $a_2^\alpha = c - \dfrac{(c-b)\alpha}{w_A}$.

So, $A_\alpha = \left[a_1^\alpha, a_2^\alpha\right] = \left[a + \dfrac{(b-a)\alpha}{w_A}, c - \dfrac{(c-b)\alpha}{w_A}\right]$

Likewise, we get β-cut for nonmembership function,

$$A_\beta = \left[a_1^\beta, a_2^\beta\right] = \left[\frac{b - au_A - \beta(b-a)}{1 - u_A}, \frac{b - cu_A + \beta(c-b)}{1 - u_A}\right],$$

where a_1, a_2 are the lower and upper levels of the cuts. The (α,β) cut set of membership and nonmembership functions are shown in Figure 2.17.

b) Trapezoidal fuzzy number: Similar to the cut set of triangular fuzzy number, (α,β) cut set of a trapezoidal intuitionistic fuzzy number, $A = [(a,b,c,d); w_A, u_A]$ is a crisp subset, which is defined as [13, 15]:

A_α is computed using Eq. (2.36) and the computation of α-cut is similar to that of α-cut of TIFN and it is written as:

$$A_\alpha = \left[a_1^\alpha, a_2^\alpha\right] = \left[a + \frac{(b-a)\alpha}{w_A}, d - \frac{(d-c)\alpha}{w_A}\right],$$

$$A_\beta = \left[a_1^\beta, a_2^\beta\right] = \left[\frac{b - au_A - \beta(b-a)}{1 - u_A}, \frac{c - du_A + (d-c)\beta}{1 - u_A}\right].$$

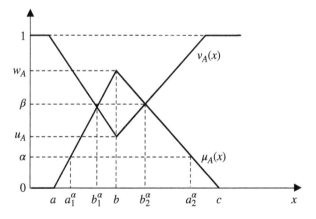

Figure 2.17 α-Cut set and β-cut set of membership and nonmembership fumction (*Source:* Adapted from [14]). α Representation of an intuitionistic fuzzy trianfular number.

2.15 Distances Between Two Intuitionistic Fuzzy Numbers

We will show a method to compute Hamming and Euclidean distances between two trapezoidal intuitionistic fuzzy numbers [16]. Consider two trapezoidal intuitionistic fuzzy numbers, $A = [(a,b,c,d); w_A, u_A]$, $B = [(a',b',c',d'); w_B, u_B]$, then Hamming distance:

$$d_h(A,B) = \frac{1}{4}[|a-a'| + |b-b'| + |c-c'| + |d-d'|] + \max\{|w_A - w_B|, |u_A - u_B|\}$$

Euclidean distance:

$$d_e(A,B) = \frac{1}{2}\sqrt{[(a-a')^2 + (b-b')^2 + (c-c')^2 + (d-d')^2 + \max\{|w_A - w_B|^2, |u_A - u_B|^2\}]}$$

Distance between TIFNs [14]: Consider two TIFNs, $A = (a,b,c; w_A, u_A)$, $B = (a',b',c'; w_B, u_B)$, Hamming distance:

$$d_h(A,B) = \frac{1}{3}[|a-a'| + |b-b'| + |c-c'|] + \max\{|w_A - w_B|, |u_A - u_B|\}$$

Euclidean distance:

$$d_e(A,B) = \sqrt{\frac{1}{3}[(a-a')^2 + (b-b')^2 + (c-c')^2 + \max\{|w_A - w_B|^2, |u_A - u_B|^2\}]}$$

2.16 Summary

This chapter provides a detailed overview of fuzzy numbers and also the operations on fuzzy numbers. Zadeh's extension principle, operation of fuzzy numbers using extension principle, fuzzy numbers with intervals and operations, operations on fuzzy numbers with alpha cut, and LR-type fuzzy numbers are explained in detail along with examples. Intuitionistic fuzzy number and especially triangular and trapezoidal intuitionistic fuzzy numbers along with operations are also discussed. Alpha–beta cut of both types of intuitionistic fuzzy numbers, and distance measures between two intuitionistic fuzzy numbers are also discussed.

References

1 Zadeh, L.A. (1975). The concept of a linguistic variable and its application to approximate reasoning. Parts 1, 2, and 3. *Information Science* 8: 199–249; 8: 301–357; 9: 43–80.

2 Zimmerman, H.J. (2002). *Fuzzy Set Theory and Its Application.* Kluwer Academic Publisher.

3 Dubois, D. and Prade, H. (1980). *Fuzzy Sets and Systems: Theory and Applications.* Academic Press Inc.

4 de Barros, L.L., Bassanezi, R.C., and Lodwick, W.A. (2016). A first course in fuzzy logic, fuzzy dynamical systems, and biomathematics. *Studies in Fuzziness and Soft Computing* 347: 23–41.

5 Nguyen, H.T. (1978). A note on the extension principle for fuzzy sets. *Journal of Mathematical Analysis and Application* 64: 369–380.

6 Digkman, J., Van Haeringen, H., and De Lange, J. (1983). Fuzzy numbers. *Journal of Mathematical Analysis and Applications* 92: 301–341.

7 Yeh, C.T. and Chu, H.M. (2014). Approximations by LR-type fuzzy numbers. *Fuzzy Sets and Systems* 257: 23–40.

8 Firozja, M.A., Fath-Tabar, G.H., and Eslampia, Z. (2012). The similarity measure of generalized fuzzy numbers based on interval distance. *Applied Mathematics Letters* 25: 1528–1534.

9 Mohedano, V. and Burillo, P.H. (1994). Some definitions on intuitionistic fuzzy number, fuzzy based expert systems. *Fuzzy Bulgarian Enthusiasts.* Sofia, Bulgaria.

10 Wang, J.Q., Hong-yu, R.N., and Zhang, X.C. (2013). New operators on triangular intuitionistic fuzzy numbers and their applications in system fault analysis. *Information Sciences* 251: 79–95.

11 Wang, S.P., Wang, F., Lin, L.-L., and Dong, J.-Y. (2016). Some new generalized aggregation operators for triangular intuitionistic fuzzy numbers and application to multi-attribute group decision making. *Computers and Industrial Engineering* 93: 286–301.

12 Li, D.F. (2008). A note on "using intuitionistic fuzzy sets for fault-tree analysis on printed circuit board assembly". *Microelectronics Reliability* 48: 1741.

13 Li, D.F. (2010). A ratio ranking method of triangular intuitionistic fuzzy numbers and its application to MADM problems. *Computers and Mathematics with Applications* 60: 1557–1570.

14 Wan, S.P., Wang, Q.Y., and Dong, J.Y. (2013). The extended VIKOR method for multi-attribute group decision making with triangular intuitionistic fuzzy numbers. *Knowledge-Based Systems* 52: 65–77.

15 Wan, S.-P. and Dong, J.-Y. (2015). Power geometric operators of trapezoidal intuitionistic fuzzy numbers and application to multi-attribute group decision making. *Applied Soft Computing* 29: 153–168.

16 Wan, S.P. (2013). Power average operators of trapezoidal intuitionistic fuzzy numbers and application to multi-attribute group decision making. *Applied Mathematical Modelling* 37: 4112–4126.

17 Wei, G. (2010). Some arithmetic aggregation operators with intuitionistic trapezoidal fuzzy numbers and their application to group decision making. *Journal of Computers* 5 (3): 345–351.

3

Similarity Measures and Measures of Fuzziness

3.1 Introduction

Since the introduction of fuzzy set, similarity measures, distance measures, measures of fuzziness of fuzzy sets, and inclusion measure have become important topics in fuzzy set theory and have successfully been applied in many fields, such as image processing, fuzzy neural networks, fuzzy reasoning, and fuzzy control. The fuzziness of a fuzzy set or fuzzy entropy describes the degree of fuzziness of a fuzzy set. The inclusion measure of fuzzy sets indicates the degree to which a fuzzy set A is contained in another fuzzy set B. The similarity measure indicates the degree of similarity of fuzzy sets A and B. Similarity measure between two fuzzy sets depends on their matched portion, that is, the greater the match between the objects the more similarity they are. Similarity measure and distance measure are closely related and these two measures are expressed using a functional relationship. The two measures are inversely related and so similarity measures can be used to define distance measures and vice versa. The study of distance measure is very much significant because if we know the distance measure between two objects, then we can find out the similarity. If the distance between the objects is more, then there is less similarity. Thus, the distance measure between fuzzy sets and fuzzy numbers gained more attention among the researchers. There has been a lot of definitions on the distance between fuzzy sets and recent research on the distance between fuzzy numbers has gained importance.

3.2 Distance and Similarity Measures

In this section we will discuss distance measure between two fuzzy sets. There are different types of distance measures suggested by different authors. Let us consider a universal set X and $F(X)$ be a fuzzy set. For two fuzzy sets, A and B

Fuzzy Set and Its Extension: The Intuitionistic Fuzzy Set, First Edition. Tamalika Chaira.
© 2019 John Wiley & Sons, Inc. Published 2019 by John Wiley & Sons, Inc.

such that $A,B \in F(X)$, the properties of similarity measure and distance measure between two fuzzy sets A and B with membership functions $\mu_A(x)$ and $\mu_B(x)$ are given as follows:

3.2.1   Distance Measure

The properties of distance measure are:

1) It is bounded, i.e. $0 \le D(A,B) \le 1$
2) It is commutative, i.e. $D(A,B) = D(B,A)$, $\forall\ A,B \in F(X)$
3) If A,B are normalized and $A = B$, then $D(A,B) = 0$
4) For three fuzzy sets, $B,C, \forall A,B,C \in F(X)$, if $A \subset B \subset C$ then $D(A,B) \le D(A,C)$ and $D(B,C) \le D(A,C)$.

3.2.2   Similarity Measure

The properties of similarity measure are:

1) It is bounded, i.e. $0 \le S(A,B) \le 1$
2) It is commutative, i.e. $S(A,B) = S(B,A)$, $\forall A,B \in F(X)$
3) If A, B are normalized and $A = B$, then $S(A,B) = 1$
4) If $A \cap B = \emptyset$, then $S(A,B) = 0$
5) For three fuzzy sets $A,B,C, \forall A,B,C \in F(X)$, if $A \subset B \subset C$ then $S(A,B) \ge S(A,C)$ and $S(B,C) \ge S(A,C)$.

As distance measure and similarity measures are complimentary concepts, distance measure and similarity measure are related as:

$$S(A,B) = 1 - D(A,B)$$

3.3   Types of Distance Measure Between Fuzzy Sets

Consider two fuzzy sets A and B in a set X with $\mu_A(x)$ and $\mu_B(x)$ be the membership functions of sets A and B, respectively. The most common distance measures are as follows:

Hamming distance: Hamming distance between two fuzzy sets A and B is given as:

$$d(A,B) = \sum_{i=1}^{n} |\mu_A(x) - \mu_B(x)|. \tag{3.1}$$

Euclidean distance: Euclidean distance between two fuzzy sets A and B is given as:

$$d(A,B) = \sqrt{\sum_{i=1}^{n} (\mu_A(x) - \mu_B(x))^2}. \tag{3.2}$$

3.4 Types of Similarity Measures Between Fuzzy Sets

Similarity measure between two fuzzy sets A and B containing 'n' number of elements with membership functions $\mu_A(x)$ and $\mu_B(x)$ are given as follows:
The simplest similarity measure by Hyung and Song [1] is:

$$S(A,B) = \max_i \min(\mu_A(x_i), \mu_B(x_i)), \quad i = 1, 2, 3, \ldots, n.$$

Few similarity measures given by Wang [39] are as follows:

i) $$S_1(A,B) = \frac{1}{n} \sum_{i=1}^{n} \frac{\min(\mu_A(x_i), \mu_B(x_i))}{\max(\mu_A(x_i), \mu_B(x_i))}. \tag{3.3}$$

When $\mu_A(x_i) = \mu_B(x_i)$, then $S_1(A,B) = 1$, implies that the two sets are exactly similar.

ii) $$S_2(A,B) = \frac{\sum_{i=1}^{n} 1 - |\mu_A(x_i) - \mu_B(x_i)|}{n}. \tag{3.4}$$

When $\mu_A(x_i) = \mu_B(x_i)$, then $S_2(A,B) = 1$, implies that the two sets are exactly similar.

iii) $$S_3(A,B) = 1 - \frac{\sum_{i=1}^{n} |\mu_A(x_i) - \mu_B(x_i)|}{\sum_{i=1}^{n} (\mu_A(x_i) + \mu_B(x_i))}.$$

When $\mu_A(x_i) = \mu_B(x_i)$, then $S_3(A,B) = 1$, implies that the two sets are exactly similar.
Pappis and Karacapilidis [3] suggested two similarity measures:

i) $$S(A,B) = \frac{\sum_{i=1}^{n} \min(\mu_A(x_i), \mu_B(x_i))}{\sum_{i=1}^{n} \max(\mu_A(x_i), \mu_B(x_i))} \tag{3.5}$$

ii) $$S(A,B) = 1 - \max_i(|\mu_A(x_i) - \mu_B(x_i)|) \tag{3.6}$$

Till now we see that distance measure basically computes the crisp distance between two fuzzy sets or fuzzy numbers. But Voxman [2] suggested that "if we are not certain about the numbers themselves how can we be 'certain' about the distances among them." Taking this as a point of view, Voxman first introduced the distance measure between two normal numbers.
Before discussing on distance measure, an overview of generalized fuzzy number is given.

3.5 Generalized Fuzzy Number

A generalized fuzzy number $A = \{p,q,r,s; w\}$, $0 < w \leq 1$ is a fuzzy subset of real line R where p, q, r, s are real numbers such that $p \leq q \leq r \leq s$ with membership function μ such that [5]:

i) μ is continuous mapping from R to a closed interval in the interval $[0,1]$.

ii) $\mu(x) = 0$ for all $x \in [-\infty,p]$

iii) μ is increasing in the interval $[p,q]$ and decreasing in the interval $[r,s]$.

iv) $\mu(x) \in w, \forall x \in (q,r)$, w is constant and $0 < w \leq 1$.

v) $\mu(x) = 0$ for all $x \in [s,\infty]$

The above fuzzy number is a generalized trapezoidal fuzzy number. Trapezoidal fuzzy number is already discussed in Chapter 2. Similar to this, a generalized trapezoidal fuzzy number is defined as $A = (p,q,r,s; w)$, where $p \leq q \leq r \leq s$ and $0 \leq w \leq 1$ whose membership function $\mu_A(x)$ is defined as:

$$\mu_A(x) = \begin{bmatrix} 0, & x < p \\ \dfrac{w(x-p)}{q-p}, & p \leq x \leq q \\ w, & q \leq x \leq r \\ \dfrac{w(x-s)}{r-s}, & r \leq x \leq s \\ 0, & x > s, \end{bmatrix}$$

where w is called the height of the fuzzy number. It is to be noted that q, r are the shoulder of the trapezium and p, s are the base of the trapezium. If $w = 1$, then generalized trapezoidal fuzzy number becomes normal trapezoidal fuzzy number, $A = (p,q,r,s)$.

Considering $w = 1$, $\mu_A{}^L(x) = \left(\dfrac{x-p}{q-p}\right)$ and $\mu_A{}^U(x) = \left(\dfrac{x-s}{r-s}\right)$ are continuously the increasing and decreasing function in the interval $[p,q]$ and $[r,s]$, respectively. If $q = r$, then it becomes generalized triangular fuzzy number.

Figure 3.1 shows two generalized trapezoidal fuzzy number with different w values, $w = 0.7, 0.8$. The trapezoidal fuzzy numbers are represented as: $P = (0.2,0.3,0.6,0.7; 0.7)$ and $Q = (0.2,0.3,0.6,0.7; 0.9)$. Figure 3.2 shows a triangular fuzzy number, when $q = r$.

The generalized fuzzy number may be written in terms of left and right spread. It is defined as $A = (a,b,c,d; w)$.

Here, we give Chen's [6] fuzzy arithmetical operators using generalized trapezoidal fuzzy numbers.

For two fuzzy numbers, $P = (p_1,p_2,p_3,p_4; w_P)$ and $Q = (q_1,q_2,q_3,q_4; w_Q)$, where $0 \leq p_1 < p_2 < p_3 < p_4 \leq 1$, $0 \leq q_1 < q_2 < q_3 < q_4 \leq 1$ and $0 \leq w_P \leq 1$, $0 \leq w_Q \leq 1$.

Fuzzy number addition: $= (p_1,p_2,p_3,p_4; w_P) \oplus (q_1,q_2,q_3,q_4; w_Q)$

$$P \oplus Q = \left[p_1 + q_1, p_2 + q_2, p_3 + q_3, p_4 + q_4; \min(w_P, w_Q)\right]$$

Fuzzy number subtraction:

$$P \ominus Q = \left[p_1 - q_4, p_2 - q_3, p_3 - q_2, p_4 - q_1; \min(w_P, w_Q)\right]$$

Figure 3.1 Two generalized trapezoidal fuzzy numbers.

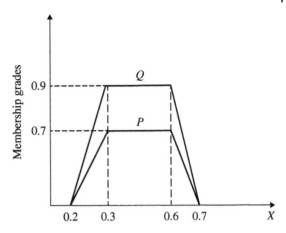

Figure 3.2 Triangular fuzzy numbers.

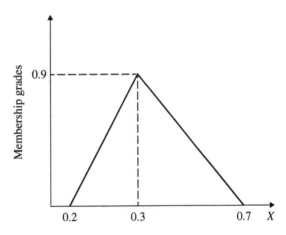

Fuzzy number multiplication:

$$P \otimes Q = \left(c, d, e, f; \min\left(w_P, w_Q\right)\right)$$

where $c = \min(p_1 \times q_1, p_1 \times q_4, p_4 \times q_1, p_4 \times q_4)$

$$d = \min\left(p_2 \times q_2, p_2 \times q_3, p_3 \times q_2, p_3 \times q_3\right)$$

$$f = \max\left(p_1 \times q_1, p_1 \times q_4, p_4 \times q_1, p_4 \times q_4\right)$$

$$e = \max\left(p_2 \times q_2, p_2 \times q_3, p_3 \times q_2, p_3 \times q_3\right)$$

If p_1, p_2, p_3, p_4, q_1, q_2, q_3, q_4 are positive real numbers, then the multiplication becomes:

$$P \otimes Q = \left[p_1 \times q_1, p_2 \times q_2, p_3 \times q_3, p_4 \times q_4; \min(w_P, w_Q) \right]$$

Fuzzy number division: $(p_1, p_2, p_3, p_4, w_P) \oslash (q_1, q_2, q_3, q_4; w_Q)$

For positive real number, $P \oslash Q = \left[\dfrac{p_1}{q_4}, \dfrac{p_2}{q_3}, \dfrac{p_3}{q_2}, \dfrac{p_4}{q_1}; \min(w_P, w_Q) \right]$

3.6 Similarity Measures Between Two Fuzzy Numbers

Yong et al. [7] suggested few similarity measures based on normal fuzzy numbers.

Assume there are two fuzzy numbers, $P = (p_1, p_2, p_3, p_4)$ and $Q = (q_1, q_2, q_3, q_4)$. The degree of similarity between two trapezoidal fuzzy numbers is:

$$S(P, Q) = \frac{\sum_{i=1}^{4} 1 - |p_i - q_i|}{4}, \ S(P, Q) \text{ lies between 0 and 1.} \tag{3.7}$$

As per Lee's [8, 9] similarity measure, similarity degree between two fuzzy numbers is:

$$S(P, Q) = 1 - \frac{\|P - Q\|_{lp}}{\|U\|} \times 4^{-1/p}, \tag{3.8}$$

where $\|P - Q\|_{lp} = \left[\sum_{i=1}^{4} (|p_i - q_i|)^p \right]^{1/p}$ and $\|U\| = \max U - \min U$, U be the universe of discourse.

Chen [10] suggested a distance measure for two trapezoidal fuzzy numbers

$$D(P, Q) = 1 - \frac{|p_1 - q_1| + |p_2 - q_2| + |p_3 - q_3| + |p_4 - q_4|}{4}. \tag{3.9}$$

For triangular numbers:

$$D(P, Q) = 1 - \frac{|p_1 - q_1| + |p_2 - q_2| + |p_3 - q_3|}{3}. \tag{3.10}$$

These measures are developed for normal fuzzy numbers and so they may not give accurate similarity measure for generalized fuzzy numbers. So, Chen and Chen [5] suggested few similarity measures based on generalized trapezoidal fuzzy numbers based on the center of gravity.

The similarity measure introduced by Chen and Chen [5, 11] considers a different type of formulation for center of gravity apart from the traditional center of gravity due to time consumption while integrating. This is because, the traditional center of gravity is given as:

$$x^* = \frac{\int x \mu_A(x) dx}{\int \mu_A(x) dx}$$ and for a trapezoidal fuzzy number with $\mu_A{}^L(x) = \left(\frac{x-p}{q-p}\right)$

and $\mu_A{}^R(x) = \left(\frac{x-s}{s-r}\right)$, the center of gravity becomes $x^* =$

$$\frac{\int x \mu_A{}^L(x) dx + \int x dx + \int x \mu_A{}^R(x) dx}{\int \mu_A{}^L(x) dx + \int dx + \int \mu_A{}^R(x) dx}$$, which is very time consuming.

Also, for crisp interval, the denominator in $\mu_A{}^L$ and $\mu_A{}^R$ becomes zero and center of gravity cannot be computed.

So, the new center of gravity method is given as follows:

Let the center of gravity of a fuzzy number, P, be $\left(x_P^*, y_P^*\right)$.

The center of gravity lies on the median curve. From Figure 3.3, we know that the center of gravity of a rectangle with base on x-axis is $w/2$ and the center of gravity of triangle is $x_P^* = (x_1 + x_2 + x_3)/3$, $y_P^* = (y_1 + y_2 + y_3)/3$. From Figure 3.3b, we see that $y_1 = y_3 = 0$, $y_2 = w$, then $y_P^* = w/3$. From Figure 3.3c, it is seen that the center of gravity of trapezium lies between $w/2$ and $w/3$.

The center of gravity of a trapezoidal fuzzy number $P = (p_1, p_2, p_3, p_4; w_P)$ denoted as $\left(x_P^*, y_P^*\right)$ is given as follows [5]:

$$y_P^* = \begin{bmatrix} \dfrac{w_P \left(\dfrac{p_3 - p_2}{p_4 - p_1} + 2\right)}{6}, & \text{if } p_1 \neq p_4 \text{ and } 0 < w_P \leq 1 \\ \dfrac{w_P}{2}, & \text{if } p_1 = p_4 \text{ and } 0 < w_P \leq 1 \end{bmatrix} \quad (3.11)$$

For finding x_P^*, the value of y_P^* is substituted in the equation of a median curve that joins at (x_5, x_5) and (x_6, x_6). Coordinate points of all the three polygons are shown in Figure 3.3.

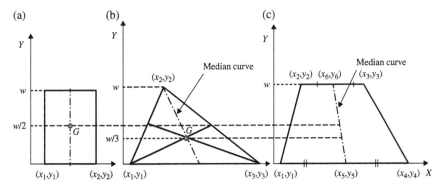

Figure 3.3 (a) Center of gravity of rectangle. (b) Center of gravity of a triangle. (c) Center of gravity of trapezium (*Source:* Adapted from [5]).

The equation of the median curve of a trapezium to compute the center of gravity, x_P^*, is:

$$\frac{y_P^* - y_5}{x_P^* - x_5} = \frac{y_6 - y_5}{x_6 - x_5}.$$

From the figure we see that $x_5 = \frac{x_4 + x_1}{2}$, $x_6 = \frac{x_2 + x_3}{2}$, $y_5 = 0$, $y_6 = w$. (Coordinate point (x_5, y_5) is the midpoint of points (x_1, y_1) and (x_4, y_4) and likewise coordinate point (x_6, y_6) is the midpoint of points (x_2, y_2) and (x_3, y_3)).

For a trapezoidal fuzzy number $P = (p_1, p_2, p_3, p_4; w_P)$, we substitute the values as: $x_1 = p_1$, $x_2 = p_2$, $x_3 = p_3$, $x_4 = p_4$, $y_6 = w_P$. So we obtain,

$$\frac{y_P^*}{x_P^* - x_5} = \frac{w_P}{x_6 - x_5} \Rightarrow x_P^* = \frac{1}{w_P}\left[y_P^*(x_6 - x_5) + w_P x_5\right],$$

$$\Rightarrow x_P^* = \frac{y_P^*(p_3 + p_2) + (p_4 + p_1)(w_P - y_P^*)}{2 w_P}. \tag{3.12}$$

A similarity measure suggested by Chen and Chen [5] for two trapezoidal fuzzy numbers, $P = (p_1, p_2, p_3, p_4)$ and $Q = (q_1, q_2, q_3, q_4)$ is:

$$S(P,Q) = \left(1 - \frac{\sum_{i=1}^{4}|p_i - q_i|}{4}\right) \times \left(1 - |x_P^* - x_Q^*|\right)^{B(S_P, S_Q)} \times \frac{\min\left(y_P^*, y_Q^*\right)}{\max\left(y_P^*, y_Q^*\right)}, \tag{3.13}$$

where S_P, S_Q are the lengths of the bases of the generalized trapezoidal fuzzy numbers which are defined as follows:

$$B(S_P, S_Q) = \begin{cases} 1, & S_P + S_Q > 0 \\ 0, & S_P + S_Q = 0 \end{cases}$$

$$S_P = p_4 - p_1,$$

$$S_Q = q_4 - q_1.$$

The larger the value of the similarity measure, the more is the similarity between the fuzzy numbers.

Wen et al. [12] suggested a similarity measure that considers the horizontal center of gravity, perimeter, height, and area of the two fuzzy numbers.

$$S(P,Q) = \left[1 - |x_P^* - x_Q^*|\right] \times \left[1 - |w_P - w_Q| \times \frac{\min(Per(P), Per(Q)) + \min(Area(P), Area(Q))}{\max(Per(P), Per(Q)) + \max(Area(P), Area(Q))}\right], \tag{3.14}$$

where x_P^*, x_Q^* are the horizontal center of gravity of trapezoidal fuzzy numbers that is given in Eqs. (3.11, 3.12). $Per(P)$ and $Per(Q)$ are the perimeters of two generalized trapezoidal fuzzy numbers which are calculated as follows using distance formula:

$$Per(P) = \sqrt{(p_1 - p_2)^2 + w_P^2} + \sqrt{(p_3 - p_4)^2 + w_P^2} + (p_3 - p_2) + (p_4 - p_1)$$

$$Per(Q) = \sqrt{(q_1 - q_2)^2 + w_Q^2} + \sqrt{(q_3 - q_4)^2 + w_Q^2} + (q_3 - q_2) + (q_4 - q_1)$$

$$(3.15)$$

$Area(P)$ and $Area(Q)$ are the areas of two generalized trapezoidal fuzzy numbers which are computed as follows:

$$Area(P) = \frac{1}{2} w_P[(p_3 - p_2) + (p_4 - p_1)],$$

$$Area(Q) = \frac{1}{2} w_Q[(q_3 - q_2) + (q_4 - q_1)]. \qquad (3.16)$$

Perimeters and area are computed using elementary mathematics.

If the similarity measure is more, then the fuzzy numbers are more similar.

A similarity measure suggested by Wei and Chen [11] using perimeter concept of generalized trapezoidal fuzzy number is given as:

$$S(P,Q) = \left(1 - \frac{\sum_{i=1}^{4} |p_i - q_i|}{4}\right) \times \frac{\min(Per(P), Per(Q)) + \min\left(w_P, w_Q\right)}{\max(Per(P), Per(Q)) + \max\left(w_P, w_Q\right)},$$

where $Per(P)$ and $Per(Q)$ are the perimeters of the two fuzzy numbers in Eq. (3.15).

Hsieh and Chen [13] introduced a similarity measure between fuzzy numbers using a concept "graded mean integration representation distance." Graded mean integration is used to measure the distance and similarity between fuzzy number with LR-type membership function. It is computed as:

$$S(P,Q) = \frac{1}{1 + d(P,Q)}, \qquad (3.17)$$

where $d(P,Q) = |m(P) - m(Q)|$, $m(P)$, $m(Q)$ are the graded means of P and Q which are computed for trapezoidal fuzzy numbers as:

$$m(P) = \frac{p_1 + 2p_2 + 2p_3 + p_4}{6} \text{ and } m(Q) = \frac{q_1 + 2q_2 + 2q_3 + q_4}{6}.$$

For triangular fuzzy numbers, $m(P) = \dfrac{p_1 + 4p_2 + p_3}{6}$ and $m(Q) = \dfrac{q_1 + 4q_2 + q_3}{6}$. Graded mean integration is computed as follows [13, 14]:

Definition 1 Let $P = (p_1, p_2, p_3, p_4; w)$ be a generalized trapezoidal fuzzy number. Let L_P and R_P be L and R functions of the fuzzy number P and $L^{-1}(h)$ and $R^{-1}(h)$ are the inverse functions of L and R of the fuzzy number P at h-level, respectively. Graded mean h-value representation of the generalized number P is: $\dfrac{h(L^{-1}(h) + R^{-1}(h))}{2}$ as shown in Figure 3.4. Graded mean integration representation of P is given as:

$$m(P) = \int_0^w \frac{h(\lambda L^{-1}(h) + (1-\lambda)R^{-1}(h))dh}{\int_0^w h\,dh}, 0 \le w \le 1 \text{ and } 0 < h < w. \quad (3.18)$$

Normally, $\lambda = 1/2$ is taken as it does not bias to left or right. With $\lambda = 1/2$, Eq. (3.18) becomes a simplified form of graded λ preference mean integration:

$$m(P) = \int_0^w \frac{h\left(\frac{L^{-1}(h) + R^{-1}(h)}{2}\right)dh}{\int_0^w h\,dh}. \quad (3.19)$$

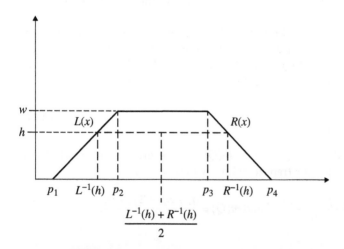

Figure 3.4 Graded mean h-level of trapezoidal fuzzy number (*Source:* Adapted from [14]).

Consider a fuzzy number $P = (p_1, p_2, p_3, p_4; w)$ with its membership function as:

$$\mu(P) = \begin{cases} \dfrac{w(x-p_1)}{p_2-p_1}, & p_1 \le x \le p_2 \\[2mm] w, & p_2 \le x \le p_3 \\[2mm] \dfrac{w(p_4-x)}{p_4-p_3}, & p_3 \le x \le p_4 \end{cases}$$

Since, $L_P(x) = L(x) = \left(\dfrac{w(x-p_1)}{p_2-p_1}\right), p_1 \le x \le p_2$ (for simplicity we assume $L_P = L$,

$R_P = R$) and $R_P(x) = R(x) = \left(\dfrac{w(p_4-x)}{p_4-p_3}\right), p_3 \le x \le p_4$, we compute at h-level,

$h = \dfrac{w(x-p_1)}{p_2-p_1} \Rightarrow x = p_1 + \dfrac{(p_2-p_1)h}{w}$ and likewise from the other one we get

$x = p_4 - \dfrac{(p_4-p_3)h}{w}$.

Thus, $L^{-1}(h) = p_1 + \dfrac{(p_2-p_1)h}{w}, 0 \le h \le w$

and $R^{-1}(h) = p_4 - \dfrac{(p_4-p_3)h}{w}, 0 \le h \le w$.

This is shown in Figure 3.4.

So, $m(P) = \displaystyle\int_0^w \dfrac{h\left(\dfrac{L^{-1}(h)+R^{-1}(h)}{2}\right)}{\int_0^w h\,dh}\,dh$

Now,

$$\dfrac{L^{-1}(h)+R^{-1}(h)}{2} = \dfrac{\left[p_1 + \dfrac{(p_2-p_1)h}{w} + p_4 - \dfrac{(p_4-p_3)h}{w}\right]}{2}$$

$$= \dfrac{p_1 + p_4 + (p_2-p_1-(p_4-p_3))h/w}{2}.$$

So,

$$m(P) = \int_0^w \dfrac{h\left(\dfrac{L^{-1}(h)+R^{-1}(h)}{2}\right)}{\int_0^w h\,dh}\,dh$$

$$= \dfrac{\displaystyle\int_0^w h\dfrac{p_1+p_4+(p_2-p_1-p_4+p_3)h/w}{2}\,dh}{\int_0^w h\,dh}$$

$$= \dfrac{\displaystyle\int_0^w h\dfrac{p_1+p_4}{2}\,dh + \int_0^w \dfrac{(p_2-p_1-p_4+p_3)h^2}{2w}\,dh}{w^2/2}$$

$$= \dfrac{\dfrac{(p_1+p_4)w^2}{4} + \dfrac{(p_2-p_1-p_4+p_3)w^3}{2w\times 3}}{w^2/2} = \dfrac{(p_1+p_4)}{2} + \dfrac{(p_2-p_1-p_4+p_3)}{3}$$

$$m(P) = \frac{3p_1 + 3p_4 + 2p_2 - 2p_1 - 2p_4 + 2p_3}{6} = \frac{p_1 + 2p_2 + 2p_3 + p_4}{6}. \tag{3.20}$$

For a generalized triangular fuzzy number $P = (p_1, p_2, p_3; w)$, which is a special case of generalized trapezoidal fuzzy number, the same definition follows when $p_2 = p_3$. Then substituting $p_2 = p_3$ in Eq. (3.20), the graded mean integration representation of triangular fuzzy number P is:

$$m(P) = \frac{p_1 + 2p_2 + 2p_3 + p_4}{6} = \frac{p_1 + 4p_2 + p_3}{6}.$$

Xu et al. [15] introduced another similarity measure:

$$S(P,Q) = 1 - \frac{\sum_{i=1}^{4} |p_i - q_i|}{8} - \frac{d(P,Q)}{2}, \tag{3.21}$$

where $d(P,Q) = \dfrac{\sqrt{\left(x^*_P{}^2 - x^*_Q{}^2\right)} + \sqrt{\left(y^*_P{}^2 - y^*_Q{}^2\right)}}{\sqrt{1.25}}.$

Few similarity measures are defined using inclusion measure.

3.7 Inclusion Measure

Zadeh extended the classical set inclusion to fuzzy set inclusion. Inclusion measure of fuzzy sets indicates the degree to which a fuzzy set A is contained in another fuzzy set B. Using the definitions of the inclusion measure and the similarity measure of fuzzy sets, Zeng and Li [16] discussed the relation between the inclusion measure and the similarity measure of fuzzy sets.

Definition 2 For two fuzzy sets, A and B such that $A, B \in F(X)$, $I(A,B)$ is an inclusion if the mapping $I : F(X) \times F(X) \to [0,1]$ satisfies the following properties:

i) $I(X, \emptyset) = 0$
ii) $I(A,B) = 1 \Longrightarrow A \subseteq B$
iii) For all $A, B, C \in F(X)$, if $A \subset B \subset C$ then $I(C,A) \leq I(B,A)$ and $I(C,A) \leq I(C,B)$.

The inclusion measure of two fuzzy sets A and B for a finite set $X = \{x_1, x_2, x_3, \dots, x_n\}$ is defined as:

$$I(A,B) = \begin{cases} \dfrac{\sum_{i=1}^{n} A(x_i) \wedge B(x_i)}{\sum_{i=1}^{n} A(x_i)}, & A \neq \emptyset \\ 1, & A = \emptyset \end{cases} \tag{3.22}$$

Then, similarity measure using the inclusion measure is given as: $S(A,B) = I(A,B) \wedge I(B,A)$

$$S(A,B) = \frac{\sum_{i=1}^{n} A(x_i) \wedge B(x_i)}{\sum_{i=1}^{n} A(x_i)} \wedge \frac{\sum_{i=1}^{n} A(x_i) \wedge B(x_i)}{\sum_{i=1}^{n} B(x_i)}$$

$$= \frac{\sum_{i=1}^{n} (A(x_i) \wedge B(x_i))}{\sum_{i=1}^{n} A(x_i) \vee \sum_{i=1}^{n} B(x_i)}. \tag{3.23}$$

3.8 Measures of Fuzziness

The measure of fuzziness provides a way to measure the fuzziness of a fuzzy set. There are many definitions that describe the measures of fuzziness in a different manner. Yager [17] suggested the measure of fuzziness as a distance between fuzzy set and its complement. Kaufmann [18] suggested a measure known as indices of fuzziness which is represented by a metric distance between its membership function and the membership function of its nearest crisp set. De Luca and Termini [19] used Shannon's entropy as a measure of fuzziness.

Three properties of the measures of fuzziness (I) were given by De Luca and Termini [19]:

1) $I(A) = 0$ or minimum iff $\mu_A(x_i) = 0$ or 1.
2) $I(A) = 1$ or maximum iff $\mu_A(x_i) = 0.5$.
3) $I(A) \geq I(B)$, if B is sharper (crisper) than A, i.e.
   $\mu_B(x_i) \leq \mu_A(x_i)$ if $\mu_A(x_i) \leq 0.5$ and
   $\mu_B(x_i) \geq \mu_A(x_i)$ if $\mu_A(x_i) \geq 0.5$.

3.8.1 Index of Fuzziness

It was introduced by Kaufmann that denotes the degree of ambiguity or fuzziness present in a set by measuring the distance between the membership values of the elements of the fuzzy set A and its nearest ordinary set $\tilde{A}$. The index of fuzziness is defined as: $I(A) = \left(\dfrac{2}{n^k}\right) \cdot d(A,\tilde{A})$, where $d(A,\tilde{A})$ denotes the distance between fuzzy set A and its nearest ordinary set $\tilde{A}$.

An ordinary set $\tilde{A}$ nearest to the fuzzy set A is defined as

$$\mu_{\tilde{A}}(x_i) = \begin{cases} 0 & \text{if } \mu_A(x_i) < 0.5 \\ 1 & \text{if } \mu_A(x_i) \geq 0.5 \end{cases}$$

There are two types of index of fuzziness – linear index of fuzziness and quadratic index of fuzziness. If $k = 1$, the distance 'd' becomes Hamming distance. Then, the linear index of fuzziness is defined as [18]:

$$I(A) = \frac{2}{n}\sum_{i=1}^{n}\left|\mu_A(x_i) - \mu_{\bar{A}}(x_i)\right| = \frac{2}{n}\sum_{i=1}^{n}\mu_{A\cap\bar{A}}(x_i)$$

$$= \frac{2}{n}\sum_{i=1}^{n}\min(\mu_A(x_i), (1-\mu_A(x_i))). \tag{3.24}$$

$\mu_{A\cap\bar{A}}(x_i)$ is the intersection or common values of the membership degree of the element x_i at ith point in the fuzzy set A and its complement, n is the number of elements in a fuzzy set.

If $k = 0.5$, then 'd' becomes Euclidean distance. Then, the quadratic index of fuzziness is defined as:

$$I(A) = \frac{2}{\sqrt{n}}\left[\sum_{i=1}^{n}\left(\mu_A(x_i) - (1-\mu_A(x_i))\right)^2\right]^{1/2} \tag{3.25}$$

3.8.2 Yager's Measure

Yager suggested that the measure of fuzziness depends on the relationship between fuzzy set and its complement. It is a measure of lack of distinction between a fuzzy set A and its complement $\bar{A}$. The distance between a fuzzy set A and its complement $\bar{A}$ is defined as:

$$D(A,\bar{A}) = \left[\sum_{i=1}^{n}\left|\mu_A(x_i) - \mu_{\bar{A}}(x_i)\right|^p\right]^{1/p}, i = 1, 2, 3, \ldots, n,$$

where $\mu_{\bar{A}}(x_i) = 1 - \mu_A(x_i)$.

Yager's measure of fuzziness is defined as : $m_Y(A) = 1 - \frac{D(A,\bar{A})}{A^{\frac{1}{p}}}$, $\tag{3.26}$

$p = 1$ for Hamming distance and $p = 2$ for Euclidean distance.

3.8.3 Fuzzy Entropy

Fuzzy entropy is a measure of information present in a fuzzy set. It is analogous to the classical Shannon's entropy in information theory, but with a slight variation. Zadeh [20] introduced fuzzy entropy as a measure of fuzzy information based on Shannon's entropy. In Shannon's entropy, there is a concept of probabilistic theory and it contains randomness uncertainty where as in fuzzy entropy, there is a concept of ambiguity and vagueness uncertainty. Fuzzy entropy is defined using the concept of membership function. For a finite set, $X = \{x_1, x_2, x_3, \dots, x_n\}$, the entropy of a fuzzy set A is defined as:

$$E(A) = \frac{1}{n \ln(2)} \sum_i [-\mu_A(x_i) \ln(\mu_A(x_i)) - (1 - \mu_A(x_i)) \ln(1 - \mu_A(x_i))],$$

(3.27)

where $i = 1, 2, 3, \dots, n$ and $\mu_A(x_i)$ is the membership degree of element x_i in A.

De Luca and Termini [19] described fuzzy entropy along with its properties. For a fuzzy entropy $E(X)$ of set A, the following properties hold:

i) $0 \le E(A) \le 1$.
ii) $E(A)$ is minimum if $\mu_A(x_{mn}) = 0$ or 1 for all m, n, i.e. there is no fuzziness.
iii) $E(A) = 1$ when $\mu_A(x_{mn}) = 0$ for all m, n, i.e. when there is maximum fuzziness.
iv) $E(A) \ge E(A^*)$, where $E(A^*)$ is the entropy of A^*, which is sharper or crisper than A.
   A^* is a sharpened version of A implies
   $\mu_{A^*}(x_i) \le \mu_A(x_i)$ if $\mu_A(x_i) \le 0.5$ and
   $\mu_{A^*}(x_i) \ge \mu_A(x_i)$ if $\mu_A(x_i) \ge 0.5$.
v) $E(A) = E(\bar{A})$, $\bar{A}$ is the complement of A.

Pal and Pal [21] introduced the concept of an exponential entropy.

$$E(A) = \frac{1}{n\sqrt{e} - 1} \left[\mu(x_i) \cdot e^{1 - \mu(x_i)} + (1 - \mu(x_i)) \cdot e^{\mu(x_i)} - 1 \right].$$

(3.28)

Zeng and Li [16] suggested a measure of fuzziness using inclusion measure. For a fuzzy set A, if $m(A)(x) = \dfrac{1 + |A(x) - A^c(x)|}{2}$, $n(A)(x) = \dfrac{1 - |A(x) - A^c(x)|}{2}$, then from the equation of inclusion measure in Eq. (3.22), $I(A,B) = \dfrac{\sum_{i=1}^{n} A(x_i) \wedge B(x_i)}{\sum_{i=1}^{n} A(x_i)}$, fuzziness of a fuzzy set is given as:

$$
\begin{aligned}
I(m(A), n(A)) &= \frac{\sum_{i=1}^{n} m(A)(x_i) \wedge n(A)(x_i)}{\sum_{i=1}^{n} m(A)(x_i)} = \frac{\sum_{i=1}^{n} n(A)(x_i)}{\sum_{i=1}^{n} m(A)(x_i)} \\
&= \frac{\sum_{i=1}^{n} 1 - |A(x_i) - A^c(x_i)|}{\sum_{i=1}^{n} 1 + |A(x_i) - A^c(x_i)|} \\
&= \frac{n - \sum_{i=1}^{n} |A(x_i) - A^c(x_i)|}{n + \sum_{i=1}^{n} |A(x_i) - A^c(x_i)|}.
\end{aligned}
$$

(3.29)

3.9 Intuitionistic Fuzzy Distance and Similarity Measures

Just like fuzzy sets, similarity and distance measures are of high importance using intuitionistic fuzzy sets (IFSs) introduced by Atanassov in 1965 [22]. They have been successfully implemented in various fields in image processing, pattern recognition, decision making, and so on. Many researchers have conducted extensive studies on similarity and distance measures between IFSs. The similarity measures are useful in real-time application where more number of uncertainties, membership, and nonmembership functions are involved.

If A and B are two IFSs and the sets are similar if $A = B$ if for all x, $\mu_A(x_i) = \mu_B(x_i)$ and $\nu_A(x_i) = \nu_B(x_i)$.

Li and Cheng [24] and Mitchell [23] suggested few definitions of similarity measures and distance measures between two IFSs. For two IFSs, A and B such that $A, B \in F(X)$, the properties of similarity measure between two IFSs A and B, $S(A, B)$ are given as follows:

i) Similarity measure $S(A,B)$ is bounded, i.e. $0 \leq S(A,B) \leq 1$,
ii) It is commutative, i.e. $S(A,B) = S(B,A), \forall A,B \in F(X)$,
iii) If A,B are normalized and $A = B$, then $S(A,B) = 1$,
iv) For three fuzzy sets $A,B,C, \forall A,B,C \in F(X)$, if $A \subset B \subset C$ then $S(A,B) \geq S(A,C)$ and $S(B,C) \geq S(A,C)$.

Wang and Xin [25] suggested a definition of distance measure $D(A, B)$ as follows:

1) It is bounded, i.e. $0 \leq D(A,B) \leq 1$,
2) It is commutative, i.e. $D(A,B) = D(B,A), \forall A,B \in F(X)$,
3) If A, B are normalized and $A = B$, then $D(A,B) = 0$,
4) For three fuzzy sets, $B,C, \forall A,B,C \in F(X)$, if $A \subset B \subset C$ then $D(A,B) \leq D(A,C)$ and $D(B,C) \leq D(A,C)$.

The most widely used distance measures are:

Let A and B be two IFSs where $A = \{(x, \mu_A(x), \nu_A(x)) \mid x \in X\}$ and $B = \{(x, \mu_B(x), \nu_B(x)) \mid x \in X\}$, $\mu_A(x)$ and $\nu_A(x)$ are the membership and nonmembership functions of sets A and B, respectively, and $\pi_A(x) = 1 - \mu_A(x) - \nu_A(x)$ and $\pi_B(x) = 1 - \mu_B(x) - \nu_B(x)$ are the hesitation degrees of A and B. Taking into account the intuitionistic-type representation of fuzzy set and normlization, Szmidt and Kacprzyk [26] introduced some distance measures between two IFSs which are given as:

Intuitionistic Euclidean Distance:

$$D_{IFS}(A,B) = \sqrt{\left(\frac{1}{2}\right)\left(\sum_{i=1}^{n}(\mu_A(x_i) - \mu_B(x_i))^2 + (\nu_A(x_i) - \nu_B(x_i))^2 + (\pi_A(x_i) - \pi_B(x_i))^2\right)}.$$

(3.30)

Intuitionistic Normalized Euclidean Distance:

$$D_{IFS}(A,B) = \sqrt{\left(\frac{1}{2n}\right)\left(\sum_{i=1}^{n}(\mu_A(x_i)-\mu_B(x_i))^2 + (\nu_A(x_i)-\nu_B(x_i))^2 + (\pi_A(x_i)-\pi_B(x_i))^2\right)}.$$

$$(3.31)$$

Intuitionistic Hamming Distance:

$$D_{IFS}(A,B) = \left(\frac{1}{2}\right)\sum_{i=1}^{n}(|\mu_A(x_i)-\mu_A(x_i)| + |\nu_A(x_i)-\nu_B(x_i)| + |\pi_A(x_i)-\pi_B(x_i)|).$$

$$(3.32)$$

Intuitionistic Normalized Hamming Distance:

$$D_{IFS}(A,B) = \left(\frac{1}{2n}\right)\sum_{i=1}^{n}(|\mu_A(x_i)-\mu_B(x_i)| + |\nu_A(x_i)-\nu_B(x_i)| + |\pi_A(x_i)-\pi_B(x_i)|).$$

$$(3.33)$$

Xu [27] suggested generalized normalized distance measure as:

$$D_{IFS}(A,B) = \left(\frac{1}{2n}\right)\sum_{i=1}^{n}[|\mu_A(x_i)-\mu_B(x_i)|^{\alpha} + |\nu_A(x_i)-\nu_B(x_i)|^{\alpha} + |\pi_A(x_i)-\pi_B(x_i)|^{\alpha}]^{\frac{1}{\alpha}},$$

$\alpha = 1,2$ for Hamming distance and Euclidean distance.

$$(3.34)$$

If w_i be the weight of element x_i, then weighted distance measure becomes:

$$D_{IFS}(A,B) = \left(\frac{1}{2n}\right)\sum_{i=1}^{n}[w_i(|\mu_A(x_i)-\mu_B(x_i)|^{\alpha} + |\nu_A(x_i)-\nu_B(x_i)|^{\alpha}$$

$$+ |\pi_A(x_i)-\pi_B(x_i)|^{\alpha})]^{1/\alpha}, \qquad 0 \le w_i \le 1 \text{ and } \sum_{i=1}^{n}w = 1.$$

$$(3.35)$$

Dengfeng and Chuntian [28] suggested two types of similarity measures

i) $S_{IFS}(A,B) = 1 - \dfrac{1}{\sqrt[p]{n}}\sqrt[p]{\sum_{i=1}^{n}|\varphi_A(x_i)-\varphi_B(x_i)|^p}, \quad 1 \le p \le \infty.$ $\qquad(3.36)$

ii) $S_{IFS}(A,B) = 1 - \sqrt[p]{\sum_{i=1}^{n}w_i|\varphi_A(x_i)-\varphi_B(x_i)|^p}, \quad 0 \le w_i \le 1 \quad \text{and} \quad \sum_{i=1}^{n}w = 1,$

$$(3.37)$$

where $\varphi_A(x_i) = \dfrac{\mu_A(x_i) + 1 - \nu_A(x_i)}{2}$ and $\varphi_B(x_i) = \dfrac{\mu_B(x_i) + 1 - \nu_B(x_i)}{2}$.

iii) Hung and Wang [29] defined some similarity measures that are the extensions of fuzzy sets which are given as:

i) $S_{IFS}(A,B) = 1 - \dfrac{1}{2}\left(\max|\mu_A(x_i) - \mu_B(x_i)| + \max|\nu_A(x_i) - \nu_B(x_i)|\right)$ (3.38)

ii) $S_{IFS}(A,B) = 1 - \dfrac{\sum_{i=1}^{n}|\mu_A(x_i) - \mu_B(x_i)| + |\nu_A(x_i) - \nu_B(x_i)|}{\sum_{i=1}^{n}|\mu_A(x_i) + \mu_B(x_i)| + |\nu_A(x_i) + \nu_B(x_i)|}$ (3.39)

iii) $S_{IFS}(A,B) = \dfrac{1}{n}\dfrac{\sum_{i=1}^{n}\min(\mu_A(x_i),\mu_B(x_i)) + \min(\nu_A(x_i),\nu_B(x_i))}{\sum_{i=1}^{n}\max(\mu_A(x_i),\mu_B(x_i)) + \max(\nu_A(x_i),\nu_B(x_i))}$ (3.40)

iv) $S_{IFS}(A,B) = \dfrac{1}{n}\sum_{i=1}^{n}\left(1 - \dfrac{1}{2}(|\mu_A(x_i) - \mu_B(x_i)| + |\nu_A(x_i) - \nu_B(x_i)|)\right)$ (3.41)

Liang and Shi [30] defined two similarity measures between two IFSs with intervals $[\mu_A(x_i), 1 - \nu_A(x_i)]$ and $[\mu_B(x_i), 1 - \nu_B(x_i)]$. The similarity measure is written as:

i) $S_{IFS}(A, B) = 1 - \dfrac{1}{\sqrt[p]{n}}\sqrt[p]{\sum_{i=1}^{n}(\varphi_A(x_i) + \varphi_B(x_i))^p}, 1 \le p \le \infty$ (3.42)

where $\varphi_A(x_i) = \dfrac{|\mu_A(x_i) - \mu_B(x_i)|}{2}$ and $\varphi_B(x_i) = \left|\dfrac{1 - \nu_A(x_i)}{2} + \dfrac{1 - \nu_B(x_i)}{2}\right|$.

ii) For IFSs A and B with intervals $[\mu_A(x_i), 1 - \nu_A(x_i)]$ and $[\mu_B(x_i), 1 - \nu_B(x_i)]$, the similarity between two IFSs is given as:

$$S_{IFS}(A,B) = 1 - \dfrac{1}{\sqrt[p]{n}}\sqrt[p]{\sum_{i=1}^{n}(\varphi_A(x_i) + \varphi_B(x_i))^p},$$ (3.43)

where $\varphi_A(x_i) = \dfrac{|m_{A1}(x_i) - m_{B1}(x_i)|}{2}$ and $\varphi_B(x_i) = \dfrac{|m_{A2}(x_i) - m_{B2}(x_i)|}{2}$ and $m_A(x_i)$ is the median value of the interval of the fuzzy set A, which is given as:

$$m_A(x_i) = \dfrac{\mu_A(x_i) + 1 - \nu_A(x_i)}{2}.$$

With this median value, the interval, $[\mu_A(x_i), 1 - \nu_A(x_i)]$, is divided into two subintervals:

$$[\mu_A(x_i), m_A(x_i)], [m_A(x_i), 1 - \nu_A(x_i)].$$

Median values of two subintervals of the fuzzy set A are computed as:

$$m_{A1}(x_i) = \dfrac{\mu_A(x_i) + m_A(x_i)}{2} \text{ and } m_{A2}(x_i) = \dfrac{m_A(x_i) + (1 - \nu_A(x_i))}{2}.$$

Likewise, the median value of the interval of the fuzzy set B, $[\mu_B(x_i), 1 - \nu_B(x_i)]$, is:

$m_B(x_i) = \dfrac{\mu_B(x_i) + 1 - \nu_B(x_i)}{2}$ and the two subintervals are $[\mu_B(x_i),\ m_B(x_i)]$, $[m_B(x_i),\ 1 - \nu_B(x_i)]$.

Then, the median values of the two subintervals of the fuzzy set B is:

$$m_{B1}(x_i) = \frac{\mu_B(x_i) + m_B(x_i)}{2} \quad \text{and} \quad m_{B2}(x_i) = \frac{m_B(x_i) + (1 - \nu_B(x_i))}{2}.$$

Wang and Xin [25] suggested a distance measure between two IFSs which is given as:

$d(A,B)$

$$= \frac{1}{n}\sum_{i=1}^{n}\left[\frac{|\mu_A(x_i) - \mu_B(x_i)|}{4} + \frac{|\mu_A(x_i) - \mu_B(x_i)|}{4} + \frac{\max\{|\mu_A(x_i) - \mu_B(x_i)|, |\nu_A(x_i) - \nu_B(x_i)|\}}{2}\right].$$

Song [31] suggested that the weights of $|\mu_A(x_i) - \mu_B(x_i)|$, $|\mu_A(x_i) - \mu_B(x_i)|$, max $\{|\mu_A(x_i) - \mu_B(x_i)|, |\nu_A(x_i) - \nu_B(x_i)|\}$ are set at $\frac{1}{4}, \frac{1}{4}, \frac{1}{2}$, respectively. But in practice, fixed weights are not convenient. So, a new type of similarity measure is suggested where a different type of weights are used. If $\alpha, \beta, \gamma \in [0,1]$ and $\alpha + \beta + \gamma = 1$, then

$d(A,B)$

$$= \sum_{i=1}^{n}[\alpha|\mu_A(x_i) - \mu_B(x_i)| + \beta|\nu_A(x_i) - \nu_B(x_i)| + \gamma \cdot \max\{|\mu_A(x_i) - \mu_B(x_i)|, |\nu_A(x_i) - \nu_B(x_i)|\}]$$

(3.44)

The decision makers can change the parameters according to their requirements.

In Eq. (3.44) any element x_i in $X = \{x_1, x_2, x_3, \ldots, x_n\}$ has the same weight parameters α, β, γ. In many applications, there may be some important elements that has an effect on the distance measure. In that situation, weights are taken into account, $w_i > 0$, $i = 1, 2, 3, \ldots, n$ along with $\alpha, \beta, \gamma \in [0,1]$ and $\alpha + \beta + \gamma = 1$. The new distance measure is given as:

$$d(A,B) = \sum_{i=1}^{n}\frac{w_i}{\sum_{i=1}^{n}w_i}$$

$$[\alpha|\mu_A(x_i) - \mu_B(x_i)| + \beta|\nu_A(x_i) - \nu_B(x_i)| + \gamma \cdot \max\{|\mu_A(x_i) - \mu_B(x_i)|, |\nu_A(x_i) - \nu_B(x_i)|\}]$$

(3.45)

Zhang and Yu [32] suggested a new similarity measure using IFS. Consider two IFSs, $P = [\mu_P(x_i), \nu_P(x_i)]$, $Q = [\mu_Q(x_i), \nu_Q(x_i)]$. For simplicity, the two IFSs P and Q are translated to interval-valued fuzzy sets with intervals $[\mu_P(x_i), 1 - \nu_P(x_i)]$ and $[\mu_Q(x_i), 1 - \nu_Q(x_i)]$. First, IFSs are transformed into symmetric triangular fuzzy numbers as: $\tilde{P} = [\mu_P(x_i), m_P, 1 - \nu_P(x_i)]$ and $\tilde{Q} = [\mu_Q(x_i), m_Q, 1 - \nu_Q(x_i)]$ where $m_P = \dfrac{\mu_P(x_i) + 1 - \nu_P(x_i)}{2}$ and $m_Q = \dfrac{\mu_Q(x_i) + 1 - \nu_Q(x_i)}{2}$.

Considering $m_P \le m_Q$, the membership function $\mu_{\tilde{P}}(x)$ of the symmetric triangular fuzzy number is given as:

$$\mu_{\tilde{P}}(x) = \begin{cases} \dfrac{x - \mu_P}{m_P - \mu_P}, & \mu_P \le x \le m_P \\ \dfrac{1 - \upsilon_P - x}{1 - \upsilon_P - m_P}, & m_P \le x \le 1 - \upsilon_P \\ 0 & \text{otherwise} \end{cases}$$

Likewise, for set Q, the membership function $\mu_{\tilde{Q}}(x)$ of the symmetric triangular fuzzy number is:

$$\mu_{\tilde{Q}}(x) = \begin{cases} \dfrac{x - \mu_Q}{m_Q - \mu_Q}, & \mu_Q \le x \le m_P \\ \dfrac{1 - \upsilon_Q - x}{1 - \upsilon_Q - m_Q}, & m_Q \le x \le 1 - \upsilon_Q \\ 0 & \text{otherwise} \end{cases}$$

Consider two figures, Figures 3.5 and 3.6. Two symmetrical triangular fuzzy numbers are shown where there is no overlap and there is an overlap between the fuzzy numbers.

The distance between two fuzzy numbers is:

$$d = A_U - A_I,$$

$$\text{where } A_I = \int_0^1 \min(\mu_{\tilde{P}}(x), \mu_{\tilde{Q}}(x))dx \text{ and} \tag{3.46}$$

$$A_U = \int_0^{m_P} \max(\mu_{\tilde{P}}(x), \mu_{\tilde{Q}}(x))dx + |m_Q - m_P| + \int_{m_Q}^1 \max(\mu_{\tilde{P}}(x), \mu_{\tilde{Q}}(x))dx. \tag{3.47}$$

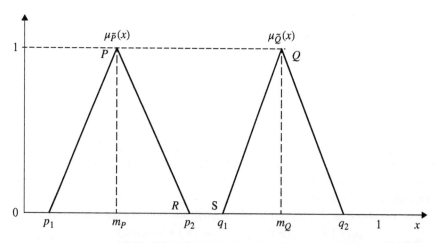

Figure 3.5 $p_2 \le q_1$ (*Source:* Adapted from [32]).

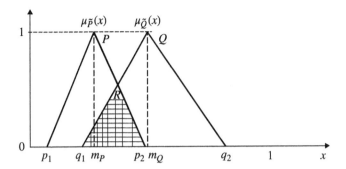

Figure 3.6 $p_1 \leq q_1 \leq p_2 \leq q_2$ (*Source*: Adapted from [32]).

These are actually the areas covered by the two fuzzy numbers.

Equations (3.46) and (3.47) show the area under the curve. Equation (3.46) is the area of intersection of the two triangular fuzzy numbers, i.e. the area of overlap. In Figure 3.5, as there is no intersection, the area is zero, i.e. $A_I = 0$ and for Figure 3.6, it is the area of overlap shown as shaded region. Equation (3.47) is the total area covered by the two regions – one region is enclosed by the horizontal axis and the outermost lines of the fuzzy number $\tilde{P}$ and $\tilde{Q}$, and the other region is the area between $\tilde{P}$ and $\tilde{Q}$, i.e. the trapezium PQRS for Figure 3.5.

This method assumes that if different points on the intervals of interval-valued fuzzy set are taken, then different amounts of information for the distance measure are obtained, i.e. a point that is closer to the midpoint provides more information.

Consider $\tilde{P} = [\mu_P(x_i), m_P, 1 - \nu_P(x_i)]$ and $\tilde{Q} = [\mu_Q(x_i), m_Q, 1 - \nu_Q(x_i)]$ as $\tilde{P} = [p_1, m_P, p_2]$ and $\tilde{Q} = [q_1, m_Q, q_2]$, respectively.

Distance measures for different conditions (position of the points) are computed, assuming $m_P \leq m_Q$:

Case i: $p_2 \leq q_1$ (Figure 3.5)

Then, A_U can be computed directly from the area under the curve, i.e.

$$A_U = \frac{1}{2}(m_P - p_1) + m_Q - m_P + \frac{1}{2}(q_2 - m_Q) = \frac{1}{2}(m_P - m_Q + q_2 - p_1) + m_Q - m_P$$

$$= \frac{m_P - m_Q + q_2 - p_1 + 2m_Q - 2m_P}{2} = \frac{m_Q - m_P + q_2 - p_1}{2}$$

and $A_I = O$.

Then, the distance

$$d = A_U - A_I = \frac{m_Q - m_P + q_2 - p_1}{2}. \tag{3.48}$$

Case ii: When $p_1 \le q_1 \le p_2 \le q_2$ (Figure 3.6)

Here, an intersecting region is present and so to find the intersection point, we will find the ordinate of point O. From the geometry of intersection of two straight lines, Pp_2 and Qq_1, we get the ordinate value of point $R = \dfrac{2(p_2 - q_1)}{(p_2 - p_1 + q_2 - q_1)}$. Using $p_1 = \mu_P(x_i)$, $p_2 = 1 - \nu_P(x_i)$, $q_1 = \mu_Q(x_i)$, $q_2 = 1 - \nu_Q(x_i)$ and the hesitation degree as $\pi_P(x_i) = 1 - \mu_P(x_i) - \nu_P(x_i)$, then the ordinate value of point $R = \dfrac{2(p_2 - q_1)}{\pi_P + \pi_Q}$, considering $\pi_P(x_i) = \pi_P$ and $\pi_Q(x_i) = \pi_Q$ for simplicity.

It is to be noted that in computation, the values of m_P and m_Q are written in terms of p_1, p_2, q_1, q_2. So, the value of $A_I = \frac{1}{2}(p_2 - q_1)\dfrac{2(p_2 - q_1)}{\pi_P + \pi_Q} = \dfrac{(p_2 - q_1)^2}{\pi_P + \pi_Q}$ and $A_U = \dfrac{m_Q - m_P + q_2 - p_1}{2}$.

Then, the distance,

$$d = A_U - A_I = \frac{m_Q - m_P + q_2 - p_1}{2} - \frac{(p_2 - q_1)^2}{\pi_P + \pi_Q}. \tag{3.49}$$

Case iii: Likewise, for the other cases, i.e. ($p_1 \ge q_1$ and $p_2 \le q_2$) or $p_1 \le q_1$ and $p_2 \ge q_2$ (Figure 3.7), A_U and A_I are computed based on the area covered.

So, A_U = area under $q_1 Q q_2$ + area under PQR.

A_I = area under SRp_1p_2 = area under Pp_2p_1 – area under PSR = area under Pp_2p_1 – area under $(PRQ - SPQ)$.

Ordinate values of S and R are required and these are obtained from geometry of straight lines.

After computation, we get:

$$d = \frac{|\pi_P - \pi_Q|}{2} + (m_Q - m_P)^2 \left[\frac{2}{|\pi_P - \pi_Q|} - \frac{1}{\pi_P + \pi_Q} \right]. \tag{3.50}$$

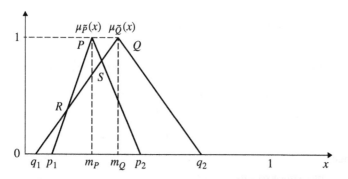

Figure 3.7 $p_1 \ge q_1$ and $p_2 \le q_2$ (*Source:* Adapted from [32]).

Cosine similarity measure between two fuzzy sets given by Bhattacharya [33] is defined as the inner product of two vectors divided by the product of their lengths or said in another way it is the cosine of the angle between the vector representations of two fuzzy sets. Similar to the fuzzy set, Ye [34] introduced cosine similarity measure between two IFSs, which is given by:

$$S(A,B) = \frac{1}{n}\sum_{i=1}^{n} \frac{\mu_A(x_i) \cdot \mu_B(x_i) + \nu_A(x_i) \cdot \nu_B(x_i)}{\sqrt{\mu_A(x_i)^2 \mu_B(x_i)^2} \sqrt{\nu_A(x_i)^2 \nu_B(x_i)^2}}. \tag{3.51}$$

If weights are considered, then weighted cosine similarity measure is given as:

$$S(A,B) = \frac{1}{n}\sum_{i=1}^{n} w_i \frac{\mu_A(x_i) \cdot \mu_B(x_i) + \nu_A(x_i) \cdot \nu_B(x_i)}{\sqrt{\mu_A(x_i)^2 \mu_B(x_i)^2} \sqrt{\nu_A(x_i)^2 \nu_B(x_i)^2}}, \tag{3.52}$$

where $\sum_{i=1}^{n} w_i = 1$, $w_i \in [0,1]$ and $i = 1, 2, 3, \ldots, n$

3.10 Intuitionistic Fuzzy Entropy

Intuitionistic fuzzy entropy (IFE) gives an amount of vagueness or ambiguity in a set. Many authors defined IFE in a different manner. Two definitions of entropy of IFS were defined by Burillo and Bustince [35] and Szmidt and Kacpryzk [36]. Burillo and Bustince defined entropy for the first time in terms of the degree of intuitionism of an IFS. Szmidt and Kacpryzk defined entropy in terms of non-probabilistic type of entropy. The properties of IFE given by Burillo and Bustince [35] are:
A real function $IFE = IFSs(X) \rightarrow [0,1]$ is called IFE on $IFSs(X)$ if:

i) $IFE(A) = 0, \forall A \in FS(X)$,
ii) $IFE(A) =$ cardinal $(X) = n$, iff $\mu_A(x_i) = \nu_A(x_i) = 0, \forall x_i$, i.e. entropy is maximum if the set is totally intuitionistic,
iii) $IFE(A) \leq IFE(B)$ if $\mu_A(x_i) \geq \mu_B(x_i)$, $\nu_A(x_i) \geq \nu_B(x_i)$. This implies $\mu_A(x_i) + \nu_A(x_i) \geq \mu_B(x_i) + \nu_B(x_i) \Longrightarrow 1 - (\mu_A(x_i) + \nu_A(x_i)) \leq 1 - (\mu_B(x_i) + \nu_B(x_i)) \Longrightarrow \pi_A(x_i) \leq \pi_B(x_i) \Longrightarrow IFE(A) \leq IFE(B)$,
iv) $IFE(A) = IFE(A^C)$.

They defined the entropy as:

$$IFE(A) = \sum_{i=1}^{n} \pi_A(x_i).$$

Szmidt and Kacpryzk [36] defined IFE in a different way.
A real function $IFE = IFSs(X) \rightarrow [0,1]$ is called an IFE on $IFSs(X)$ if

1) $IFE(A) = 0$ if A is a crisp set, i.e. $\mu_A(x_i) = 0$ or $\mu_A(x_i) = 1$ for all $x_i \in X$,
2) $IFE(A) = 1$, if $\mu_A(x_i) = \nu_A(x_i)$ for all $x_i \in X$,
3) $IFE(A) \leq IFE(B)$ if A is less fuzzy than B, i.e. $\mu_A(x_i) \leq \mu_B(x_i)$ and $\nu_A(x_i) \geq \nu_B(x_i)$, for $\mu_B(x_i) \leq \nu_B(x_i)$, or $\mu_A(x_i) \geq \mu_B(x_i)$ and $\nu_A(x_i) \leq \nu_B(x_i)$, for $\mu_B(x_i) \geq \nu_B(x_i)$ for any $x_i \in X$.
4) $IFE(A) = IFE(A^C)$.

3.11 Different Types of Intuitionistic Fuzzy Entropies

i) Entropy by Chaira [37]: For a probability distribution, $p = p_1, p_2, \ldots, p_n$, the exponential entropy is defined as: $H = \sum_{i=1}^{n} p_i \cdot e^{(1-p_i)}$.

If $\mu_A(x_i)$, $\nu_A(x_i)$, $\pi_A(x_i)$ are the membership, nonmembership, and hesitation degrees of the elements of the set $X = \{x_1, x_2, \ldots, x_n\}$, then IFE that denotes the degree of intuitionism in an IFS is given as:

$$IFE(A) = \sum_{i=1}^{n} \pi_A(x_i) \cdot e^{[1-\pi_A(x_i)]} \tag{3.53}$$

where $\pi_A(x_i) = 1 - \mu_A(x_i) - \nu_A(x_i)$.

ii) Szmidt and Kacpryzk formulated a similar type of entropy which is given as:

$$IFE(A) = \frac{1}{n}\sum_{i=1}^{n} \frac{\min\{\mu_A(x_i), \nu_A(x_i)\} + \pi_A(x_i)}{\max\{\mu_A(x_i), \nu_A(x_i)\} + \pi_A(x_i)}. \tag{3.54}$$

iii) Huang and Liu [38] suggested an entropy using vague sets and later he extended to IFSs, which is defined as:

$$IFE(A) = \frac{1}{n}\sum_{i=1}^{n} \frac{1 - |\mu_A(x_i) - \nu_A(x_i)| + \pi_A(x_i)}{1 + |\mu_A(x_i) - \nu_A(x_i)| + \pi_A(x_i)}. \tag{3.55}$$

iv) Vlachos and Sergiadis [4] introduced an entropy which is defined as:

$$IFE(A) = \frac{1}{n}\sum_{i=1}^{n} \frac{2\mu_A(x_i) + \nu_A(x_i) + \pi_A^2(x_i)}{\mu_A^2(x_i) + \nu_A^2(x_i) + \pi_A^2(x_i)}. \tag{3.56}$$

They also defined an entropy measure based on Shannon's entropy [36]:

$$IFE(A) = \frac{1}{n\ln 2}\sum_{i=1}^{n}$$

$$[\mu_A(x_i)\ln\mu_A(x_i) + \nu_A(x_i)\ln\nu_A(x_i) - (1-\pi_A(x_i))\ln(1-\pi_A(x_i)) - \pi_A(x_i)\ln 2].$$

Ye [34] introduced two different kinds of entropy measures using Ifss, which are given as:

$$IFE(A) = \frac{1}{n}\sum_{i=1}^{n}\left[\left\{\sin\frac{\pi \times [1+\mu_A(x_i)-\nu_A(x_i)]}{4} + \sin\frac{\pi \times [1-\mu_A(x_i)+\nu_A(x_i)]}{4} - 1\right\} \times \frac{1}{\sqrt{2}-1}\right],$$

$$IFE(A) = \frac{1}{n}\sum_{i=1}^{n}\left[\left\{\cos\frac{\pi \times [1+\mu_A(x_i)-\nu_A(x_i)]}{4} + \cos\frac{\pi \times [1-\mu_A(x_i)+\nu_A(x_i)]}{4} - 1\right\} \times \frac{1}{\sqrt{2}-1}\right].$$

(3.57)

3.12 Summary

This chapter provides a clear overview and definition of similarity measure, distance measure, inclusion measure, and measures of fuzziness between fuzzy sets and fuzzy numbers. Different types of similarity measures, distance measures, and entropy using fuzzy sets are given. A detailed discussion on different types of similarity measures using triangular and trapezoidal fuzzy numbers are also given. Distance and similarity measures between two IFSs are also discussed and also few recent distance measures that use center of gravity, perimeter, area, and graded mean integration are provided along with explanation of the geometrical terms. At the end, measures of fuzziness and entropy are explained along with different measures of fuzziness.

References

1 Hyung, L.K. and Song, Y.S. (1994). Similarity measure between fuzzy sets and between elements. *Fuzzy Sets and Systems* 62: 291–293.

2 Voxman, W. (1998). Some remarks on distances between fuzzy numbers. *Fuzzy Sets and Systems* 100: 353–365.

3 Pappis, C.P. and Karacapilidis, N.I. (1993). A comparative assessment of measures of similarity of fuzzy values. *Fuzzy Sets and Systems* 56: 171–174.

4 Vlachos, I.K. and Sergiadis, G.D. (2006). Inner product based entropy in the intuitionistic fuzzy setting. *International Journal of Uncertainty. Fuzziness Knowledge based Systems* 14: 351–366.

5 Chen, S.J. and Chen, S.M. (2003). Fuzzy risk analysis based on similarity measures of generalized fuzzy numbers. *IEEE Transaction on Fuzzy Systems* 11: 45–56.

6 Chen, S.M. and Chen, J.H. (2009). Fuzzy risk analysis based on ranking generalized fuzzy numbers with different heights and different spreads. *Expert Systems with Applications* 36: 6833–6842.

7 Yong, D., Wenkang, S., Feng, D., and Qi, L. (2004). A new similarity measure of generalized fuzzy numbers and its application to pattern recognition. *Pattern Recognition Letters* 25: 875–883.

8 Lee, H.S. (2002). Optimal consensus of fuzzy opinions under group decision making environment. *Fuzzy Sets and Systems* 132: 303–315.

9 Lee, H.S. (1999). An optimal aggregation method for fuzzy opinions of group decision. *IEEE International Conference on Systems, Man, and Cybernetics* (12–15 October 1999), Japan.

10 Chen, S.M. (1996). New methods for subjective mental workload assessment and fuzzy risk analysis. *Cybernetics and Systems* 27: 449–472.

11 Wei, S.H. and Chen, S.M. (2009). A new approach for fuzzy risk analysis based on similarity measures of generalized fuzzy numbers. *Expert Systems and Applications* 36: 589–598.

12 Wen, J., Fan, X., Duanmu, D., and Yong, D. (2011). A modified similarity measure of generalized fuzzy numbers. *Procedia Engineering* 15: 2773–2777.

13 Hsieh, C.H. and Chen, S.H. (1999). Similarity of generalized fuzzy numbers with graded mean integration representation. *Eighth International Fuzzy Systems Association World Congress*, vol. 2, pp. 551–558 (17–20 August), Taipei, Taiwan, Republic of China.

14 Chou, C.C. (2003). The canonical representation of multiplication operation on triangular fuzzy numbers. *Computers and Mathematics with Applications* 45: 1601–1610.

15 Xu, Z., Shang, S., Qian, W., and Shu, W. (2010). A method for fuzzy risk analysis based on the new similarity of trapezoidal fuzzy numbers. *Expert Systems and Applications* 37 (3): 1920–1927.

16 Zeng, W. and Li, H. (2006). Inclusion measures, similarity measures, and the fuzziness of fuzzy sets and their relations. *International Journal of Intelligent Systems* 21: 639–653.

17 Yager, R.R. (1979). On the measure of fuzziness and negation. Part I: Membership in the unit interval. *International Journal of General Systems* 5: 189–200.

18 Kaufmann, A. (1975). *Introduction to the Theory of Fuzzy Subsets, Vol. 1, Fundamental Theoretical Elements*. New York: Academic Press.

19 De Luca, A. and Termini, S. (1972). A definition of non-probabilistic fuzziness in the setting of fuzzy sets theory. *Information Control* 20: 301–312.

20 Zadeh, L.A. (1965). Fuzzy sets. *Information and Control* 8: 338–353.

21 Pal, N.K. and Pal, S.K. (1999). Entropy: a new definitions and its applications. *IEEE Transactions on Systems, Man and Cybernetics* 21 (5): 1260–1270.

22 Atanassov, K. and Gargov, G. (1989). Interval-valued intuitionistic fuzzy sets. *Fuzzy Sets and Systems* 31: 343–349.

23 Mitchell, H.B. (2003). On the Dengfeng–Chuntian similarity measure and its application to pattern recognition. *Pattern Recognition Letters* 24: 3101–3104.

24 Li, D. and Cheng, C. (2002). New similarity measures of intuitionistic fuzzy sets and application to pattern recognition. *Pattern Recognition Letters* 23: 221–225.

25 Wang, W. and Xin, X. (2005). Distance measure between intuitionistic fuzzy sets. *Pattern Recognition Letters* 26: 2063–2069.

26 Szmidt, E. and Kacprzyk, J. (2000). Distances between intuitionistic fuzzy set. *Fuzzy Sets and Systems* 114 (3): 505–518.

27 Xu, Z.S. (2007). Some similarity measures of intuitionistic fuzzy sets and their applications to multiple attribute decision making. *Fuzzy Optimization and Decision Making* 6: 109–121.

28 Dengfeng, L. and Chuntian, C. (2002). New similarity measure of intuitionistic fuzzy sets and application to pattern recognitions. *Pattern Recognition Letters* 23: 221–225.

29 Hung, W.L. and Yang, M.S. (2008). On similarity measures between intuitionistic fuzzy sets. *International Journal of Intelligent Systems* 23: 364–383.

30 Liang, D. and Shi, C. (2002). New similarity measures of intuitionistic fuzzy sets and application to pattern recognition. *Pattern Recognition Letters* 23: 221–225.

31 Song, Y.Y. and Zhou, X.G. (2009). New properties and measures of distance measure between intuitionistic fuzzy sets. *Sixth IEEE International Conference on Fuzzy Systems and Knowledge Discovery* (14–16 August), Tianjin, China.

32 Zhang, H. and Yu, L. (2013). New distance measures between intuitionistic fuzzy sets and interval-valued fuzzy sets. *Information Sciences* 245: 181–196.

33 Bhattacharya, A. (1946). On a measure of divergence of two multinomial populations. *Sankhya* 7: 401–406.

34 Ye, J. (2011). Cosine similarity measures for intuitionistic fuzzy sets and their applications. *Mathematical and Computer Modelling* 53: 91–97.

35 Burillo, P. and Bustince, H. (1996). Entropy on intuitionistic fuzzy sets and on interval-valued fuzzy sets. *Fuzzy Sets and Systems* 78: 305–316.

36 Szmidt, E. and Kacprzyk, J. (2001). Entropy for intuitionistic fuzzy set. *Fuzzy Sets and Systems* 118: 467–477.

37 Chaira, T. (2011). A novel intuitionistic fuzzy clustering algorithm and its application to medical images. *Applied Soft Computing* 11 (2): 1711–1717.

38 Huang, G.S. and Liu, Y.S. (2005). The fuzzy entropy of vague sets based on non-fuzzy sets. *Computer Applications and Software* 22 (6): 16–17.

39 Wang, W.J. (1997). New similarity measures on fuzzy sets and on elements. *Fuzzy Sets and Systems* 85: 305–309.

4

Fuzzy/Intuitionistic Fuzzy Measures and Fuzzy Integrals

4.1 Introduction

The problem of defining and measuring fuzziness in a fuzzy set is an important part in fuzzy mathematics. In mathematics, fuzzy measure theory is considered as a generalized measure and it was introduced by Choquet in 1953 and defined by Sugeno in 1974 in the context of fuzzy integrals. Fuzzy set deals with membership grades whereas fuzzy measure deals with measures of fuzzy set. It considers degree of possibility that a given element belongs to a fuzzy set or a non-fuzzy set. They have many applications in engineering and their main characteristic is additivity. Classical measure holds additive property. Additive property can be very useful in many applications but may not be adequate in real-time world problems such as in approximate reasoning, fuzzy logic, artificial intelligence, data mining, etc. For example, if we measure the efficiency of a set of students and the efficiency of the same set of students working in a group, it will be seen that the efficiencies are not equal. This is due to the reasons that the efficiency of the students working in a team is not the addition of the efficiency of each students working on their own. The concept of fuzzy measure does not require additivity. It requires monotonicity.

4.2 Definition of Fuzzy Measure

Let $X = \{x_1, x_1, x_3, \ldots, x_n\}$ be a finite set of reference. A fuzzy measure on X is a mapping $\mu : P(X) \longrightarrow [0, 1]$, where $P(X)$ is a power set of X (2^X) that fulfills the following conditions [1]:

i) $\mu(0) = 0$ and $\mu(X) = 1$. This is the boundary condition,
ii) μ is said to be monotone, i.e. if $A \subseteq B$, then $\mu(A) \leq \mu(B)$, A and B belong to $P(X)$,

Fuzzy Set and Its Extension: The Intuitionistic Fuzzy Set, First Edition. Tamalika Chaira.
© 2019 John Wiley & Sons, Inc. Published 2019 by John Wiley & Sons, Inc.

iii) For a monotone and convergence sequence, A_n, such that $A_1 \subseteq A_2 \subseteq \ldots$
$\subseteq A_n$ then $\lim_{n \to \infty} \mu(A_n) = \mu(\lim_{n \to \infty} A_n)$,
iv) $\mu(X) = 1$.

This fuzzy measure is a monotone and continuous function.

There are *some* properties of fuzzy measures [2, 3]:

Let $\mu : P(X) \to [0,1]$ be the fuzzy measure on finite set X, and $(A \cap B) = \emptyset, \forall A, B \in P(X)$, then

v) μ is said to be subadditive if $\mu(A \cup B) \leq \mu(A) + \mu(B)$,
vi) μ is said to be additive if $\mu(A \cup B) = \mu(A) + \mu(B)$ for any $A, B \in P(X)$,
vii) μ is said to be superadditive if $\mu(A \cup B) > \mu(A) + \mu(B)$ for any $A, B \in P(X)$,
viii) μ is said to be super subtractive if $\mu(A \cap B^c) \geq \mu(A) - \mu(B)$ for any $A, B \in P(X)$,
ix) If $|A| = |B|$ implies $\mu(A) = \mu(B)$, then μ is symmetric.

The above measures will be useful in defining fuzzy integrals.

4.3 Fuzzy Measures

In this section we will present different types of fuzzy measures: Sugeno measure, belief measure, plausibility measure, possibility measure, and necessity fuzzy measure.

4.3.1 Sugeno λ-Fuzzy Measure

Michio Sugeno introduced a special class of fuzzy measure known as Sugeno fuzzy λ-measure [1, 4]. It is very much close to probability measure. It is defined on monotone family and a nonadditive fuzzy measure.

Let us take an example: Consider two disjoint groups A and B and they work to produce goods. Let $\mu(A)$ be the number of goods produced by A in per hour. If they work separately, then the total number of goods they produce is $\mu(A \cup B) = \mu(A) + \mu(B)$. Here, additive property holds. But, if they work together and on interacting with each other, this equality may not hold. Their efficiency may be increased due to the efficient cooperation of the members and this leads to the inequality, i.e. $\mu(A \cup B) > \mu(A) + \mu(B)$. Again, sometimes it may happen that on interacting, due to insufficient space and/or equipment, their productivity is decreased, then the combined work will be $\mu(A \cup B) < \mu(A) + \mu(B)$.

From the definition of fuzzy measure for monotonicity in Section 4.2(ii), it is observed that the property of monotonicity is weaker than that of additivity considering for a probability measure [5]. By relaxing the additive property of probability such that $A \cap B = \emptyset$ implies there is no interaction between A and B, and $\forall A, B \in P(X)$, then λ-Sugeno measure for $\lambda \in [-1, \infty]$ is formulated as:

$$\mu(A \cup B) = \mu(A) + \mu(B) + \lambda \cdot \mu(A) \cdot \mu(B). \tag{4.1}$$

It is observed that if $\lambda = 0$, additivity is maintained and if $\lambda \neq 0$, then additive property is not maintained.

If $\lambda > 0$, then $\mu(A \cup B) > \mu(A) + \mu(B)$. This implies some sort of multiplicative effect on $\{A,B\}$ is present.

So, Sugeno λ-fuzzy measure, which is a monotone fuzzy measure, is defined as:

If $X = \{x_1, x_2, ..., x_n\}$ be a universe of discourse and let g_λ be a λ-fuzzy measure in the set X, then

$$\text{i) } g_\lambda(X) = 1,$$
$$\text{ii) } g_\lambda(A \cup B) = g_\lambda(A) + g_\lambda(B) + \lambda \cdot g_\lambda(A) \cdot g_\lambda(B),$$

(4.2)

with $A \cap B = \emptyset$ and $\lambda > -1$.

Before proceeding further to compute Sugeno fuzzy measure, we define fuzzy density function [5].

Fuzzy density function of fuzzy measure g_λ defined on $X = \{x_1, x_2, ..., x_n\}$ is a function $g : x_i \in X \longrightarrow [0,1]$ such that

$$g(x_i) = g_\lambda(\{x_i\}), \, i = 1, 2, 3, ..., k \quad \text{for all } x \in X.$$

The fuzzy measure of a finite set is obtained from a set of values of fuzzy density function g as follows [5]:

Let us consider a set $X = \{x_1, x_2\}$ consists of two elements, then λ fuzzy measure of a finite set, X, is written as:

$$g_\lambda(\cup_{i=1}^n x_i) = g_\lambda(x) = g_\lambda(\{x_1, x_2\}) = g(x_1) + g(x_2) + \lambda \cdot g(x_1) \cdot g(x_2).$$

Likewise, $g_\lambda(\{x_1, x_2, x_3\}) = g(x_1) + g(x_2) + g(x_3) + \lambda \cdot (g(x_1) \cdot g(x_2) + g(x_2) \cdot g(x_3) + g(x_1) \cdot g(x_3)) + \lambda^2 \cdot g(x_1) \cdot g(x_2) \cdot g(x_3).$

For the sake of simplicity, we can write: $g(x_i) = g_i$, for $i = 1, 2, 3, ..., k$.

Extending the set X consists of n elements, $X = \{x_1, x_2, ..., x_n\}$, the general formulation of fuzzy measure $g(X) = g_\lambda(\{x_1, x_2, ..., x_n\})$ can be written as:

$$g_\lambda(\{x_1, x_2, ..., x_n\}) = \sum_{i=1}^n g_i + \lambda \sum_{i=1}^{n-1} \sum_{j=i+1}^n g_i g_j + ... + \lambda^{n-1} g_1 \cdot g_2 \cdot g_3 ... g_n, \quad (4.3)$$

where $i, j = 1, 2, 3, ..., n$.

We see that if $\sum_{i=1}^n g_i = 1$, then we get $\lambda = 0$.

For $\lambda \neq 0$, we can write

$$g_\lambda(\{x_1, x_2, ..., x_n\}) = \frac{1}{\lambda} \left[\prod_{i=1}^n (1 + \lambda g_i) - 1 \right], \lambda \in [-1, \infty]$$

$$\Rightarrow g_\lambda(\{x_1, x_2, ..., x_n\}) \cdot \lambda = \prod_{i=1}^n (1 + \lambda g_i) - 1$$

$$\Rightarrow g_\lambda(X) = \frac{1}{\lambda} \left[\prod_{i=1}^n (1 + \lambda g_i) - 1 \right], X = \{x_1, x_2, ..., x_n\}.$$

We know from the definition of Sugeno measure, that $g_\lambda(X) = 1$.

$$\text{So, } \lambda + 1 = \prod_{i=1}^{n} (1 + \lambda g_i). \tag{4.4}$$

Now, we state some following properties:

i) From Eq. (4.3), when $\lambda = 0$, the equation turns to additive measure corresponding to probability measure $g(A \cup B) = g(A) + g(B)$.

From the value of λ, we can come to know how close the Sugeno g_λ measure is to the probability measure. The smaller the value of $|\lambda|$, the closer the Sugeno g_λ measure is.

ii) When $\lambda = -1$, then

$$g_\lambda(A \cup B) = g_\lambda(A) + g_\lambda(B) - \lambda \cdot g_\lambda(A) \cdot g_\lambda(B).$$

This is similar to T-conorm $- S(a, b) = a + b - ab$ which is the algebraic sum.

iii) From Eq. (4.3), $g_\lambda(A \cup B) = g_\lambda(A) + g_\lambda(B) + \lambda \cdot g_\lambda(A) \cdot g_\lambda(B)$.

As $g_\lambda(A \cup B) = 1$, so

$$1 = g_\lambda(A) + g_\lambda(B)(1 + \lambda g_\lambda(A))$$

$$\Rightarrow g_\lambda(B) = \frac{1 - g_\lambda(A)}{1 + \lambda g_\lambda(A)}. \tag{4.5}$$

If we take $B = \bar{A}$, then Eq. (4.5) becomes

$$g_\lambda(\bar{A}) = \frac{1 - g_\lambda(A)}{1 + \lambda g_\lambda(A)}.$$

This is also called Sugeno fuzzy complement.
If $\lambda = 0$, then
$g(\bar{A}) = 1 - g(A)$ is a standard fuzzy complement.
Let us take an example on Sugeno λ-fuzzy measure:

Example 1 Consider a set $X = \{x_1, x_2, x_3\}$. Fuzzy densities values are: $g_\lambda(\{x_1\}) = 0.3$, $g_\lambda(\{x_2\}) = 0.2$, $g_\lambda(\{x_3\}) = 0.1$. Compute λ, $g_\lambda(\{x_1, x_2\})$, $g_\lambda(\{x_2, x_3\})$, $g_\lambda(\{x_1, x_3\})$ and $g(X)$.

Solution

First, we need to compute λ:
From Eq. (4.3) and considering $g(x_i) = g_i$, we have

$$\lambda + 1 = \prod_{i=1}^{n} (1 + \lambda g_i)$$
$$= (1 + 0.3\lambda)(1 + 0.2\lambda)(1 + 0.1\lambda),$$
$$= (1 + 0.5\lambda + 0.06\lambda^2)(1 + 0.1\lambda)$$

$$\Rightarrow 0.006\lambda^2 + 0.11\lambda - 0.4 = 0$$

$$\Rightarrow \lambda = \{0, 3.109, -21.44\}$$

As $\lambda > -1$, $\lambda = 3.109$ is selected.
At $\lambda = 3.109$,

$$g_\lambda(\{x_1, x_2\}) = g_\lambda(x_1) + g_\lambda(x_2) + \lambda \cdot g_\lambda(x_1) \cdot g_\lambda(x_2)$$

$$= 0.3 + 0.2 + 3.109 \times 0.3 \times 0.2$$

$$= 0.686,$$

Likewise, $g_\lambda(\{x_2, x_3\}) = g_\lambda(x_2) + g_\lambda(x_3) + \lambda \cdot g_\lambda(x_2) \cdot g_\lambda(x_3)$

$$= 0.2 + 0.1 + 3.109 \times 0.2 \times 0.1 = 0.3622,$$

$$g_\lambda(\{x_1, x_3\}) = g_\lambda(x_1) + g_\lambda(x_3) + \lambda \cdot g_\lambda(x_1) \cdot g_\lambda(x_3)$$

$$\Rightarrow g_\lambda(\{x_1, x_3\}) = 0.4 + 3.109 \times 0.03 = 0.493$$

$$g(X) = g_\lambda(\{x_1, x_2\}) + g_\lambda(x_3) + \lambda \cdot g_\lambda(\{x_1, x_2\}) \cdot g_\lambda(x_3)$$

$$= 0.686 + 0.1 + 3.109(0.686)(0.1) = 1.$$

4.3.2 Belief Measure

Theory of belief function is also known as evidence theory or Dempstar–Shafer theory. Belief function is a fuzzy measure. After the introduction of fuzzy sets, Shafer [6] introduced the concept of belief measure. Belief is actually the amount or degree of belief that a given element in X belongs to set A as well as to any of the subsets of A. Degree of belief is the judgement based on evidence that how strongly we consider the evidence to be correct.

Belief measure (*Bel*) is a function on X such that [6, 7]

i) Boundary condition – $Bel(0) = 0$, $Bel(X) = 1$,
ii) For all $A \in P(X)$, $0 \le Bel(A) \le 1$,
iii) For all $A_1, A_2, \dots, A_n \in P(X)$,

$$Bel(A_1 \cup A_2 \cup \dots \cup A_n)$$

$$\ge \sum_{i=1}^{n} Bel(A_i) - \sum_{i < j} Bel(A_i \cap A_j) + \dots + (-1)^{n+1} Bel(A_1 \cap A_2 \cap \dots \cap A_n),$$

iv) Monotonicity – If $A \subseteq B$, then $Bel(A) \le Bel(B)$.

Condition (i) shows that zero means no belief and one means total belief.

In condition (iii), if we put $n = 2$, then we get

$$Bel(A_1 \cup A_2) \geq Bel(A_1) + Bel(A_2) - Bel(A_1 \cap A_2).$$

Likewise, for $n = 3$, we get

$$Bel(A_1 \cup A_2 \cup A_3)$$
$$\geq Bel(A_1) + Bel(A_2) + Bel(A_3) - Bel(A_1 \cap A_2) - Bel(A_2 \cap A_3)$$
$$- Bel(A_1 \cap A_3) + Bel(A_1 \cap A_2 \cap A_3).$$

If A_1 and A_2 are pairwise disjoint, i.e. if $(A_1 \cap A_2) = \emptyset$, then $Bel(A_1 \cup A_2) \geq Bel(A_1) + Bel(A_2)$.

This inequality replaces the additivity requirement of the probability measures and it shows that the belief measures are called superadditive.

Again, $Bel(A) + Bel(\bar{A}) \leq 1$.

The above equation implies that if there is a strong belief in $x \in A$, then it may not be the case that there is lack of belief in $x \in \bar{A}$.

A λ-fuzzy measure is a belief measure iff $\lambda \geq 0$ [8].

4.3.3 Plausibility Measure

Plausibility measure, defined by Shafer [6] in 1976, is a dual of belief measure which is given as [3]:

$\forall A \in P(X), Pl(A) = 1 - Bel(\bar{A})$, where Bel is a belief function.

$Bel(A) = 1 - Pl(\bar{A})$, where $\bar{A}$ is complement of A.

The following properties are:

i) For all $A \in P(X)$, $0 \leq Pl(A) \leq 1$,
ii) For all $A_1, A_2, \ldots, A_n \in X$,
   $Pl(A_1 \cap A_2 \cap \ldots \cap A_n) \leq$
   $\sum_{i=1}^{n} Pl(A_i) - \sum_{i<j} Pl(A_i \cup A_j) + \ldots + (-1)^{n+1} Pl(A_1 \cup A_2 \cup \ldots \cup A_n)$,
iii) If $A \subseteq B$, then $Pl(A) \leq Pl(B)$.

Similar to belief measure, if $(A_1 \cap A_2) = \emptyset$, then for $n = 2$,
$Pl(A_1 \cup A_2) \leq Pl(A_1) + Pl(A_2)$, which is a subadditive.

Again, $Pl(A) + Pl(\bar{A}) \geq 1$.

A λ-fuzzy measure is a plausibility measure iff $-1 < \lambda \leq 0$ [8].

4.3.4 Possibility Measure and Necessity Measure

Possibility and necessity measures play a key role in decision making and approximate reasoning. In applications where these measures are used, selection of the measures affects the results and inferences. Possibility measure and necessity measure can be defined in many ways. It can be defined using conjunction and implication operator [9]. Since there are many conjunction and implication operators, there are also a great number of possibility and necessity measures. But, selection of conjunction and implication operators for particular application is still a difficult problem.

4.3.4.1 Possibility Measure

A possibility measure evaluates to what extent a fact that when we know that x is in a fuzzy set B is in a fuzzy set A is possible. It is a mathematical theory that deals with uncertainty and is an alternative to probability theory. Possibility measure for a fuzzy set is given by Zadeh [10]. But, probability is a measure of frequency of occurrence of an event but possibility is a measure to quantify the meaning of an event. A possibility measure evaluates to what extent a fact is possible that a variable x, present in a fuzzy set A, is in a fuzzy set B. In a finite case, a fuzzy measure is called possibility measure when it is maxitive, i.e.

$$\mu(A \cup B) = \mu(A) \vee \mu(B).$$

A fuzzy measure is called necessity measure, which is a dual of possibility measure, when it is minitive, i.e.

$\mu(A \cap B) = \mu(A) \wedge \mu(B)$, respectively, where μ is a fuzzy measure but is still less developed than probability measure. It is a class of nonadditive measure.

Definition 1 A possibility measure is a function $\pi : P(X) \rightarrow [0,1]$, $P(X)$ is a power set of X, has the following properties [8, 11, 12]:

i) $\pi(0) = 0$, $\pi(X) = 1$,
ii) If $A \subseteq B$, then $\pi(A) \leq \pi(B)$, $A, B \in P(X)$,
iii) $\pi \cup_{j \in J} A_j = sup_{j \in J} \pi(A_j)$, $j \in J$, J is an index set.

When A is a crisp subset of X, possibility measure can be determined using the possibility distribution function $p : X \longrightarrow [0,1]$ by

$$\pi(A) = \sup_{x \in A} p(x), \text{ for all, } A \subset X. \tag{4.6}$$

It is clear that p is defined by $p(x) = \pi(\{x\})$, $\forall x \in X$ [13].
But possibility measure is not always a fuzzy measure [11]. It is a fuzzy measure if X is finite and if the possibility distribution is normal, then it is a mapping into [0,1].

Example 2 Consider a set of integers, $X = \{1,2,3,4,5,6,7,8,9,10\}$ and the possibility that X is close to 7 is

$$\pi(\{x\}) = \{0,0,0,0.1,0.6,0.8,1,0.7,0.5,0.3, \}.$$

Let us take a set crisp set $A = \{3,6,8\}$.

Then, the possibility that A contains an integer close to 7 is:

$$\pi(A) = \sup_{x \in A} \pi(\{x\})$$
$$= \sup\{\pi(\{3\}), \pi(\{6\}), \pi(\{8\})\}$$
$$= \sup\{0, 0.8, 0.7\}$$
$$= 0.8.$$

So, the possibility of set A contains an integer close to 7 is 0.8.

For a fuzzy set, the definition of possibility measure does not take the same form, which is given by Zadeh [10]. Let F be a fuzzy set in the universe of discourse, U, and X be a variable taking values in U. Let Π_x be the possibility distribution associated with the variable X. Then, possibility measure, $\pi_x(F)$, of F, is defined as:

$$(\text{Poss } X \text{ is } F) = \pi(F) = \sup_{u \in U} \min(\mu_F(u), \pi_x(u)), u \in U \qquad (4.7)$$

where $\pi_x(u)$ is the possibility distribution function of the possibility distribution, Π_x and $\mu_F(u)$ is the membership function of fuzzy set F.

Possibility distribution function, $\pi_x(u)$, of a possibility distribution, Π_x, is defined to be numerically equal to the membership function $\mu_F(u)$ of F, i.e. $\pi_x(u)$, the possibility that $X = u$ is equal to $\mu_F(u)$.

The same may be expressed as:

$$\Pi_x = F.$$

For a non-fuzzy case [10] similar to that defined above by [8, 12], let A be a non-fuzzy set in the universe of discourse U and X be a variable taking values in U. Also, let $\pi_x(u)$ be the possibility distribution function of the possibility distribution Π_x associated with the variable X. Then, possibility measure, $\pi_x(A)$, is defined as [10]:

$$\text{Poss}(X \in A) = \pi(A) = \sup_{u \in A}(\pi_x(u)).$$

Let us take an example. Let U be a universe of discourse and F be a fuzzy set of large integers:

$$F = \{(4,0.1), (5,0.3), (7,0.6), (8,0.7), (9,0.8), (10,0.9), (11,0.9), (12,1)\}.$$

Then, the proposition "X is a large integer" has a possibility distribution $\Pi_x = F$.

So, $\Pi_x = \{(4,0.1), (5,0.3), (7,0.6), (8,0.7), (9,0.8), (10,0.9), (11,0.9), (12,1)\}$,

where the term (8,0.7) signifies that the possibility of X is 8 when X is a large integer is 0.7.

In limiting case, the possibility measure is defined as:
If $A \subseteq X$, $A \subseteq X$ and $A \cap B = 0$, then

$$\pi(A \cup B) = \max(\pi(A) \cdot \pi(B)),$$
$$\pi(A \cap B) = \min(\pi(A) \cdot \pi(B)).$$

Example 3   [12] Consider a possibility distribution with proposition "X is a small integer"

$$\Pi_x = \left\{ \frac{1}{1}, \frac{1}{2}, \frac{0.9}{3}, \frac{0.7}{4}, \frac{0.4}{5}, \frac{0.3}{6}, \frac{0.2}{7} \right\}.$$

i) Consider a fuzzy set F= "integers which are large"

$$F = \left\{ \frac{0.1}{1}, \frac{0.2}{2}, \frac{0.3}{3}, \frac{0.5}{4}, \frac{0.7}{5}, \frac{0.9}{6}, \frac{1}{7}, \dots \right\}.$$

Compute the possibility measure for fuzzy and crisp cases.

Solution

Possibility measure of "X is a large integer" = $\sup_{u \in U} \min(\mu_F(u), \pi_x(u))$
= sup(0.1,0.2,0.3,0.5,0.4,0.3,0.2) = 0.5.
ii) If A is a crisp set, $A = \{3,5,6\}$, then
possibility measure, $\pi(A) = \sup_{x \in A} \pi_x(u)$
$\qquad\qquad\qquad\qquad = \sup\{0.9, 0.4, 0.3\}$
$\qquad\qquad\qquad\qquad = 0.9.$

Possibility measure can also be defined using conjunction (or equivalently intersection) operator which is explained in the next section.

4.3.4.2 Necessity Measure
A necessity measure $N_A(B)$ evaluates to what extent a fact that when we know that x is in a fuzzy set A is in a fuzzy set B, is certain. Just like possibility measure, necessity measure can also be defined using implication (or equivalently union) operator. Necessity measure can be said as dual of possibility measure.

$$N(A \cap B) = \min(N(A) \cdot N(B)), \text{ for all } A, B \in P(X) \cdot P(X)$$
$$\text{is all the subsets of } X,$$

where N is a necessity measure.
It satisfies the following properties:
Boundary condition – For all $A \in P(X)$, $N(0) = 0$, $N(A) = 1$,
$N(A) = 1 - \pi(A^c)$, A^c is the complement of A and π is a possibility measure. This is the dual relationship of possibility and necessity measure, $N(A) = 1$ means that A is necessarily true.

The definition of possibility and necessity measure as given by Inuiguchi et al. [9] using conjunction and implication operator, respectively, is given as:

$$\text{Possibility measure}: \pi_A(B) = \sup_x \left(T(\mu_A(x), \mu_B(x)) \right), \tag{4.8}$$

where T is a conjunction function with $T: [0,1] \times [0,1] \rightarrow [0,1]$ such that $T(0,0) = T(0,1) = T(1,0) = 0$ and $T(1,1) = 1$ and $\mu_A(x)$ and $\mu_B(x)$ are the membership functions of fuzzy sets A and B.
Different types of T norms are explained in Chapter 5.

$$\text{Necessity measure}: N_A(B) = \inf_x(I(\mu_A(x), \mu_B(x))),$$

where $\mu_A(x)$ and $\mu_B(x)$ are the membership functions of fuzzy set A and B. I is an implication function with $I: [0,1] \times [0,1] \rightarrow [0,1]$ such that

$$I(0,0) = I(0,1) = I(1,1) = 1 \text{ and } I(1,0) = 0.$$

Different types of implication operators are discussed in Chapter 5.

Since there are many conjunction and implication operators, there are a great number of possibility and necessity measures.

Necessity measures are chosen in such a way that it should be adequate with decision maker's preference. When the uncertainty is expressed in a qualitative way (relating to description, measure. etc.), Dubois et al. [14] had given a different type of definition for possibility and necessity measures. These are defined as:

$$\text{Possibility measure}: \pi_A(B) = \sup_x \min(\mu_A(x), \mu_B(x)). \tag{4.9}$$

$$\text{Necessity measure}: N_A(B) = \inf_x \max(C(\mu_A(x)), \mu_B(x)), \tag{4.10}$$

where $C: [0,1] \rightarrow [0,1]$ is an order-reversing function which is a strictly decreasing bijection (one-to-one correspondence) such that if $a < b$ then $C(a) > C(b)$.

Possibility and necessity measures are a particular case of belief and plausibility measures. It is used in approximate reasoning, information retrieval, fuzzy mathematical programming, and so on.

The relationships between possibility measures and necessity measures are given as follows [8]:

$$\pi(A) \geq N(A),$$

$$N(A) > 0 \Rightarrow \pi(A) = 1,$$

$$\pi(A) < 1 \Rightarrow N(A) = 0.$$

4.4 Fuzzy Integrals

The concept of integral was defined by Murofushi and Sugeno [4] and it provides a useful way of aggregating multiple information. There are many types of fuzzy integrals – the Choquet integral, the Šipoš integral, the Sugeno integral, the T-conorm integral, etc. Fuzzy integral is replaced by a fuzzy measure with additivity replaced by monotonicity. The basic property of Sugeno and Choquet integrals, in particular, is that they are extensively used as an aggregation operator. They are used in different types of decision-making applications [7, 15] to model situations in which sources are independent and also situations where the sources are not independent. As the interpretation is not straight forward, their use is not easy. Fuzzy integrals combine the data obtained from different information sources based to a fuzzy measure. This fuzzy measure is a set function from the set of information sources and it denotes the background knowledge on the information sources. During aggregation, this fuzzy measure represents the importance of the sources.

Let g be a fuzzy measure on X and f be a function on X. The integral $\int f dg$ of $f : X \rightarrow [0,1]$ with respect to fuzzy measure g is defined as [4]:

$$\int f dg = \sum_{x \in X} f(x) \cdot g(\{x\})$$

The integral may be written as (shown in Figure 4.1) follows where different regions are shown in different types of dashes:

$$\int f dg = f(x_1) \cdot g(\{x_1\}) + f(x_2) \cdot g(\{x_2\}) + f(x_3) \cdot g(\{x_3\}) + f(x_4) \cdot g(\{x_4\}).$$

$$(4.11)$$

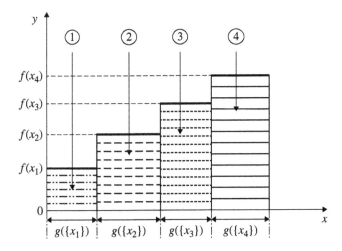

Figure 4.1 Integral of f (*Source:* Adapted from [4]).

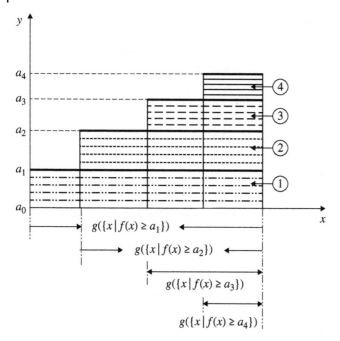

Figure 4.2 Integral of f (*Source:* Adapted from [4]).

The integral can also be computed as shown in Figure 4.2:

$$\int fdg = \sum_{i=1}^{n} (a_i - a_{i-1}) \cdot g(\{x \mid f(x) \geq a_i\}),$$ (4.12)

where a_i is the range $f(x)$ of f and $a_1 < a_2 < a_3 \ldots < a_n$ and $a_0 = 0$.

4.4.1 Sugeno Integral

Sugeno integral is a type of fuzzy integral with respect to a fuzzy measure, which is nonadditive. It takes the range $[0,1]$.

Let g be a fuzzy measure on X and f be a real-valued function $f: X \to [0,1]$. Then, Sugeno integral of f with respect to a fuzzy measure g is defined as [15]:

$$S_g(f) = \int_X f(x) \circ g = \vee_{i=1}^{n} \left(f(x_{(i)}) \wedge g(A_{(i)}) \right)$$
$$= \max_{1 \leq i \leq n} \left(\min \left(f(x_{(i)}), g(A_{(i)}) \right) \right).$$ (4.13)

where (i) indicates that the indices have been permuted so that

$$0 \leq f(x_{(1)}) \leq f(x_{(2)}) \leq f(x_{(3)}) \leq \ldots \leq f(x_{(n)}) \leq 1$$

and $A_{(i)} = \{x_{(i)}, x_{(i+1)}, \ldots, x_{(n)}\}$, $A_{(n+1)} = \emptyset$.

It should be noted that both $f(x)$ and $g(A)$ should be in the same domain, otherwise, minimum cannot be applied to combine the two.

Example 4   This is an example of Sugeno integral for λ-fuzzy measure. Consider a set $X = \{x_1, x_2, x_3\}$ and the function $f(x)$ or the range is defined as: $f(x_1) = 0.5, f(x_2) = 0.3, f(x_3) = 0.2$. The fuzzy densities g are given as: $g(\{x_1\}) = 0.2, g(\{x_2\}) = 0.4, g(\{x_3\}) = 0.6$. Compute Sugeno integral.

Solution
First, let us calculate the value of λ from eq. (4.4):

From Eq. (4.3),
$$\lambda + 1 = \prod_{i=1}^{n}(1 + \lambda g_i)$$
$$= (1 + 0.2\lambda)(1 + 0.4\lambda)(1 + 0.6\lambda)$$
$$= (1 + 0.6\lambda + 0.08\lambda^2)(1 + 0.6\lambda)$$
$$\Longrightarrow \lambda(0.048\lambda^2 + 0.44\lambda + 0.2) = 0$$
$$\lambda = \{0, -0.4796, -8.687\}.$$

As $\lambda > -1$, we select $\lambda = -0.4796$.
At $\lambda = -0.4796$, we will compute

$$g_\lambda(x_1, x_2) = g_\lambda(x_1) + g_\lambda(x_2) + \lambda \cdot g_\lambda(x_1) \cdot g_\lambda(x_2)$$
$$= 0.2 + 0.4 - 0.4796 \times 0.2 \times 0.4$$
$$= 0.6 - 0.0384 = 0.5616.$$

Likewise,
$$g_\lambda(\{x_1, x_3\}) = g_\lambda(x_1) + g_\lambda(x_3) + \lambda \cdot g_\lambda(x_1) \cdot g_\lambda(x_3)$$
$$= 0.8 - 0.0576 = 0.7424,$$
$$g_\lambda(\{x_2, x_3\}) = g_\lambda(x_2) + g_\lambda(x_3) + \lambda \cdot g_\lambda(x_2) \cdot g_\lambda(x_3)$$
$$= 1.0 - 0.1151 = 0.8849,$$

and
$$g_\lambda(X) = g_\lambda(x_1, x_2) + g_\lambda(x_3) + g_\lambda(x_1, x_2).g_\lambda(x_3)$$
$$= 0.5616 + 0.6 - 0.4796 \times 0.5616 \times 0.6 = 1.$$

Hence, g_λ is a fuzzy measure. For simpicity, we remove the subscript λ from g_λ. As the orders of $f(x_i)$ are not arranged in ascending order, so after reordering we get:

$$f(x_3) = 0.2$$
$$f(x_2) = 0.3$$
$$f(x_1) = 0.5$$

We see that the index is changed as: $f(x_{(1)}) = 0.2, f(x_{(2)}) = 0.3, f(x_{(3)}) = 0.5$.

From this we see that: $(1) = 3$, $(2) = 2$, $(3) = 1$. We will use (i) to find $A_{(i)}$.

Now Sugeno integral, $S_g(f) = \vee_{i=1}^{n} \left(f\left(x_{(i)}\right) \wedge g(A_i) \right)$, $A_{(i)} = \left\{ x_{(i)}, x_{(i+1)}, \ldots, x_{(n)} \right\}$

$$= \max_{1 \leq i \leq 3} (\min(f(x_i), g(A_i)))$$

$$\Longrightarrow S_g(f) = \max\left(\min\left(f\left(x_{(1)}\right), g\left(A_{(1)}\right) \right), \min\left(f\left(x_{(2)}\right), g\left(A_{(2)}\right) \right), \min\left(f\left(x_{(3)}\right), g\left(A_{(3)}\right) \right) \right)$$

Here,

$$A_{(1)} = \left\{ x_{(1)}, x_{(2)}, x_{(3)} \right\} = \left\{ x_3, x_2, x_1 \right\},$$

$$A_{(2)} = \left\{ x_{(2)}, x_{(3)} \right\} = \left\{ x_2, x_1 \right\},$$

$$A_{(3)} = \left\{ x_{(3)} \right\} = \left\{ x_1 \right\}.$$

Then,

$$S_g(f) = \max\left(\min\left(f\left(x_{(1)}\right), g(\{x_3, x_2, x_1\}) \right), \min\left(f\left(x_{(2)}\right), g(\{x_2, x_1\}) \right), \right.$$
$$\left. \min\left(f\left(x_{(3)}\right), g(\{x_1\}) \right) \right)$$
$$= \max(\min(0.2, 1), \min(0.3, 0.5616), \min(0.5, 0.2))$$
$$= \max(0.2, 0.3, 0.2) = 0.3.$$

Example 5 This is a practical example on a washing machine where Sugeno integral combines the importance or reliability of the machine [7]. Let $f(x_i)$ be the reliability of a machine according to the expert x_i. At the same time, we also consider the reliability of subset of experts. To deal with Sugeno integral, both reliability of the machine and the reliability of experts are required. Let X be a set of experts $X = \{x_1, x_2, x_3\}$ that evaluates the reliability of a washing machine. Let the reliability of three experts on individual basis are

$$g(\{x_1\}) = 0.2, g(\{x_2\}) = 0.4, g(\{x_3\}) = 0.6.$$

After computing λ using Sugeno measure from Eq. (4.3), we get the reliability of the group of experts:

$$g(\{x_1, x_2\}) = 0.5616, g(\{x_1, x_3\}) = 0.7424, g(\{x_2, x_3\}) = 0.8849 \text{ and}$$
$$g(\{x_1, x_2, x_3\}) = 1 \text{ implies the reliability of the experts all together.}$$

It seems that the expert x_3 is more reliable than expert x_2 and x_1. But, when the expert x_3 joins with x_1, the reliability is more than when x_1 joins with x_2. But, when x_2 joins with x_3, the reliability is much better than the other combinations.

Let $f(x_i)$ denotes the experts' opinion on the reliability of the washing machine. Let $f(x_1) = 0.4$, $f(x_2) = 0.6$, $f(x_3) = 0.7$. As $f(x_i)$ is already arranged in ascending order, so reordering is not required. Then, the overall reliability of the machine using Sugeno integral is:

$$S_g(f) = \vee_{i=1}^{3} \left(f\left(x_{(i)}\right) \wedge g(A_i) \right), i = 1,2,3$$

$$= \max_{1 \le i \le 3}(\min(f(x_i), g(A_i)))$$

$$= \max(\min(f(x_1), g(\{x_1, x_2, x_3\}))), \min(f(x_2), g(\{x_2, x_3\})),$$

$$\min(f(x_3), g(\{x_3\})))$$

$$= \max(\min(0.4, 1), \min(0.6, 0.88), \min(0.7, 0.6))$$

$$= \max(0.4, 0.6, 0.6) = 0.6.$$

So, Sugeno integral leads to 0.6 that corresponds to the value of the most reliable experts in this case.

4.4.2 Choquet Integral

Let $X = \{x_1, x_2, \dots, x_n\}$. The Choquet integral of a function $f : X \to [0,1]$ with respect to fuzzy measure g is defined as [15]:

$$C_g(f) = \int f dg = \sum_{i=1}^{n} \left(f\left(x_{(i)}\right) - f\left(x_{(i-1)}\right) \right) g\left(A_{(i)}\right), \tag{4.14}$$

where $A_{(i)} \subset X$ for $i = 1, 2, 3, \dots, n$,
$f(x_0) = 0$ and (i) indicates that the indices that have been permuted so that

$$0 \le f\left(x_{(1)}\right) \le f\left(x_{(2)}\right) \le f\left(x_{(3)}\right) \le \dots \le f\left(x_{(n)}\right) \le 1,$$

i.e. $f(x_{(i)})$ indicates that the indices are permuted.

$$A_{(i)} = \left\{x_{(i)}, x_{(i+1)}, \dots, x_{(n)}\right\}, A_{(n+1)} = \varnothing.$$

It is noted that Eq. (4.14) is similar to Eq. (4.12)
$C_g = \int f dg = \sum_{i=1}^{n} (a_i - a_{i-1}).g(\{x \,|\, f(x) \ge a_i\})$ (a_i is the range $f(x)$ of f and $a_1 < a_2 < a_3 \dots < a_n$ and $a_0 = 0$).
Equation (4.14) can equally be written as [15]:

$$C_g = \int f dg = \sum_{i=1}^{n} f\left(x_{(i)}\right) \left(g\left(A_{(i)}\right) - g\left(A_{(i+1)}\right)\right),$$

and follows similar notations as for Eq. (4.14)

Example 6 A practical workshop example is given to explain Choquet integral [16]. Suppose there are n number of workers, $X = \{x_1, x_2, x_3, \dots, x_n\}$, working in a factory to produce same goods. Let $A \in P(X)$, P be the power set of X. Group A has various ways to work whether jointly or separately. Each group works in an efficient way. So, for each $A \subset X$, we consider a situation, when the members of group A work in the workshop. Each worker x_i works for $f(x_i)$ hours from the opening hour and they are labeled in ascending order as: $f(x_1) \leq f(x_2) \leq f(x_3) \leq f(x_4) \leq \ldots \leq f(x_n)$.

Then, for $i = 2$, we have

$$f(x_2) - f(x_1) \geq 0 \quad \text{and} \quad \text{for } i \geq 2$$

$$f(x_i) = f(x_1) + [f(x_2) - f(x_1)] + [f(x_3) - f(x_2)] + \cdots + [f(x_i) - f(x_{i-1})].$$

Now, aggregation of working hours of all the workers may be written in the following way.

Initially, the entire group X consists of n workers worked for $f(x_1)$ hours,
Next, the group with $\{x_2, x_3, \ldots, x_n\}$ workers work for $f(x_2) - f(x_1)$ hours,
Next, the group with $\{x_3, \ldots, x_n\}$ workers work for $f(x_3) - f(x_2)$ hours,
In this way at last, $\{x_n\}$ worker works for $f(x_n) - f(x_{n-1})$ hours.

Suppose $g(A)$ be the number of the products made by group A in one hour or the productivity in one hour where $A \subset X$.
Then, the total number of products produced by the workers will be:

$$\int f dg = f(x_1) \cdot g(X) + (f(x_2) - f(x_1)) \cdot g(\{x_2, x_3, \ldots, x_n\})$$

$$+ (f(x_3) - f(x_2)) \cdot g(\{x_3, \ldots, x_n\}) + \cdots + (f(x_n) - f(x_{n-1})) \cdot g(\{x_n\})$$

$$= \sum_{i=1}^{n} (f(x_i) - f(x_{i-1})) \cdot g(A_i).$$

Let us take an example:

Example 7 Consider a set $X = \{x_1, x_2, x_3\}$ and the function $f(x)$ or the range is defined as:
$f(x_1) = 0.5$, $f(x_2) = 0.3$, $f(x_3) = 0.2$. Find the Choquet integral for λ-fuzzy measure.
The fuzzy density values are given as:

$$g(\{x_1\}) = 0.2, g(\{x_2\}) = 0.5, g(\{x_3\}) = 0.6.$$

Solution

First we calculate the value of λ using Eq. (4.4) to find $g_\lambda(\{x_1, x_2\})$, $g_\lambda(\{x_1, x_3\})$, $g_\lambda(\{x_2, x_3\})$.
From Eq. (4.4), $\lambda + 1 = \prod_{i=1}^{n}(1 + \lambda g_i)$, considering

$$g_i = g(\{x_i\})$$

$$= (1 + 0.2\lambda)(1 + 0.5\lambda)(1 + 0.6\lambda)$$

$$= (1 + 0.7\lambda + 0.1\lambda^2)(1 + 0.6\lambda).$$

$$\Longrightarrow \lambda(0.06\lambda^2 + 0.52\lambda + 0.3) = 0$$

$$\Longrightarrow \lambda = \{0, -0.6215, -8.0462\}.$$

A $\lambda > -1$, so $\lambda = -0.6215$.

At $\lambda = -0.6215$, we will compute

$$g_\lambda(\{x_1, x_2\}) = g_\lambda(x_1) + g_\lambda(x_2) + \lambda \cdot g_\lambda(x_1) \cdot g_\lambda(x_2)$$

$$= 0.2 + 0.5 - 0.6215 \times 0.2 \times 0.5$$

$$= 0.7 - 0.0622 = 0.6378,$$

Likewise, $g_\lambda(\{x_1, x_3\}) = g_\lambda(x_1) + g_\lambda(x_3) + \lambda \cdot g_\lambda(x_1) \cdot g_\lambda(x_3)$

$$= 0.8 - 0.0746 = 0.7254,$$

$$g_\lambda(\{x_2, x_3\}) = g_\lambda(x_2) + g_\lambda(x_3) + \lambda \cdot g_\lambda(x_2) \cdot g_\lambda(x_3)$$

$$= 1.1 - 0.1865 = 0.9135,$$

$$g_\lambda(X) = g_\lambda(x_1, x_2) + g_\lambda(x_3) + g_\lambda(\{x_1, x_2\}) \cdot g_\lambda(x_3) = 1.$$

g_λ is a monotonic fuzzy measure. For simpicity, we remove the subscript λ from g_λ.

As the orders of $f(x_i)$ are not arranged in ascending order, we get permuted $f(x_{(i)})$ by arranging in ascending order $f(x_{(1)}) \le f(x_{(2)}) \le f(x_{(3)})$.

So, we get $f(x_{(1)}) = 0.2, f(x_{(2)}) = 0.3, f(x_{(3)}) = 0.5$.

We see that $(1) = 3, (2) = 2, (3) = 1$.

Then, Choquet integral $C_g = \sum_{i=1}^{3} \left(f(x_{(i)}) - f(x_{(i-1)}) \right) g(A_{(i)})$.

Now, we will use (i) to find $A_{(i)}$.

Here, $g(A_{(1)}) = g(\{x_{(1)}, x_{(2)}, x_{(3)}\}) = g(\{x_3, x_2, x_1\})$,

Likewise, $g(A_{(2)}) = g(\{x_{(2)}, x_{(3)}\}) = g(\{x_2, x_1\})$,

$$g(A_{(3)}) = g(\{x_{(3)}\}) = g(\{x_1\}).$$

$$C_g(f) = \left(f(x_{(1)}) - f(x_{(0)}) \right) g(A_{(1)}) + \left(f(x_{(2)}) - f(x_{(1)}) \right) g(A_{(2)})$$

$$+ \left(f(x_{(3)}) - f(x_{(2)}) \right) g(A_{(3)})$$

$$= f(x_{(1)}) \cdot g(\{x_1, x_2, x_3\}) + \left(f(x_{(2)}) - f(x_{(1)}) \right) \cdot g(\{x_2, x_1\})$$

$$+ \left(f(x_{(3)}) - f(x_{(2)}) \right) \cdot g(\{x_1\})$$

$$= 0.2 \times 1.0 + (0.3 - 0.2) \times 0.6378 + (0.5 - 0.3) \times 0.2$$

$$= 0.2 + 0.0638 + 0.04 = 0.3038.$$

This is Choquet integral for a fuzzy measure g with respect to function f.

It is to be noted that Sugeno integral and Choquet integral are different. Sugeno integral depends on nonlinear operator, i.e. max–min function and Choquet integral is based on linear operator.

A practical example on Choquet integral is given, based on the performance of school students in three subjects.

Example 8 Consider three students and these students are evaluated on the basis of their the performance in three subjects – physics, mathematics, and literature. Normally, students who are good in physics are also good in mathematics. The normalized performance/ marks is given in tabular form (marks obtained by the students are on a 20-point scale).

Students	Mathematics	Physics	Literature
A	0.85	0.75	0.55
B	0.7	0.75	0.75
C	0.55	0.65	0.9

As the school is more of scientific oriented, weights of physics and mathematics are assumed to be similar. School wants to evaluate a student whose performance is equally good in all the subjects.

Solution

Let us assume x_1= mathematics, x_2= physics, and x_3= literature.
Marks obtained by students in three subjects – mathematics, physics, and literature are written as follows.

$$A = \{0.85, 0.75, 0.55\}, \ B = \{0.70, 0.75, 0.75\}, \ C = \{0.55, 0.65, 0.9\}.$$

As the school is more science based, so fuzzy measures of mathematics and physics are considered to be similar. Fuzzy measures on three subjects [15] on $X = \{x_1, x_2, x_3\}$ are

$$g(\{x_1\}) = 0.45, \ g(\{x_2\}) = 0.45, \ g(\{x_3\}) = 0.3, \ g(\{x_1, x_2\})$$
$$= 0.5, \ g(\{x_2, x_3\}) = 0.9, \ g(\{x_1, x_3\}) = 0.9.$$

Marks of student A: $a_1 = 0.85$, $a_2 = 0.75$, $a_3 = 0.55$. As these are not arranged in ascending, we get permuted $a_{(i)}$ by arranging a_i as: $a_{(1)} \le a_{(2)} \le a_{(3)}$.
So, $a_{(1)} = 0.55$, $a_{(2)} = 0.75$, $a_{(3)} = 0.85$.
We see that (1) = 3, (2) = 2, (3) = 1.
Aggregation of student A: $C_g = \sum_{i=1}^{3} \left(a_{(i)} - a_{(i-1)} \right) g\left(A_{(i)} \right)$, $A_{(i)} = \{x_{(i)}, \dots, x_{(3)}\}$, i = 1, 2, 3
Here,

$$g\left(A_{(1)} \right) = g\left(A_{(1)} \right) = \{x_{(1)}, x_{(2)}, x_{(3)}\} = g(\{x_3, x_2, x_1\}),$$
$$g\left(A_{(2)} \right) = g\left(\{x_{(2)}, x_{(3)}\} \right) = g(\{x_2, x_1\}),$$
$$g\left(A_{(3)} \right) = g\left(\{x_{(3)}\} \right) = g(\{x_1\}).$$

So, $C_g = a_{(1)}g(\{x_1,x_2,x_3\}) + (a_{(2)} - a_{(1)}) \cdot g(\{x_2,x_1\}) + (a_{(3)} - a_{(2)}) \cdot g(\{x_1\})$

$= 0.55 \times 1.0 + 0.2 \times 0.5 + 0.1 \times 0.45$

$= 0.55 + 0.1 + 0.045 = 0.695,$

Likewise, marks for student B: $b_1 = 0.7$, $b_2 = 0.75$, $b_3 = 0.75$. As these are already arranged in ascending order, so, $b_{(1)} = 0.7$, $b_{(2)} = 0.75$, $b_{(3)} = 0.75$

Aggregation of student B: $C_g = \sum_{i=1}^{3} (b_{(i)} - b_{(i-1)}) g(A_{(i)})$

$= b_{(1)} \cdot g(\{x_1,x_2,x_3\}) + (b_{(2)} - b_{(1)}) \cdot g(\{x_2,x_3\}) + (b_{(3)} - b_{(2)}) \cdot g(\{x_3\})$

$= 0.7 \times 1.0 + 0.05 \times 0.9 + 0 \times 0.3$

$= 0.7 + 0.045 + 0 = 0.745,$

Likewise, marks for student C: $c_1 = 0.55$, $c_2 = 0.65$, $c_3 = 0.9$. As these are already arranged in ascending order, so, c_i is $c_{(1)} = 0.55$, $c_{(2)} = 0.65$, $c_{(3)} = 0.9$.

Aggregation of student C: $C_g = \sum_{i=1}^{3} (c_{(i)} - c_{(i-1)}) g(A_{(i)})$

$= 0.55 \times 1.0 + 0.1 \times 0.9 + 0.25 \times 0.3$

$= 0.55 + 0.09 + 0.075 = 0.715.$

So the evaluations of students are: $A = 0.695$, $B = 0.745$, $C = 0.715$. So, student B is the best who performs almost equally well in all the subjects. Then comes student C and then student A.

4.4.3 Šipoš Integral

This integral follows the principle of fuzzy integral. Let g be a fuzzy measure on X and f be a real-valued function $f: X \rightarrow [0,1]$ with range $\{a_1, a_2, a_3, \ldots, a_m, b_1, b_2, b_3, \ldots, b_n\}$ where $b_n \leq b_{n-1} \leq \ldots \leq b_1 \leq 0 \leq a_1 \leq a_2 \leq \ldots \leq a_m$. Then Šipoš integral of f with respect to a fuzzy measure g is defined as [2]:

$$\check{S}_g \int f dg = \sum_{i=1}^{n} (a_i - a_{i-1}) g(\{x \mid f(x) \geq a_i\}) + \sum_{i=1}^{n} (b_i - b_{i-1}) g(\{x \mid f(x) \leq b_i\}),$$

$A_i \subset X$ for $i = 1,2,3,\ldots,n$

$$(4.15)$$

where $a_0 = 0$ and $b_0 = 0$

$0 \leq a_1 \leq a_2 \leq a_3 \ldots \leq a_n \leq 1,$

$0 \geq b_1 \geq b_2 \geq b_3 \ldots \geq b_n \geq 1$

4.5 Intuitionistic Fuzzy Integral

For real-time decision-making problems, fuzzy integral-based aggregation operators are generally used. But, in real-time decision-making problems, there is always some degree of inter-dependent characteristics between the attributes as there is an interaction among preferences of decision makers. To overcome this limitation, intuitionistic fuzzy integrals are used.

4.5.1 Intuitionistic Fuzzy Choquet Integral

Intuitionistic fuzzy Choquet integral was introduced by Atanassov et al. [17]. This has been used as an aggregation operator [7, 18–20] to aggregate decision-making problems. It is similar to fuzzy integral, the only point that is considered is the nonmembership degree.

If $p_i = (\mu_{p_i}, \nu_{p_i})$, with $i = 1, 2, 3, \dots, n$ be a collection of intuitionistic fuzzy values associated with space X, then intuitionistic fuzzy Choquet integral of p_i with fuzzy measure m on X is written as:

$$
\begin{aligned}
IFC_m(p_1,p_2,p_3,\dots,p_n) \\
= p_{(1)}\left(m\left(A_{(1)}\right)-m\left(A_{(2)}\right)\right) \oplus p_{(2)}\left(m\left(A_{(2)}\right)-m\left(A_{(3)}\right)\right) \\
\oplus \cdots \oplus p_{(n)}\left(m\left(A_{(n)}\right)-m\left(A_{(n+1)}\right)\right) \qquad (4.16) \\
= \sum_{i=1}^{n} \oplus p_{(i)}\left(m\left(A_{(i)}\right)-m\left(A_{(i+1)}\right)\right),
\end{aligned}
$$

where (i) indicates that the indices have been permuted such that

$$
p_{(1)} \leq p_{(2)} \leq p_{(3)} \leq \dots \leq p_{(n)}
$$

and $A_{(i)} = \{x_{(i)}, x_{(i+2)}, x_{(i+3)}, \dots, x_{(n)}\}$, $i = 1, 2, 3, \dots, n$ and $A_{(n+1)} = \emptyset$.
It can also be defined as [21]:

$$
\begin{aligned}
IFC_m(p_1,p_2,p_3,\dots,p_n) = p_{(1)}\left(m\left(A_{(1)}\right)-m\left(A_{(0)}\right)\right) \oplus \\
p_{(2)}\left(m\left(A_{(2)}\right)-m\left(A_{(1)}\right)\right) \oplus \cdots \oplus \\
p_{(n)}\left(m\left(A_{(n)}\right)-m\left(A_{(n-1)}\right)\right) \qquad (4.17) \\
= \sum_{i=1}^{n} \oplus p_{(i)}\left(m(A_i)-m(A_{i-1})\right)
\end{aligned}
$$

such that $p_{(1)} \geq p_{(2)} \geq p_{(3)} \geq \dots \geq p_{(n)}$
where $A_{(i)} = \{x_{(1)}, x_{(2)}, x_{(3)}, \dots, x_{(i)}\}$, $i = 1, 2, 3, \dots, n$ and $A_{(0)} = \emptyset$.

The aggregated value of p_i, which is an intuitionistic fuzzy value, is given as:

$$IFC_m(p_1, p_2, p_3, \ldots, p_n)$$

$$= \left(1 - \prod_{i=1}^{n}\left(1 - \mu_{p(i)}\right)^{m\left(A_{(i)}\right) - m\left(A_{(i-1)}\right)}, \prod_{i=1}^{n}\nu_{p(i)}{}^{m\left(A_{(i)}\right) - m\left(A_{(i-1)}\right)}\right).$$

Fuzzy and intuitionistic fuzzy integrals are used as aggregation operators which are used in decision-making problems. There are different types of aggregation operators using fuzzy/intuitionistic fuzzy integrals and these are explained with examples in the next chapter.

4.6 Summary

This chapter provides an overview of different types of fuzzy measures such as possibility measure, belief measure, plausibility measures, and necessity measures. Fuzzy integrals that include Sugeno, Choquet, and Šipoš integrals are also explained along with examples. Practical examples related to factory or day-to-day work are also presented. Also, Choquet integral operator using intuitionistic fuzzy set is discussed with examples.

References

1 Sugeno, M. (1974). Theory of fuzzy integrals and its applications. PhD thesis, Tokyo Institute of Technology.

2 Murofushi, T. and Sugeno, M. (1991). Theory of fuzzy measures: representation, the Choquet integral and null sets. *Journal of Mathematical Analysis and Applications* 159: 532–549.

3 Turksen, I.B. (2004). Belief, plausibility, and probability measures on interval-valued type 2 fuzzy sets. *International Journal of Intelligent Systems* 19: 681–699.

4 Murofushi, T. and Sugeno, M. (2000). Fuzzy measures and fuzzy integrals. In: *Fuzzy Measures and Integrals: Theory and Applications* (ed. M. Grabisch, T. Murofushi and M. Sugeno), 3–41. Physica-Verlag.

5 Leszczyński, K., Penczek, P., and Grochulski, W. (1985). Sugeno fuzzy measure and fuzzy clustering. *Fuzzy Sets and Systems* 15: 147–158.

6 Shafer, G. (1976). *Mathematical Theory of Evidence*. Princeton, NJ: Princeton University Press.

7 Torra, V. and Narikawa, Y. (2006). Interpretation to fuzzy integrals and application to fuzzy systems. *International Journal of Approximate Reasoning* 41: 43–58.

8 Dubois, D. and Prade, H. (1980). *Fuzzy Sets and Systems: Theory and Application*. Boston, MA: Academic Press.

9 Inuiguchi, M., Greco, S., Słowinski, R., and Tanino, T. (2003). Possibility and necessity measure specification using modifiers for decision making under fuzziness. *Fuzzy Sets and Systems* 137: 151–175.

10 Zadeh, L.A. (1978). Fuzzy sets as a basis for the theory of possibility. *Fuzzy Sets and Systems* 1: 3–28.

11 Puri, M.L. and Ralescvu, D. (1982). Possibility measure is not a fuzzy measure. *Fuzzy sets and Systems* 7: 311–313.

12 Zimmerman, H.J. (2001). *Fuzzy Set Theory and Its Application.* Kluwer Academic Publishers.

13 Wang, Z. and Klir, G.J. (1992). *Fuzzy Measure Theory,* 1992. New York: Plenum Press.

14 Dubois, D., Prade, H., and Sabbadin, R. (2001). Decision-theoretic foundations of qualitative possibility theory. *European Journal of Operational Research* 128: 459–478.

15 Grabisch, M. (1996). The application of fuzzy integrals in multicriteria decision making. *European Journal of Operational Research* 89: 445–456.

16 Murofushi, T., Sugeno, M., and Mahida, M. (1994). Non-monotonic fuzzy measures and the Choquet integral. *Fuzzy Sets and Systems* 64: 73–86.

17 Atanassov, K., Vassilev, P., and Tsvetkov, R. (2013). *Intuitionistic Fuzzy Sets, Measures and Integrals.* Sofia: "Prof. M. Drinov" Academic Publishing House.

18 Melin, P., Martinez, G.E., and Tsvetkov, R. (2017). Choquet and Sugeno integrals and intuitionistic fuzzy integrals as aggregation operators. *Notes on Intuitionistic Fuzzy Sets* 23 (1): 95–99.

19 Narukawa, Y. and Murofushi, T. (2003). Choquet integral and Sugeno integral as aggregation function. In: *Information Fusion in Data Mining* (ed. V. Torra), 27–39. Berlin Heidelberg: Springer-Verlag.

20 Tan, C. and Chen, X. (2010). Intuitionistic fuzzy Choquet integral operator for multi-criteria decision making. *Expert Systems with Applications* 37 (2010): 149–157.

21 Tan, C. and Chen, X. (2010). Induced Choquet ordered averaging operator and its application to group decision making. *International Journal of Intelligent Systems* 25: 59–82.

5

Operations on Fuzzy/Intuitionistic Fuzzy Sets and Application in Decision Making

5.1 Introduction

A fuzzy set may be viewed as an extension of an ordinary set where operations like union, intersection, and complement have been extended to logical operations. There are lot of operators on fuzzy sets implementing basic operations of union and intersection. Among the well-known operations, which are performed on fuzzy sets are fuzzy union, fuzzy intersection, fuzzy complement, fuzzy algebraic sum, and algebraic product. Intuitionistic fuzzy set, which is an extension of fuzzy set, is very much useful when dealing with real-time application that takes into account more uncertainty. Like fuzzy set, there are also few operators on intuitionistic fuzzy set. These operators are used in almost all types of applications such as fuzzy/intuitionistic fuzzy decision making, image processing, pattern recognition, linguistics, information retrieval, and so on. In addition, Zadeh [1] introduced bounded sum and bounded difference, which are used in fuzzy reasoning, decision making, pattern recognition, information retrieval, and so on.

5.2 Fuzzy Operations

In this section we will define various types of fuzzy operations. Among many operations that are defined in this section, fuzzy union, intersection, and fuzzy complements are the most common operations that are used.

Fuzzy Set and Its Extension: The Intuitionistic Fuzzy Set, First Edition. Tamalika Chaira.
© 2019 John Wiley & Sons, Inc. Published 2019 by John Wiley & Sons, Inc.

5.2.1 Fuzzy Union

Union of two fuzzy sets A and B is given as:

$$A \cup B = \{x, \mu_{A \cup B}(x) \mid x \in X\},$$

where $\mu_{A \cup B} = \mu(A) \vee \mu(B)$.

5.2.2 Fuzzy Intersection

Intersection of two fuzzy sets A and B is given as:

$$A \cap B = \{\mu_{A \cap B}(x) \mid x \in X\},$$

where $\mu_{A \cap B} = \mu(A) \wedge \mu(B)$.

5.2.3 Fuzzy Complements

Let A be a fuzzy set on a universe of discourse X and $\mu_A(x)$ be the membership function. Fuzzy complement is a fuzzy negation. Fuzzy complement is defined as a function $c(\mu_A(x)) : [0,1] \rightarrow [0,1]$ that assigns a value to each membership degree $\mu_A(x)$ of any given set A. It is defined as:

$$c(\mu_A(x)) = 1 - \mu_A(x), \text{ for all } x \in A.$$

Properties of fuzzy complement:

i) c is involutive implies $c(c(a)) = a$, $a \in [0,1]$,
ii) c is continuous,
iii) Boundary condition – $c(1) = 0$, $c(0) = 1$,
iv) Monotonicity – if $a \leq b$, then $c(a) \geq c(b)$, for all $a,b \in [0,1]$.

This is the simplest and extensively used in fuzzy set theory. There are different types of fuzzy complements in the literature.

Fuzzy complement is computed from fuzzy complement functional which is defined as:

$$N(\mu(x)) = f^{-1}(f(1)) - f(\mu(x)), \tag{5.1}$$

where $f(\cdot)$ is an increasing function with $f(0) = 0$.

Some of the fuzzy complements suggested by different authors are as follows:

a) Sugeno-type fuzzy complement [2] is generated using fuzzy complement function using an increasing function which is given as:

$$f(\mu(x)) = \frac{1}{\alpha}\log(1 + \alpha\mu(x)) \tag{5.2}$$

Inverse function of $f(\mu(x))$ is

$$f^{-1}(\mu(y)) = \frac{e^{\alpha\mu(y)} - 1}{\alpha}. \tag{5.3}$$

Substituting the value $f(\mu(x))$ in Eq. (5.2), we get

$$N(\mu(x)) = f^{-1}\left[\frac{1}{\alpha}\log(1+\alpha) - \frac{1}{\alpha}\log(1+\alpha\cdot\mu(x))\right]$$

$$\Rightarrow N(\mu(x)) = f^{-1}\left[\frac{1}{\alpha}\log\left(\frac{1+\alpha}{1+\alpha\cdot\mu(x)}\right)\right]$$

By induction method, Eq. (5.3) becomes

$$f^{-1}\left[\frac{1}{\alpha}\log\left(\frac{1+\alpha}{1+\alpha\cdot\mu(x)}\right)\right] = \frac{e^{\alpha\left(\frac{1}{\alpha}\log\frac{1+\alpha}{1+\alpha\cdot\mu(x)}\right)} - 1}{\alpha}$$

$$\frac{\dfrac{1+\alpha}{1+\alpha\cdot\mu(x)} - 1}{\alpha} = \frac{1-\mu(x)}{1+\alpha\cdot\mu(x)}$$

Hence,

$$N(\mu(x)) = \frac{1-\mu(x)}{1+\alpha\cdot\mu(x)}, \text{ with } \alpha \in [-1, \infty]. \tag{5.4}$$

When $\alpha = 0$, Eq. (5.4) reduces to a standard fuzzy complement: $N(\mu(x)) = 1 - \mu(x)$.
b) Yager [3] proposed another type of fuzzy complement function using a function:

$$f(\mu(x)) = \mu(x)^{\beta}.$$

Fuzzy complement is given as:

$$N(\mu(x)) = \left[1 - \mu(x)^{\beta}\right]^{\frac{1}{\beta}}, \beta \in (0, \infty). \tag{5.5}$$

When $\beta = 1$, Eq. (5.5) reduces to $N(\mu(x)) = 1 - \mu(x)$.
c) Roychowdhury and Wang [4] suggested a different type of fuzzy complement function

$$N(\mu(x)) = f^{-1}\left(\frac{-f(\mu(x))}{1 + f(\mu(x))}\right),$$

where $f: [0,1] \rightarrow (-\infty, -1)$ is a continuous function with $f(0) = -\infty$ and $f(1)$ = -1, if it is strictly increasing and $f(0) = 1$ and $f(1) = -\infty$, if it is strictly decreasing.

Different types of fuzzy complements can be obtained using different functions, $f(\mu(x))$.

d) Chaira [5] introduced another type of fuzzy complement. Let the function be

$$f(\mu(x)) = \frac{1}{\lambda}\ln\left[1 + \mu(x)\left(1 - e^{-\lambda}\right)\right], \ \mu(x) \in [0,1]$$

Using the inverse function and induction method, the fuzzy complement is given as:

$$N(\mu(x)) = \frac{1 - \mu(x)}{1 - (1 - e^{\lambda})\mu(x)} = \frac{1 - \mu(x)}{1 + (e^{\lambda} - 1)\mu(x)}, \ \lambda > 0. \tag{5.6}$$

When $\lambda = 0$, Eq. (5.6) reduces to standard fuzzy complement.

e) Another form of increasing function is given an author:

$$f(\mu(x)) = \frac{\mu(x)}{\lambda + (1 - \lambda)\mu(x)}$$

and its complement function is given as:

$$N(\mu(x)) = \frac{\lambda^2(1 - \mu(x))}{\lambda^2(1 - \mu(x)) + \mu(x)}, \ \lambda > 0 \tag{5.7}$$

At $\lambda = 1$, Eq. (5.7) reduces to a standard fuzzy complement.

There is another generating function which is a dual generator suggested by Klir and Yuan [6].

$$N(\mu(x)) = f^{-1}(f(0)) - f(\mu(x)), \tag{5.8}$$

where $f(\cdot)$ is a decreasing function.

Using this complement equation, we can also find fuzzy complement.

5.2.4 Algebraic Product

Product of two fuzzy sets A and B is given in terms of membership function as:

$$P = A \cdot B = \{x, \mu_{A \cdot B}(x) \mid x \in X\},$$

where

$$= \mu_{A \cdot B} = \mu(A) \cdot \mu(B). \tag{5.9}$$

5.2.5 Algebraic Sum

Algebraic sum of two fuzzy sets A and B is given as:

$$P = A + B = \{x, \mu_{A+B}(x) \mid x \in X\},$$

where

$$\mu_{A+B} = \mu(A) + \mu(B) - \mu(A) \cdot \mu(B). \tag{5.10}$$

5.2.6 Simple Difference

Algebraic sum of two fuzzy sets A and B is given as:

$$P = A - B = \{x, \mu_{A-B}(x) \mid x \in X\},$$

where

$$\mu_{A-B} = \min(\mu(A), 1 - \mu(B)). \tag{5.11}$$

5.2.7 Bounded Sum

Bounded sum of two fuzzy sets A and B is given as:

$$P = A \oplus B = \{x, \mu_{A \oplus B}(x) \mid x \in X\},$$

where

$$\begin{aligned}
\mu_{A \oplus B} &= \min(1, \mu(A) + \mu(B)) \\
&= 1 \wedge (\mu(A) + \mu(B)).
\end{aligned} \tag{5.12}$$

5.2.8 Bounded Difference

Bounded difference of two fuzzy sets A and B is given as:

$$P = A \ominus B = \{x, \mu_{A \oplus B}(x) \mid x \in X\},$$

where

$$\mu_{A \ominus B} = \max(0, \mu(A) - \mu(B)). \tag{5.13}$$

5.2.9 Bounded Product

$$P = A \odot B = \{x, \mu_{A \odot B}(x) \mid x \in X\},$$

where

$$\mu_{A \odot B} = 0 \vee (\mu(A) + \mu(B) - 1). \tag{5.14}$$

Example 1   Consider two fuzzy sets
Let

$A = \{(3,0.4),(5,0.7),(7,0.2),(9,0.6)\}$
$B = \{(5,1.0),(9,0.9)\}$,
$A \cup B = \{(5,1),(9,0.9)\}$,
$A \cap B = \{(5,0.7),(9,0.6)\}$,
$A + B = \{(3,0.4),(5,1),(7,0.2),(9,0.96)\}$,
$\bar{B} = \{(0.5,0.0),(9,0.1)\}$,
$A - B = \{(0.5,0),(9,0.1)\}$,
$A \oplus B = \{(3,0.4),(5,1),(7,0.2),(9,1)\}$,
$A \ominus B = \{(5,0),(9,0)\}$,
$A \cdot B = \{(5,0.7),(9,0.54)\}$.

5.3   Fuzzy Other Operators: Fuzzy T-Norms and T-Conorms

Triangular norms and triangular conorms originated from probabilistic metric spaces in which triangular inequalities are extended using the theory of T-norms and T-conorms. T-norm and a T-conorm are generalizations of the classical conjunction and disjunction. Later, many authors [7–12] introduced T-norms and T-conorms in fuzzy set theory and these norms can be used as intersection and union of fuzzy sets, respectively. T-norms and T-conorms are used to find the membership function of union and intersection of fuzzy sets, respectively. They are a kind of binary operation used in the framework of fuzzy logic and probabilistic metric space.

5.3.1   Definition of T-Norm

Let T be a T-norm such that $T : [0,1] \times [0,1] \to [0,1]$ that represents intersection in fuzzy set theory or an "anding" operator. It satisfies the following properties, where for all $x,y,z \to [0,1]$ [13, 14]:

1) $T(1,1) = 1$ and $T(0,1) = T(1,0) = T(0,0) = 0$ (Boundary condition),
2) $T(x,y) = T(y,x)$ (Commutativity),
3) $T(T(x,y),z) = T(x,T(y,z))$ (Associativity),
4) $T(x,y) \leq T(x,z)$ if $y \leq z$ (Monotonicity),
5) $T(x,1) = x$ (One-identity).

A T-norm, T is called Archimedian iff

$T(x,y)$ is continuous,
$T(x,x) < x$ for all $x \in [0,1]$.

An Archimedian T-norm is strict iff

$T(x_1,y_1) < T(x,y)$ if $x_1 < x,\ y_1 < y\ \forall x,y,x_1,y_1 \in [0,1]$.

T is called an Archimedian T norm iff there exists a decreasing and continuous function $f: [0,1] \rightarrow [0,\infty]$ with $f(1) = 0$ so that,

$$T(x,y) = f^{-1}(f(x) + f(y)), \tag{5.15}$$

where function $T: [0,1] \times [0,1] \rightarrow [0,1]$ and f is an additive generator.

5.3.2 Definition of T-Conorm

Let T^* be a T-conorm $T^*: [0,1] \times [0,1] \rightarrow [0,1]$. It represents union in fuzzy set theory or an "oring" operator [4, 14]. It satisfies the following properties, where for all $x,y,z \rightarrow [0,1]$:

1) $T^*(0,0) = 0$ and $T^*(0,1) = T^*(1,0) = T^*(1,1) = 1$ (Boundary condition),
2) $T^*(x,y) = T^*(y,x)$ (Commutativity),
3) $T^*(x,y) \leq T^*(x,z)$ if $y \leq z$ (Monotinicity),
4) $T^*(T^*(x,y),z) = T^*(x,T^*(y,z))$ (Associativity),
5) $T^*(x,0) = x$ (Zero-identity).

A T-conorm, T^* is called Archimedian iff

$T^*(x,y)$ is continuous
$T^*(x,x) > x$ for all $x \in [0,1]$.

An Archimedian T-conorm is strict iff

$T^*(x_1,y_1) < T^*(x,y)$ if $x_1 < x$, $y_1 < y \; \forall x,y,x_1,y_1 \in [0,1]$.

A function $T^*: [0,1] \times [0,1] \rightarrow [0,1]$ is called an Archimedian T-conorm, iff there exists an increasing and continuous function $g: [0,1] \rightarrow [0,\infty]$ with $g(0) = 0$, so that

$$T^*(x,y) = g^{-1}(g(x) + g(y)), \tag{5.16}$$

where g is an additive generator.

There are various types of triangular operators suggested by different authors and these are classified as a class of (i) conditional operators and (ii) algebraic operators [4, 12, 14]. Conditional operators contain max or min operators or the combination of min. and max. operators. T-norms and T-conorms that belong to conditional class are as follows:

i) Zadeh's T operators are the most popular ones and are defined as follows:

$T_z(x,y) = \min(x,y)$.

$T_Z^*(x,y) = \max(x,y)$.

ii) Yager [3]

$$\text{T-norm}: \quad T_Y(x,y) = 1 - \min\left(\left[(1-x)^p + (1-y)^p\right]^{\left(\frac{1}{p}\right)}, 1\right), \tag{5.17}$$

with decreasing generator $f(x) = (1 - x)^p$ and $f^{-1}(y) = 1 - y^{1/p}$, $p > 0$.

$$\text{T-conorm}: \quad T_Y^*(x,y) = \min\left([x^p + y^p]^{1/p}, 1\right), \tag{5.18}$$

with increasing generator $g(x) = x^p$ and $g^{-1}(y) = y^{1/p}$.

iii) Weber operator [14], T-norm:

$$\text{T-norm}: \quad T_W(x,y) = \max\left(\frac{x + y - 1 + \gamma xy}{1 + \gamma}, 0\right), \gamma > -1, \tag{5.19}$$

with decreasing generator $f(x) = 1 - \dfrac{\ln(1 + \gamma x)}{\ln(1 + \gamma)}$ and $f^{-1}(y)$

$= \dfrac{1}{\gamma}\left[(1 + \gamma)^{1-y} - 1\right], y \leq 1$.

$$\text{T-conorm}, \quad T_W^*(x,y) = \min((x + y + \gamma xy), 1), \tag{5.20}$$

with increasing generator as $g(x) = \dfrac{\ln(1 + \gamma x)}{\ln(1 + \gamma)}$ and $g^{-1}(y) = \dfrac{1}{\gamma}\left[(1 + \gamma)^y - 1\right], y \leq 1$.

iv) Using Lukasiewicz logic, Giles [11] suggested a T operator as:

$$\text{T-norm}: \quad T_L(x,y) = \max(x + y - 1, 0). \tag{5.21}$$

$$\text{T-conorm}: \quad T_L^*(x,y) = \min(x + y, 1). \tag{5.22}$$

v) Dubois and Prade [15] suggested a T operator as:

$$\text{T-norm}, \quad T_D(x,y) = \frac{xy}{\max(x,y,c)}. \tag{5.23}$$

$$\text{T-conorm}, \quad T_D^*(x,y) = 1 - \frac{(1-x)(1-y)}{1 - \max(1-x, 1-y, c)} \quad \text{and} \quad c \in (0,1). \tag{5.24}$$

vi) Another T operator [2] is given as follows:

$$\text{T-norm}, \quad T_S(x,y) = \max(0, (1 + b)(x + y - 1) - bxy). \tag{5.25}$$

$$\text{T-conorm}, \quad T_S^*(x,y) = \min(1, x + y - bxy) \quad \text{and} \quad b > -1. \tag{5.26}$$

Algebraic operators do not contain conditional operators and are purely algebraic in nature. This class of operators does not have conditional function rather, they simply follow arithmetic operations.

Chaira [5] suggested a type of algebraic T operator that does not contain min or max operators. The generating function for T-norm is a decreasing function with $f(1) = 0$, which is given as:

$$f(x) = \ln\left(\frac{1 + (1 - \gamma)(1 - x)}{x}\right), \quad 0 \leq \gamma \leq 1,$$

with $f^{-1}(y) = \dfrac{2 - \gamma}{e^y + 1 - \gamma}$.

With the generating function, T-norm is computed as:

$$T_C(x,y) = \frac{xy}{(\gamma-1)(x+y-xy)+(2-\gamma)}.$$

(5.27)

Generating function for T-conorm $g(\cdot)$ is an increasing function such that $g(0) = 0$ is given as:

$$g(x) = \ln\left(\frac{1+(1+\gamma)xy}{1-x}\right), \quad 0 \le \gamma \le 1,$$

with $g^{-1}(y) = \dfrac{e^y-1}{e^y+1+\gamma}$.

Similar to T-norm, T-conorm is given as:

$$T_C^*(x,y) = \frac{x+y+\gamma xy}{(1+\gamma)xy+1}.$$

(5.28)

i) Bandler and Kohout [8] suggested T operators which are given as:

T-norm-$T_B(x,y) = x \cdot y.$

(5.29)

T-conorm-$T_B^*(x,y) = x + y - x \cdot y.$

(5.30)

ii) Hamacher [16] proposed T operators that are algebraic in nature.

$$\text{T-norm-}T_H(x,y) = \frac{x \cdot y}{\gamma + (1-\gamma) \cdot (x+y-x \cdot y)}, \quad \gamma > 0,$$

(5.31)

with decreasing generator

$$f(x) = \frac{1}{\gamma}\ln\frac{\gamma+(1-\gamma)x}{x} \text{ and } f^{-1}(y) = \frac{\gamma \cdot e^{-\gamma y}}{1-(1-\gamma) \cdot e^{-\gamma y}}.$$

$$\text{T-conorm, } T_H^*(x,y) = \frac{x+y-x \cdot y-(1-\gamma)xy}{1-(1-\gamma) \cdot xy},$$

(5.32)

with increasing generator is given as:

$$g(x) = \frac{1}{\gamma}\ln\frac{\gamma+(1-\gamma) \cdot (1-x)}{1-x} \text{ and } g^{-1}(y) = \frac{1-e^{-\gamma y}}{1-(1-\gamma) \cdot e^{-\gamma y}}.$$

iii) Dombi [9] also proposed T-norm and T-conorm that are algebraic in nature

$$\text{T-norm-}T_D(x,y) = \frac{1}{1+\left(\left(\frac{1}{x}-1\right)^\gamma+\left(\frac{1}{y}-1\right)^\gamma\right)^{1/\gamma}}.$$

(5.33)

with $f(x) = \dfrac{1}{(1+x)^{1/\gamma}}$ and $f^{-1}(y) = \left(\dfrac{1}{x}-1\right)^\gamma$.

$$\text{T-conorm}: T_D^*(x,y) = \frac{1}{1 + \left(\left(\frac{1}{x}-1\right)^{-\gamma} + \left(\frac{1}{y}-1\right)^{-\gamma}\right)^{-\frac{1}{\gamma}}}, \quad \gamma > 0, \quad (5.34)$$

with $g(x) = \dfrac{1}{(1+x)^{-1/\gamma}}$ and $g^{-1}(y) = \left(\dfrac{1}{x}-1\right)^{-\gamma}$.

iv) Frank [10] proposed logarithmic T-norm and T-conorms:

$$T_F(x,y) = \log_a\left(1 + \frac{(a^x-1)(a^y-1)}{a-1}\right). \quad (5.35)$$

$$T_B^*(x,y) = 1 - \log_a\left(1 - \frac{(a^{1-x}-1)(a^{1-y}-1)}{a-1}\right), \quad a \in (0,\infty). \quad (5.36)$$

The most basic T operators that are normally used are as follows:

T-norms:

i) Zadeh's intersection: $T_M(x,y) = \min(x,y)$.
ii) Product intersection: $T_P(x,y) = x \cdot y$.
iii) Lukasiewicz intersection: $T_L(x,y) = \max(x+y-1,0)$,
iv) Nilpotent T-norm: $T_N(x,y) = \min(x,y)$, if $(x+y) > 1$

$$= 0, \qquad \text{otherwise.}$$

T co-norms:

i) Zadeh's union: $T_M^*(x,y) = \max(x,y)$.
ii) Product union: $T_P^*(x,y) = x+y-x \cdot y$,
iii) Lukasiewicz union: $T_L^*(x,y) = \min(x+y,1)$.
iv) Nilpotent T-conorm: $T_N^*(x,y) = \max(x,y)$, if $(x+y) < 1$

$$= 0, \qquad \text{otherwise.}$$

5.4 Implication Operator

Just like T-norms and T-conorms are generalization of fuzzy conjunction and disjunction, respectively, fuzzy implication is a generalization of the classical Boolean implication to fuzzy logic. The basic interpretation of implication is that $p \rightarrow q$ is true when q is as true as p. It is false when q is as false as p is true. The overall interpretation is that the degree of truth of $p \rightarrow q$ expresses the degree by which q is at least as true as p. This is essential in many different fields. To name a few, its application ranges from approximate reasoning and fuzzy control to fuzzy mathematical morphology and image processing. There exist many families of implication operator, the most basics are S-implication (S stands for T-conorm), R-implication (residual implication), and Q–L (quantum logic) implication. Yager [17] suggested f-generated implication operators.

Definition

I is an implication function with $I: [0,1] \times [0,1] \to [0,1]$, such that [18, 19]:

i) $I(1,x) = x$,
ii) If $x_1 \le x_2$ then $I(x_1,y) \ge I(x_2,y)$, i.e. $I(.,y)$ is decreasing,
iii) If $y_1 \le y_2$ then $I(x,y_1) \le I(x,y_2)$, i.e. $I(x,.)$ is increasing,
iv) $I(0,y) = 1$
v) $I(x,1) = 1$
vi) $I(1,0) = 0$.

Some of the desirable properties are mentioned below [18].

$I(1,y) = y$
$I(x,(y,z)) = I(y,(x,z))$,
$I(x,y) = 1$, iff $x \le y$,
$I(x,0) = 1 - x$,
$I(x,y) \ge y$,
$I(x,x) = 1$.

S-implications are based on classical operator, which is given as:

$I(x,y) = S(n(x),y)$, n is a strong negation and S is a strong T-conorm.

Some S-implication operators are:

Lukasiewicz implication function $- I_{LK}(x,y) = \min(1, 1 - x + y)$.
Kleene–Dienes implication function $- I_{KD}(x,y) = \max(1 - x, y)$.
Reichenbach implication function $- I_{LK}(x,y) = 1 - x + xy$.

Some examples of R-implications are:

$$\text{Godel implication-}I_1(x,y) = \begin{cases} 1, & \text{if } x \le y \\ y, & \text{if } x > y \end{cases}$$

$$\text{Goguen-}I_1(x,y) = \begin{cases} 1, & \text{if } x \ge y \\ \dfrac{y}{x}, & \text{if } x > y \end{cases}$$

$Q - L$ implication is given as:

$$I(x,y) = S(1 - x, T(x,y)), \text{ where T is a T} - \text{norm}$$

But these basic implication operators sometimes do not follow the condition (ii) [18] and so these are not often used.

Yager suggested f-generated implication operator.

An f-generator is a function $f: [0,1] \to [0,\infty]$, which is a strictly decreasing and continuous function such that $f(1) = 0$. f is a type of additive generator used in T-norms as mentioned earlier.

f-Generated implication operator is given as:

$$I(x,y) = f^{-1}(x \cdot f(y)), \, x, y \in [0,1] \text{ and } I(x,y) \in [0,1].$$

This operator satisfies all the implication properties.
Given below are some generating functions by Yager [19].

i) Let $f(x) = -\ln(x)$ and $f^{-1}(z) = e^{-z}$.
   Then, $I(x,y) = f^{-1}(x(-\ln(y))) = f^{-1}(-x\ln(y)) = e^{x\ln(y)} = e^{\ln(y^x)} = y^x$.
   So, $I(x,y) = y^x$.

ii) $f(x) = 1 - x$ and $f^{-1}(z) = \max(1 - z, 0)$.
   Then, $I(x,y) = f^{-1}(x \cdot (1 - y)) = f^{-1}(x - xy) = \max(1 - x + xy, 0)$.
   Hence, $I(x,y) = \max(1 - x + xy, 0) = 1 - x + xy$, as $x,y \in [0,1]$.

iii) Yager's additive generator for T-norm (discussed earlier) is given as:

$$f(x) = (1-x)^{\omega}, \, \omega > 0 \text{ and } f^{-1}(z) = \max\left(1 - z^{\frac{1}{\omega}}, 0\right).$$

Then, $I(x,y) = f^{-1}(x \cdot (1-y)^{\omega}) = \max\left(1 - x^{\frac{1}{\omega}}(1-y), 0\right) = 1 - x^{\frac{1}{\omega}}(1-y)$ as $x,y \in [0,1]$.

Here, ω shows the strength of implication. One can also use another class of generator where g is an increasing and continuous function with $g(0) = 0$. This is related to a type of generator that represents T-conorm.

5.5 Aggregation Operator with Application in Decision Making

Aggregation operators are mathematical functions that combine different pieces of information. They aggregate all the objects into a single object of a same set so that the final result of aggregation takes into account all the individual values.

These are generally used in computational intelligence where linguistically expressed pieces of information are fused together. Some well-known examples are (i) arithmetic mean, (ii) weighted minimum and maximum, (iii) weighted sum, (iv) median, and (v) ordered weighted averaging (OWA) operators. Also, fuzzy integrals such as Choquet and Sugeno integrals are used as aggregation operator. We will discuss some of the well-known fuzzy aggregating operators.

5.5.1 Fuzzy Weighted Averaging Operator (FWA)

Let X be a set of fuzzy numbers and a_i $(i = 1, 2, 3, \ldots, n)$ be a collection of n numbers or argument variables, which has an associated weighing vector $W = \{w_1, w_2, w_3, \ldots, w_n\}$ of dimension n with $\sum_{i=1}^{n} w_i = 1$. The averaging operator is written as:

$$FWA(a_1, a_2, a_3, \ldots, a_n) = \sum_{j=1}^{n} w_i a_i, \tag{5.37}$$

where a_i is the ith element of the a_i.

5.5.2 Fuzzy Ordered Weighted Averaging Operator (FOWA)

This operator was suggested by Yager [20]. Let a_i, $(i = 1, 2, 3, \ldots, n)$ be a collection of n variables called argument variables and has an associated weighing vector $W = \{w_1, w_2, w_3, \ldots, w_n\}$ of dimension n with $\sum_{j=1}^{n} w_j = 1$. The ordered averaging operator is written as:

$$OWA(a_1, a_2, a_3, \ldots, a_n) = \sum_{j=1}^{n} w_j b_j, \tag{5.38}$$

where $(b_1, b_2, b_3, \ldots, b_n)$ are similar to $(a_1, a_2, a_3, \ldots, a_n)$ but ordered from largest to smallest.

It should be noted that b_j is the jth largest element in a_i. This will become more clear in Example 2.

The reason for using ordered weighting operator is explained using an example.

Example 2 Let us take a general equation of $n = 2$ and the associated weights are [0.4 0.6]. Then

$$OWA(a_1, a_2) = \sum_{i=1}^{2} w_i a_i$$

$$= 0.4 a_1 + 0.6 a_2.$$

If $a_1 = 0$, $a_2 = 1$, then $OWA(a_1, a_2) = 0.6$.
If $a_1 = 1$, $a_2 = 0$, then $OWA(a_1, a_2) = 0.4$.
It is seen that $OWA(0, 1) \neq OWA(1, 0)$.

Example 3 An example of computing ordered weighing averaging operator of four variables $F(0.6, 0.7, 0.2, 0.4)$ whose associated weighting vector is $\begin{bmatrix} 0.3 \\ 0.1 \\ 0.2 \\ 0.4 \end{bmatrix}$.

Solution

The four argument variables are: $a_i = \{0.6, 0.7, 0.2, 0.4\}$.

The order of a_i is rearranged from largest to smallest, so $b_j = \{0.7, 0.6, 0.4, 0.2\}$. So $b_j \neq a_i$.

Then,

$$OWA(a_1, a_2, a_3, a_4) = [0.3\ 0.1\ 0.2\ 0.4] \begin{bmatrix} 0.7 \\ 0.6 \\ 0.4 \\ 0.2 \end{bmatrix}$$

$$= 0.21 + 0.06 + 0.08 + 0.08 = 0.43.$$

5.5.3 Fuzzy Generalized Ordered Weighted Averaging Operator (GOWA)

This operator was suggested by Yager [17, 21]. It is similar to OWA where a parameter $\lambda \in [-\infty, \infty]$. The operator of dimension n with an associated weighting vector W of dimension n and $\sum_{i=1}^{n} w_j = 1$ is written as:

$$GOWA(a_1, a_2, a_3, ..., a_n) = \left(\sum_{j=1}^{n} w_j b_j^{\lambda} \right)^{\frac{1}{\lambda}}, \quad \lambda > 0 \tag{5.39}$$

where $(b_1, b_2, b_3, ..., b_n)$ are similar to argument variables $(a_1, a_2, a_3, ..., a_n)$ but ordered from largest to smallest. When $\lambda = 1$, it becomes OWA.

5.5.4 Fuzzy Hybrid Averaging Operator (FHA)

It was suggested by Merigo and Casanovas [13]. It uses weighted averaging operator and also ordered average weighted operator together. Let a_i, $(i = 1, 2, 3, ..., n)$ be a collection of n numbers and has an associated weighing vector $W = \{w_1, w_2, w_3, ..., w_n\}$ of dimension n with $\sum_{j=1}^{n} w_j = 1$. The averaging operator is written as:

$$OWA(\tilde{a}_1, \tilde{a}_2, \tilde{a}_3, ..., \tilde{a}_n) = \sum_{j=1}^{n} w_j b_j,$$

where b_j is the jth largest element in $\tilde{a}_i$ where $\tilde{a}_i = n\omega_i a_i$, $j = 1, 2, 3, ..., n$ and $\omega = (\omega_1, \omega_2, \omega_3, ..., \omega_n)^T$ is the weighting factor of $\tilde{a}_i$. ω_i lies between 0 and 1 and $\sum_{i=1}^{n} \omega_i = 1$.

5.5.5 Fuzzy Quasi-Arithmetic Weighted Averaging Operator

This was introduced by Yang [22] and Merigo and Casanovas [23]. It is similar to weighted averaging operator where a_i is replaced by a function $f(a_i)$ which is given as:

$$QWA(a_1, a_2, a_3, ..., a_n) = f^{-1} \left(\sum_{i=1}^{n} w_i f(a_i) \right), \tag{5.40}$$

where a_i are the arguments, which are fuzzy numbers and $f(a)$ is a strictly continuous monotonic function.

Similar to quasi-weighted averaging operator and OWA operator, a fuzzy quasi-OWA operator of dimension n has an associated weighting vector W of dimension n with $\sum_{j=1}^{n} w_j = 1$ is written as:

$$QOWA(a_1, a_2, a_3, ..., a_n) = f^{-1} \left(\sum_{j=1}^{n} w_j f(b_j) \right), \tag{5.41}$$

where $f(b)$ is a strictly continuous monotonic function and b_j is the largest of a_i, i.e. $(b_1,b_2,b_3, \ldots ,b_n)$ are similar to $(a_1,a_2,a_3, \ldots ,a_n)$ but ordered from largest to smallest.

If we choose $f(b) = b^\lambda$, we get $f^{-1}(c) = c^{\frac{1}{\lambda}}$.

Then, $QOWA(a_1,a_2,a_3,\ldots,a_n) = f^{-1}\left(\sum_{j=1}^{n} w_j b_j^\lambda\right) = \left(\sum_{j=1}^{n} w_j b_j^\lambda\right)^{\frac{1}{\lambda}}$.

This becomes generalized OWA.

If we choose $f(b) = b$, then $(a_1, a_2, a_3, \ldots, a_n) = f^{-1}\left(\sum_{j=1}^{n} w_j b_j\right) = \sum_{j=1}^{n} w_j b_j$.

This is simple OWA.

5.5.6 Induced Generalized Fuzzy Averaging Operator (IGOWA)

This operator was suggested by Yager [17] and further developed by Merigo and Casanovas [23] which was then later extended by Xu and Wang [24] using intuitionistic fuzzy sets. It is an extension of OWA operator. The difference with respect to OWA is that in OWA, reordering is carried out with the values of arguments a_i, i.e. from largest to smallest, but in induced OWA, reordering is carried out using another variable u_i, where the ordered position of a_i depends on u_i. Just like OWA, if the operator of dimension n has an associated weighting vector W of dimension n with $\sum_{j=1}^{n} w_j = 1$, a parameter $\lambda \in [-\infty,\infty]$ and an order-inducing variable u_i, then IGOWA is written as:

$$I - GOWA(\langle u_1,a_1\rangle,\langle u_2,a_2\rangle,\ldots,\langle u_n,a_n\rangle) = \left(\sum_{j=1}^{n} w_j b_j^\lambda\right)^{\frac{1}{\lambda}}, \tag{5.42}$$

where b_j is the a_i value of the pair $\langle u_i,a_i\rangle$ having the jth largest value u_i of the set $\{u_1,u_2,u_3, \ldots ,u_n\}$, i.e. $(b_1,b_2,b_3, \ldots ,b_n)$ are similar to $(a_1,a_2,a_3, \ldots ,a_n)$ but reordered in decreasing order of the values of the u_i, i.e. from largest to smallest. An example will explain the definition clearly.

Example 4 Let the argument variables along with order-induced variables be: $\langle u_i,a_i\rangle = \langle 8,0.6\rangle,\langle 4,0.7\rangle,\langle 5,0.2\rangle,\langle 3,0.4\rangle$. The associated weighting vector

$W = \begin{bmatrix} 0.3 \\ 0.1 \\ 0.2 \\ 0.4 \end{bmatrix}$. Compute the aggregation of the four variables using $I - GOWA$.

Solution

As per the order-induced variable, reordering of u_i is done as follows:

$$\langle u_i,a_i\rangle = \langle 8,0.6\rangle,\langle 5,0.2\rangle,\langle 4,0.7\rangle,\langle 3,0.4\rangle$$

So, the vector $B = [0.6, 0.2, 0.7, 0.4]$.
With $\lambda = 1$,

$$I - GOWA(\langle u_1, a_1 \rangle, \langle u_2, a_2 \rangle, ..., \langle u_n, a_n \rangle) = \left(\sum_{j=1}^{n} w_j b_j \right)$$

$$= 0.3 \times 0.6 + 0.1 \times 0.2 + 0.2 \times 0.7 + 0.4 \times 0.4$$

$$= 0.18 + 0.02 + 0.14 + 0.16 = 0.50.$$

Sometimes it may happen that there is a tie in ordering operation [25]. In that case aggregations of the tied values are performed. Let us consider an example.

Example 5 Let the argument variables along with order-induced variables be: $\langle u_i, a_i \rangle = \langle 6, 0.3 \rangle, \langle 4, 0.4 \rangle, \langle 6, 0.7 \rangle, \langle 3, 0.2 \rangle$. The associated weighting vector is $W = \begin{bmatrix} 0.3 \\ 0.1 \\ 0.2 \\ 0.4 \end{bmatrix}$.

It is seen that u_i is same for two argument variables. There is a tie between $\langle 6, 0.3 \rangle$ and $\langle 6, 0.7 \rangle$.

As per the order-induced variable, reordering is done as follows:

$$\langle u_i, a_i \rangle = \langle 6, 0.3 \rangle, \langle 6, 0.7 \rangle, \langle 4, 0.4 \rangle, \langle 3, 0.2 \rangle$$

or

$$\langle u_i, a_i \rangle = \langle 6, 0.7 \rangle, \langle 6, 0.3 \rangle, \langle 4, 0.4 \rangle, \langle 3, 0.2 \rangle$$

So, the vector $B = \begin{bmatrix} 0.3 \\ 0.7 \\ 0.4 \\ 0.2 \end{bmatrix}$ or $B = \begin{bmatrix} 0.7 \\ 0.3 \\ 0.4 \\ 0.2 \end{bmatrix}$.

Assuming $\lambda = 1$, $IGOWA_1$ is computed with $B = \begin{bmatrix} 0.3 \\ 0.7 \\ 0.4 \\ 0.2 \end{bmatrix}$ and we get

$$IGOWA_1(\langle u_1, a_1 \rangle, \langle u_2, a_2 \rangle, ..., \langle u_n, a_n \rangle) = 0.3 \times 0.3 + 0.1 \times 0.7 + 0.2 \times 0.4 + 0.4 \times 0.2$$

$$= 0.09 + 0.07 + 0.08 + 0.08 = 0.32.$$

With $B = \begin{bmatrix} 0.7 \\ 0.3 \\ 0.4 \\ 0.2 \end{bmatrix}$, we compute

$$I - GOWA_{2(\langle u_1,a_1 \rangle, \langle u_2,a_2 \rangle, ..., \langle u_n,a_n \rangle)} = 0.3 \times 0.7 + 0.1 \times 0.3 + 0.2 \times 0.4 + 0.4 \times 0.2$$

$$= 0.21 + 0.03 + 0.08 + 0.08 = 0.40.$$

So, $I - GOWA\_1 \neq IGOWA\_2$.

In that case, the arguments of the tied pairs, i.e. $\langle 6,0.3 \rangle$, $\langle 6,0.7 \rangle$, are replaced by the average of the arguments of the tied pairs, i.e. $(0.3 + 0.7)/2= 0.5$. So, the final

ordered argument matrix $B = \begin{bmatrix} 0.5 \\ 0.5 \\ 0.4 \\ 0.2 \end{bmatrix}$

$$I - GOWA(\langle u_1,a_1 \rangle, \langle u_2,a_2 \rangle, ..., \langle u_n,a_n \rangle) = 0.3 \times 0.5 + 0.1 \times 0.5 + 0.2 \times 0.4 + 0.4 \times 0.2$$

$$= 0.15 + 0.05 + 0.08 + 0.08 = 0.36.$$

It is observed that $IGOWA = \dfrac{IGOWA_1 + IGOWA_2}{2} = 0.36$.

5.5.7 Choquet Aggregation Operator

This was initially introduced by Yager [17]. Choquet integral with respect to fuzzy measure is often used in fuzzy decision making, information fusion, and in many other applications. It is an adequate aggregation operator that extends the weighted arithmetic mean to aggregate a collection of arguments $(a_1,a_2,a_3, ... ,a_n)$. Choquet measure is a type of fuzzy measure, which is described in Chapter 4.

Let $X = \{x_1,x_2,x_2, ... ,x_2\}$ be a collection of objects and $\{a_1,a_2,a_3, ... ,a_n\}$ be a finite set of fuzzy values on X, that is a_i is an argument variable associated with X. Then, fuzzy Choquet integral of a_i ($i = 1, 2, 3, ... , 4$) with fuzzy measure μ on $P(X)$ is written as:

$$C_\mu(a_1,a_2,a_3,...,a_n) = \sum_{j=1}^{n} w_j b_j,$$

where b_j is the jth largest component of a_i, i.e. $(b_1,b_2,b_3, ... ,b_n)$ are similar to $(a_1, a_2,a_3, ... ,a_n)$ but reordered in decreasing order.

The components of weighing vector w_j is computed as follows:

For $j = 1, 2, 3, \ldots, n$, $w_j = \mu(A_j) - \mu(A_{j-1})$, where A_j is the subset of X and μ is a fuzzy measure.

Let $a - index(j)$ denote the jth largest component in a_i, so that we have $b_j = a_{a - index(j)}$.

$$A_j = \left\{ x_{a-\text{index}(1)}, x_{a-\text{index}(2)}, \ldots, x_{a-\text{index}(j)} \right\} \text{ and } A_{(0)} = \varnothing.$$

A_j is the subset of X that consists of 1 to j largest argument values.

It is to be noted that for simplicity $a - index(j)$ is taken as (j) and in that case we write

$$A_j = \left\{ x_{(1)}, x_{(2)}, \ldots, x_{(j)} \right\}.$$

5.5.8 Induced Choquet Ordered Aggregation Operator

Similar to Choquet aggregation operator, Choquet ordered aggregation operator was developed by Yager [17] where the ordering is done based on the argument variable a_i associated with X using ordered induced variable, u_i, $i = 1, 2, 3, \ldots, n$ similar to IOWA.

$$C_\mu(\langle u_1, a_1 \rangle, \langle u_2, a_2 \rangle, \ldots, \langle u_n, a_n \rangle) = \sum_{j=1}^{n} w_j b_j,$$

where b_j is the jth largest component of a_i reordered in decreasing order of the values of the u_i of the set $\{u_1, u_2, u_3, \ldots, u_n\}$, i.e. $(b_1, b_2, b_3, \ldots, b_n)$ are similar to $(a_1, a_2, a_3, \ldots, a_n)$ but reordered in decreasing order of the values of the u_i.

The components of weighing vector w_j are computed as follows:

For $j = 1, 2, 3, \ldots, n$, $w_j = \mu(A_j) - \mu(A_{j-1})$, where A_j is the subset of X and μ is a fuzzy measure. The components of weighing vector w_j are computed as follows:

Let $u - index(j)$ is the index of the object which is the jth largest value of u_i so that we have the argument value $b_j = a_{u - index(j)}$.

$$A_j = \left\{ x_{u-\text{index}(1)}, x_{u-\text{index}(2)}, \ldots, x_{u-\text{index}(j)} \right\} \text{ and } A_{(0)} = \varnothing.$$

A_j is the subset of X that consists of 1 to j largest argument values.

It is to be noted that for simplicity $u - index(j)$ may be taken as (j), then

$$A_j = \left\{ x_{(1)}, x_{(2)}, \ldots, x_{(j)} \right\}.$$

An example will explain the aggregation operator clearly.

Example 6   Let μ be a fuzzy measure on space $X = \{x_1, x_2, x_3\}$ in which

$\mu(0) = \varnothing$, $\mu(x_1) = 0.2$, $\mu(x_2) = 0.5$, $\mu(x_3) = 0.6$

$\mu(x_1, x_2) = 0.62$, $\mu(x_1, x_3) = 0.73$, $\mu(x_2, x_3) = 0.91$, $\mu(x_1, x_2, x_3) = 1$.

Assume that we have three pairs of fuzzy values (arguments) associated with $X = \{x_1, x_2, x_3\}$, respectively:

$$a_1 = \langle 3, 0.2 \rangle, a_2 = \langle 7, 0.3 \rangle, a_3 = \langle 6, 0.5 \rangle,$$

where the first component is the order-inducing variable. Compute aggregation of fuzzy values using fuzzy-induced Choquet ordered operator.

Solution

Initially it is required to order the three pairs based on their order-inducing variable

$$a_2 = \langle 7, 0.3 \rangle$$
$$a_3 = \langle 6, 0.5 \rangle$$
$$a_1 = \langle 3, 0.2 \rangle$$

So, it is seen that the initial *u-index* is reordered.

u-index(1) = 2, *u-index*(2) = 3, *u-index*(3) = 1. On the basis of order-inducing

variable, ordered vector $B = \begin{bmatrix} a_{u-index(1)} \\ a_{u-index(2)} \\ a_{u-index(3)} \end{bmatrix} = \begin{bmatrix} a_{(1)} \\ a_{(2)} \\ a_{(3)} \end{bmatrix} = \begin{bmatrix} a_2 \\ a_3 \\ a_1 \end{bmatrix} = \begin{bmatrix} 0.3 \\ 0.5 \\ 0.2 \end{bmatrix}$

It is to be noted that for simplicity $u - index(j)$ is taken as (j), which will be used in computing A_j.

Now, we will compute A_j, $A_j = \{x_{(1)}, x_{(2)}, x_{(3)}, \ldots, x_{(j)}\}$.

We get $\mu(A_1) = \mu(\{x_{(1)}\}) = \mu(\{x_2\}) = 0.5$.

Likewise, $\mu(A_2) = \mu(\{x_{(1)}, x_{(2)}\}) = \mu\{x_2, x_3\} = 0.91$, $\mu(A_3) = \{x_{(1)}, x_{(2)}, x_{(3)}\} = \mu(x_2, x_3, x_1) = 1$.

Using the formula $w_j = \mu(A_j) - \mu(A_{j-1})$, we get

$$w_1 = \mu(A_1) = 0.5,$$

$$w_2 = \mu(A_2) - \mu(A_1) = 0.91 - 0.5 = 0.41,$$

$$w_3 = \mu(A_3) - \mu(A_2) = 1 - 0.91 = 0.09.$$

$$C_\mu(\langle u_1, a_1 \rangle, \langle u_2, a_2 \rangle, \langle u_3, a_3 \rangle, \langle u_4, a_4 \rangle) = w^T B = \sum_{j=1}^{4} w_j b_j$$

$$= [0.5 \; 0.41 \; 0.09] \begin{bmatrix} 0.3 \\ 0.5 \\ 0.2 \end{bmatrix}$$

$$= 0.5 \times 0.3 + 0.41 \times 0.5 + 0.09 \times 0.2$$

$$= 0.15 + 0.205 + 0.018 = 0.373.$$

If the order-induced variable, i.e. "*u*" is not there, then the induced Choquet ordered aggregation operator becomes simple Choquet ordered aggregation operator where ordering of arguments is done on the basis of argument values (largest to smallest).

5.6 Intuitionistic Fuzzy Operators

There are different types of operators using intuitionistic fuzzy set that uses membership, nonmembership, and hesitation degrees, i.e. these operators consider more number of uncertainties. In real-time applications, there are uncertainties in decision making, pattern recognition, information retrieval, and so on. Logically, when we say about the degree of belongingness or presence of any element in a set, that does not mean that the degree of non-belongingness is the complement of the degree of belongingness. There may be some kind of hesitation present while selecting membership function. So, nonmembership degree will not be the complement of membership degree. In such type of applications, intuitionistic fuzzy set is used that takes into account both membership and nonmembership degree.

There are few operations on intuitionistic fuzzy sets that are required in aggregation operators.

Let A be an intuitionistic fuzzy set $A = \{X, \mu_A(x), \nu_A(x) \mid x \in A\}$, where $\mu_A(x)$ and $\nu_A(x)$ are the membership and nonmembership functions, respectively, with the conditions $\mu_A(x) + \nu_A(x) + \pi_A(x) = 1$ and $\mu_A(x), \nu_A(x) \in [0,1]$. Then, grade of membership is the interval $[\mu_A(x), 1 - \nu_A(x)]$.

Score function by Chen and Tan [26] is used to evaluate the degree of suitability in decision maker's requirement. It is mainly used in problems based on intuitionistic fuzzy aggregation operators to aggregate a collection of intuitionistic fuzzy values.

If $p = (\mu_p, \nu_p)$ is an intuitionistic fuzzy value, the score function or the score on p is given as: $S(p) = \mu_p - \nu_p$, which is used to find the deviation between membership and nonmembership degrees and $S(p) \in [-1,1]$. It represents the difference of membership and nonmembership values.

Accuracy function by Hong and Choi [11] is used to evaluate the degree of accuracy of intuitionistic fuzzy value $p = (\mu_p, \nu_p)$, which is given as:

$$H(p) = \mu_p + \nu_p, \quad \text{where} \quad H(p) \in [0,1].$$

Accuracy function represents the sum of membership and nonmembership values. The larger the value of $H(p)$, the higher the degree of accuracy of the degree of membership of the intuitionistic fuzzy value p.

Xu [27] suggested few laws of intuitionistic fuzzy values that are required to define aggregate operator. Consider two IFSs p_1, p_2, and p_3 and let the intuitionistic fuzzy values be $p_1 = (\mu_{p_1}, \nu_{p_1})$, $p_2 = (\mu_{p_2}, \nu_{p_2})$, $p_3 = (\mu_{p_3}, \nu_{p_3})$ where $\mu_A(x) + \nu_A(x) \leq 1$, then the following properties hold:

$$p_1 \oplus p_2 = \left(\mu_{p_1} + \mu_{p_2} - \mu_{p_1} \cdot \mu_{p_2}, \nu_{p_1} \cdot \nu_{p_2}\right) \tag{5.43}$$

$$\lambda p = \left(1 - \left(1 - \mu_p\right)^\lambda, \nu_p^\lambda\right), \ \lambda > 0 \tag{5.44}$$

$$p^\lambda = \left(\mu_p^\lambda, 1 - \left(1 - \nu_p\right)^\lambda\right), \ \lambda > 0 \tag{5.45}$$

$$p_1 \otimes p_2 = \left(\mu_{p_1} \cdot \mu_{p_2}, \nu_{p_1} + \nu_{p_2} - \nu_{p_1} \cdot \nu_{p_2}\right) \tag{5.46}$$

$$\lambda_1 p \oplus \lambda_2 p = (\lambda_1 + \lambda_2) p, \ \lambda_1, \lambda_2 > 0 \tag{5.47}$$

$$(p_1 \oplus p_2) \oplus p_3 = p_1 \oplus (p_2 \oplus p_3) \tag{5.48}$$

$$\left(p^{\lambda_1}\right)^{\lambda_2} = p^{\lambda_1 \lambda_2} \tag{5.49}$$

5.7 Intuitionistic Fuzzy Aggregation Operator

5.7.1 Generalized Intuitionistic Fuzzy Aggregation Operator

Zhao et al. [28] introduced few intuitionistic fuzzy aggregation operators. Let $p_j = \left(\mu_{p_j}, \nu_{p_j}\right)$ with $j = 1, 2, 3, \ldots, n$ be a collection of intuitionistic fuzzy values, then

$$GIFWA(p_1, p_2, p_3, \ldots, p_n) = \left(w_1 p_1^\lambda \oplus w_2 p_2^\lambda \oplus w_2 p_3^\lambda \oplus \cdots \oplus w_n p_n^\lambda\right)^{1/\lambda} \tag{5.50}$$

$w = (w_1, w_2, w_3, \ldots, w_n)^T$ is a weight vector associated with the operator and $\sum_{i=1}^{n} w_j = 1$, $w_j \in [0,1]$, $j = 1, 2, 3, \ldots, n$.
Considering first two terms in the bracket, i.e.

$$GIFWA(p_1, p_2) = w_1 p_1^\lambda \oplus w_2 p_2^\lambda$$

and using the operations in Eqs. (5.42–5.45), we get

$$w_1 p_1^\lambda = \left(1 - \left(1 - \mu_{p_1}^\lambda\right)^{w_1}, \left(1 - \left(1 - \nu_{p_1}\right)^\lambda\right)^{w_1}\right)$$

$$w_2 p_2^\lambda = \left(1 - \left(1 - \mu_{p_2}^\lambda\right)^{w_2}, \left(1 - \left(1 - \nu_{p_2}\right)^\lambda\right)^{w_2}\right)$$

$$w_1 p_1^\lambda = \left(1 - \left(1 - \mu_{p_1}^\lambda\right)^{w_1}, \left(1 - \left(1 - \nu_{p_1}\right)^\lambda\right)^{w_1}\right)$$

$$w_2 p_2^\lambda = \left(1 - \left(1 - \mu_{p_2}^\lambda\right)^{w_2}, \left(1 - \left(1 - \nu_{p_2}\right)^\lambda\right)^{w_2}\right)$$

So,

$$w_1 p_1^\lambda \oplus w_2 p_2^\lambda = \left(1-\left(1-\mu_{p_1}^\lambda\right)^{w_1}, \left(1-\left(1-\nu_{p_1}\right)^\lambda\right)^{w_1}\right) \oplus \left(1-\left(1-\mu_{p_2}^\lambda\right)^{w_2}, \left(1-\left(1-\nu_{p_2}\right)^\lambda\right)^{w_2}\right)$$

$$= \left(1-\left(1-\mu_{p_1}^\lambda\right)^{w_1}\right) + \left(1-\left(1-\mu_{p_2}^\lambda\right)^{w_2}\right)$$

$$- \left(1-\left(1-\mu_{p_1}^\lambda\right)^{w_1}\right) \cdot \left(1-\left(1-\mu_{p_2}^\lambda\right)^{w_2}\right), \left(1-\left(1-\nu_{p_1}\right)^\lambda\right)^{w_1} \cdot \left(1-\left(1-\nu_{p_2}\right)^\lambda\right)^{w_2}$$

$$= 1-\left(1-\mu_{p_1}^\lambda\right)^{w_1} \cdot \left(1-\mu_{p_2}^\lambda\right)^{w_2}, \left(1-\left(1-\nu_{p_1}\right)^\lambda\right)^{w_1} \cdot \left(1-\left(1-\nu_{p_2}\right)^\lambda\right)^{w_2}$$

$$= 1-\prod_{j=1}^{2}\left(1-\mu_{p_j}^\lambda\right)^{w_j}, \prod_{j=1}^{2}\left(1-\left(1-\nu_{p_2}\right)^\lambda\right)^{w_j}.$$

For n terms,

$$w_1 p_1^\lambda \oplus w_2 p_2^\lambda \oplus w_2 p_3^\lambda \oplus \cdots \oplus w_n p_4^\lambda = \left[1-\prod_{j=1}^{n}\left(1-\mu_{p_j}^\lambda\right)^{w_j}, \prod_{j=1}^{n}\left(1-\left(1-\nu_{p_2}\right)^\lambda\right)^{w_j}\right].$$

So,

$$GIFWA(p_1,p_2,p_3,...,p_n) = \left(w_1 p_1^\lambda \oplus w_2 p_2^\lambda \oplus w_2 p_3^\lambda \oplus \cdots \oplus w_n p_4^\lambda\right)^{1/\lambda}$$

$$= \left(1-\prod_{j=1}^{n}\left(1-\mu_{p_j}^\lambda\right)^{w_j}, \prod_{j=1}^{n}\left(1-\left(1-\nu_{p_j}\right)^\lambda\right)^{w_j}\right)^{1/\lambda}$$

$$= \left(1-\prod_{j=1}^{n}\left(1-\mu_{p_j}^\lambda\right)^{w_j}\right)^{1/\lambda}, 1-\left(1-\prod_{j=1}^{n}\left(1-\left(1-\nu_{p_j}\right)^\lambda\right)^{w_j}\right)^{\frac{1}{\lambda}}.$$

$$(5.51)$$

If $\lambda = 1$, then GIFWA reduces to intuitionistic fuzzy aggregating operator (IFWA)

$$GIFWA(p_1,p_2,p_3,...,p_n) = w_1 p_1 \oplus w_2 p_2 \oplus w_2 p_3 \oplus \cdots \oplus w_n p_n$$

$$= \left(1-\prod_{j=1}^{n}\left(1-\mu_{p_j}\right)^{w_j}, \prod_{j=1}^{n}\left(1-\left(1-\nu_{p_j}\right)\right)^{w_j}\right)$$

$$= \left(1-\prod_{j=1}^{n}\left(1-\mu_{p_j}\right)^{w_j}, \prod_{j=1}^{n}\nu_{p_j}^{w_j}\right).$$

$$(5.52)$$

An example is given to calculate *GIFWA* of the four intuitionistic fuzzy values.

Example 7 Suppose there are four experts who are invited to evaluate some decision alternative. Their evaluation are expressed with intuitionistic fuzzy sets. These are given as:

$$p_1 = (0.2,0.7), \ p_2 = (0.3,0.4), \ p_3 = (0.7,0.2), \ p_4 = (0.2,0.5)$$

The weight vector of the four experts is given as:

$$w = (0.3, 0.4, 0.2, 0.1)^T \text{ and } \lambda = 2, \ p_j(j = 1, 2, 3, 4),$$

then compute the evaluation of the four experts using generalized intuitionistic FWA operator.

Solution

From the intuitionistic fuzzy values, we have

$$\mu_{p_1} = 0.2, \ \mu_{p_2} = 0.3, \ \mu_{p_3} = 0.7, \ \mu_{p_4} = 0.2$$

$$\nu_{p_1} = 0.7, \ \nu_{p_2} = 0.4, \ \nu_{p_3} = 0.2, \ \nu_{p_4} = 0.5$$

$$GIFWA(p_1, p_2, p_3, \dots, p_n) = \left[\left(1 - \prod_{j=1}^{n} \left(1 - \mu_{p_j}^{\lambda} \right)^{w_j} \right)^{1/\lambda}, 1 - \left(1 - \prod_{j=1}^{n} \left(1 - \left(1 - \nu_{p_j} \right)^{\lambda} \right)^{w_j} \right)^{1/\lambda} \right]$$

$$= \left[\left(1 - \left[(1 - 0.2^2)^{0.3} \times (1 - 0.3^2)^{0.4} \times (1 - 0.7^2)^{0.2} \times (1 - 0.2^2)^{0.1} \right] \right)^{1/2}, \right.$$

$$\left. 1 - \left(1 - \left[\begin{array}{c} \left(1 - (1 - 0.7)^2 \right)^{0.3} \times \left(1 - (1 - 0.4)^2 \right)^{0.4} \times \left(1 - (1 - 0.2)^2 \right)^{0.2} \\ \times \left(1 - (1 - 0.5)^2 \right)^{0.1} \end{array} \right] \right)^{\frac{1}{2}} \right]$$

$$= (0.4147, 0.4034).$$

This shows that the satisfaction degree is: 0.4147 and dissatisfaction degree of the four experts on the decision is 0.4034. The hesitation degree is $1 - (0.4147 + 0.4034) = 0.1819$.

5.7.2 Generalized Intuitionistic Fuzzy Ordered Weighting Operator (GIFOWA)

Just like fuzzy ordered weighting operator as discussed in Section 5.5.3, GIFOWA is written as follows [28]:

If $p_j = \left(\mu_{p_j}, \nu_{p_j} \right)$, with $j = 1, 2, 3, \dots, n$ be a collection of intuitionistic fuzzy values, then

$$GIFOWA(p_1, p_2, p_3, \dots, p_n) = \left(w_1 p_{\sigma(1)}^{\lambda} \oplus w_2 p_{\sigma(2)}^{\lambda} \oplus w_2 p_{\sigma(3)}^{\lambda} \oplus \cdots \oplus w_n p_{\sigma(n)}^{\lambda} \right)^{1/\lambda},$$

$w = (w_1, w_2, w_3, \dots, w_n)^T$ is a weight vector of p_j and $\sum_{i=1}^{n} w_j = 1, j = 1, 2, 3, \dots, n$.

$$GIFOWA(p_1, p_2, p_3, \dots, p_n)$$

$$= \left[\left(1 - \prod_{j=1}^{n} \left(1 - \mu_{p_{\sigma(j)}}^{\lambda} \right)^{w_j} \right)^{1/\lambda}, 1 - \left(1 - \prod_{j=1}^{n} \left(1 - \left(1 - \nu_{p_{\sigma(j)}} \right)^{\lambda} \right)^{w_j} \right)^{1/\lambda} \right],$$

$$(5.53)$$

where $(\sigma(1),\sigma(2),\sigma(3), \dots ,\sigma(n))$ is a permutation of $(1,2,3, \dots ,n)$ such that $p_{\sigma(j)} \geq p_{\sigma(j-1)}$ for all $j = 1, 2, 3, \dots , n$, i.e. $p_{\sigma(j)}$ is the jth largest value in the set $(p_1,p_2,p_3, \dots ,p_n)$.

The following cases hold for GIFOWA:

If $\lambda = 1$, then GIOFWA reduces to IFOWA. So from Eq. (5.53)

$$GIFOWA(p_1,p_2,p_3, \dots ,p_n) = w_1 p_{\sigma(1)} \oplus w_2 p_{\sigma(2)} \oplus w_2 p_{\sigma(3)} \oplus \cdots \oplus w_n p_{\sigma(n)}$$

$$\left(1 - \prod_{j=1}^{n}\left(1 - \mu_{p_{\sigma(j)}}\right)^{w_j}, \prod_{j=1}^{n}\left(1 - \left(1 - \nu_{\sigma(j)}\right)\right)^{w_j}\right)$$

$$= \left[1 - \prod_{j=1}^{n}\left(1 - \mu_{p_{\sigma(j)}}\right)^{w_j}, \prod_{j=1}^{n}\nu_{\sigma(j)}^{w_j}\right].$$

$$(5.54)$$

Example 8 Suppose there are four experts who are invited to take some decision alternative. Their evaluations are expressed in intuitionistic fuzzy values,

$$p_1 = (0.1,0.5),\; p_2 = (0.4,0.3),\; p_3 = (0.6,0.2),\; p_4 = (0.2,0.7)$$

with position weight vector associated with the operator, $w = (0.3,0.2,0.1,0.4)^T$ and $\lambda = 1$, $p_j(j = 1, 2, 3, 4)$. Compute the evaluation of the four experts on decision alternative using intuitionistic fuzzy ordered weighting aggregation operator.

Solution

From the intuitionistic fuzzy values, we have

$$\mu_{p_1} = 0.1,\; \mu_{p_2} = 0.4,\; \mu_{p_3} = 0.6,\; \mu_{p_4} = 0.2,$$

$$\nu_{p_1} = 0.5,\; \nu_{p_2} = 0.3,\; \nu_{p_3} = 0.2,\; \nu_{p_4} = 0.7.$$

Now scores of $p_j(j = 1, 2, 3, 4)$ are computed as:

$$s(p_1) = 0.1 - 0.5 = -0.4,\; s(p_2) = 0.4 - 0.3 = 0.1,\; s(p_3) = 0.6 - 0.2 = 0.4,\; s(p_4) = 0.2 - 0.7 = -0.5.$$

It is observed that

$$s(p_3) > s(p_2) > s(p_1) > s(p_4)$$

so, $p_{\sigma(1)} = (0.6,0.2),\; p_{\sigma(2)} = (0.4,0.3),\; p_{\sigma(3)} = (0.1,0.5),\; p_{\sigma(4)} = (0.2,0.7).$

From Eq. (5.54)

$$GIFOWA(p_1,p_2,p_3,...,p_n) = \left[1 - \prod_{j=1}^{n}\left(1-\mu_{p_{\sigma(j)}}\right)^{w_j}, \prod_{j=1}^{n}\nu_{\sigma(j)}^{w_j}\right]$$

$$= \left[\begin{array}{l} 1 - \left[(1-0.6)^{0.3} \times (1-0.4)^{0.2} \times (1-0.1)^{0.1} \times (1-0.2)^{0.4}\right], \\ 0.2^{0.3} \times 0.3^{0.2} \times 0.5^{0.1} \times 0.7^{0.4} \end{array}\right]$$

$$= \left[\begin{array}{l} 1 - (0.7597 \times 0.9029 \times 0.9779 \times 0.9587), \\ 0.6170 \times 0.7860 \times 0.8513 \times 0.8670 \end{array}\right]$$

$$= [0.3793, 0.3923].$$

This shows that the satisfaction degree is: 0.3793 and dissatisfaction degree of the four experts on the decision is 0.3923. The hesitation degree is $1 - (0.3793 + 0.3923) = 0.2284$.

5.7.3 Generalized Intuitionistic Fuzzy Hybrid Operator

Till now we know that GIFWA operator weighs only intuitionistic fuzzy values and GIFOWA weighs only the ordered positions of intuitionistic fuzzy values. So, we require two operators to handle a problem. To overcome the limitation, generalized intuitionistic FHA operator is introduced that weighs both intuitionistic fuzzy values and its ordered position [28].

If $p_j = \left(\mu_{p_j}, \nu_{p_j}\right)$, with $j = 1, 2, 3, \dots, n$ be a collection of intuitionistic fuzzy values, GIFHA operator of dimension "n" with an associated vector $w = (w_1, w_2, w_3, \dots, w_n)^T$ and $\sum_{j=1}^{n} w_j = 1$, is given as:

$$GIFWA_{w,\omega}(p_1,p_2,p_3,...,p_n)$$

$$= \left(w_1\dot{p}_{\sigma(1)}^{\lambda} \oplus w_2\dot{p}_{\sigma(2)}^{\lambda} \oplus w_2\dot{p}_{\sigma(3)}^{\lambda} \oplus \cdots \oplus w_n\dot{p}_{\sigma(n)}^{\lambda}\right)^{1/\lambda}$$

$$= \left[\left(1 - \prod_{j=1}^{n}\left(1-\mu_{\dot{p}_{\sigma(j)}}^{\lambda}\right)^{w_j}\right)^{1/\lambda}, 1 - \left(1 - \prod_{j=1}^{n}\left(1-\left(1-\nu_{\dot{p}_{\sigma(j)}}\right)^{\lambda}\right)^{w_j}\right)^{1/\lambda}\right],$$

$$(5.55)$$

where $\dot{p}_{\sigma(j)} = \left(\mu_{\dot{p}_{(j)}}, \nu_{\dot{p}_{(j)}}\right)$, $\dot{p}_{(j)}$ is the jth largest weighted intuitionistic fuzzy values $\dot{p}_j$, $\dot{p}_j = n\omega_j p_j$ where $\omega_j = (\omega_1,\omega_2,\omega_3, \dots ,\omega_n)^T$ is the weight vector of p_j $(j = 1,2,3, \dots ,n)$ and $\sum_{j=1}^{n}\omega_j = 1$, $\lambda > 0$.

With $\lambda = 1$, GIFHA reduces to intuitionistic FHA operator:

$GIFWA_{w,\omega}(p_1,p_2,p_3,...,p_n)$

$$= \left[1 - \prod_{j=1}^{n}\left(1-\mu_{\dot{p}_{\sigma(j)}}\right)^{w_j}, 1-\left(1-\prod_{j=1}^{n}\left(1-\left(1-\nu_{\dot{p}_{\sigma(j)}}\right)\right)^{w_j}\right)\right]$$

$$= \left[1 - \prod_{j=1}^{n}\left(1-\mu_{\dot{p}_{\sigma(j)}}\right)^{w_j}, \sum_{j=1}^{n}\nu_{\dot{p}_{\sigma(j)}}^{w_j}\right].$$

An example will show the computation of IFHA for the five intuitionistic values.

Example 9 Let us consider five intuitionistic fuzzy values $p_1 = (0.2, 0.6)$, $p_2 = (0.4, 0.3)$, $p_3 = (0.6, 0.2)$, $p_4 = (0.1, 0.7)$, $p_5 = (0.5, 0.3)$ and the weight vector of $p_j (j = 1,2,3,4)$, $\omega = (0.25, 0.20, 0.15, 0.18, 0.22)^T$ and $\lambda = 1$. Compute the aggregation of intuitionistic fuzzy values using intuitionistic fuzzy hybrid operator.

Solution

First, weighted intuitionistic fuzzy values are computed as follows:
Using $\dot{p}_j = n\omega_j p_j = 5\omega_j p_j$, we get using Eq. (5.44)

$$\dot{p}_1 = \left[1-(1-0.2)^{5*0.25}, 0.6^{5*0.25}\right] = [0.2434, 0.5281],$$

$$\dot{p}_2 = \left[1-(1-0.4)^{5*0.2}, 0.3^{5*0.2}\right] = [0.4, 0.3],$$

$$\dot{p}_3 = \left[1-(1-0.6)^{5*0.15}, 0.2^{5*0.15}\right] = [0.497, 0.299],$$

$$\dot{p}_4 = \left[1-(1-0.1)^{5*0.18}, 0.7^{5*0.18}\right] = [0.0905, 0.7254],$$

$$\dot{p}_5 = \left[1-(1-0.5)^{5*0.22}, 0.3^{5*0.22}\right] = [0.5335, 0.2660].$$

Now scores of $\dot{p}_j$ $(j = 1,2,3,4,5)$ are computed as:

$s(\dot{p}_1) = 0.2434 - 0.5281 = -0.2847,$

$s(\dot{p}_2) = 0.4 - 0.3 = 0.1,$

$s(\dot{p}_3) = 0.497 - 0.299 = 0.198,$

$s(\dot{p}_4) = 0.0905 - 0.7254 = -0.6349,$

$s(\dot{p}_5) = 0.5335 - 0.2660 = 0.2675.$

It is seen that $s(\dot{p}_5) > s(\dot{p}_3) > s(\dot{p}_2) > s(\dot{p}_1) > s(\dot{p}_4)$.
So,

$\dot{p}_{\sigma(1)} = [0.5335, 0.2660], \dot{p}_{\sigma(2)} = [0.497, 0.299], \dot{p}_{\sigma(3)} = [0.4, 0.3]$

$\dot{p}_{\sigma(4)} = [0.2434, 0.5281], \dot{p}_{\sigma(5)} = [0.0905, 0.7254].$

Then,

$$\mu_{\dot{p}_{\sigma(1)}} = 0.5335, \mu_{\dot{p}_{\sigma(2)}} = 0.497, \mu_{\dot{p}_{\sigma(3)}} = 0.4, \mu_{\dot{p}_{\sigma(4)}} = 0.2434, \mu_{\dot{p}_{\sigma(5)}} = 0.0905.$$

$$\nu_{\dot{p}_{\sigma(1)}} = 0.2660, \nu_{\dot{p}_{\sigma(2)}} = 0.299, \nu_{\dot{p}_{\sigma(3)}} = 0.3, \nu_{\dot{p}_{\sigma(4)}} = 0.5281, \nu_{\dot{p}_{\sigma(5)}} = 0.7254.$$

For computation of weight vector of the operator, there are many methods for computing the weights of the operator, but normal distribution method is the most commonly used method as it models many natural processes [21, 29]. Let $w = (w_1, w_2, w_3, \ldots, w_n)^T$ be a weight vector ($j = 1, 2, 3, \ldots, n$) and $\mu_n = \dfrac{1}{n}\dfrac{n(n+1)}{2} = \dfrac{n+1}{2}$ is the mean of the collection of $1, 2, 3, \ldots, n$.

$\sigma_n = \sqrt{\dfrac{1}{n}\sum_{j=1}^{n}(j-\mu_n)^2}$ is the standard deviation of the collection of $1, 2, 3, \ldots, n$.

The weight vector is computed as:

$$w_j = \frac{\dfrac{1}{\sqrt{2\pi}\sigma_n}e^{-\left[\frac{(j-\mu_n)^2}{2\sigma_n^2}\right]}}{\sum_{i=1}^{n}\dfrac{1}{\sqrt{2\pi}\sigma_n}e^{-\left[\frac{(i-\mu_n)^2}{2\sigma_n^2}\right]}} = \frac{e^{-\left[\frac{(j-\mu_n)^2}{2\sigma_n^2}\right]}}{\sum_{i=1}^{n}e^{-\left[\frac{(i-\mu_n)^2}{2\sigma_n^2}\right]}}.$$

Substituting the values of μ_n and σ_n, the value of $w_j = \dfrac{e^{-\left[\frac{\left(j-\frac{n+1}{2}\right)^2}{2\sigma_n^2}\right]}}{\sum_{i=1}^{n}e^{-\left[\frac{\left(i-\frac{n+1}{2}\right)^2}{2\sigma_n^2}\right]}}.$

In the problem, number of collection is 5, i.e. $n = 5$.

So, we get $\mu_n = 3, \sigma_n = \sqrt{\dfrac{1}{5}\sum_{i=1}^{n}(i-\mu_n)^2} = \sqrt{\dfrac{1}{5}(4+1+0+1+4)} = \sqrt{2}.$

So,

$$w_j = \frac{e^{-\left[\frac{\left(j-\frac{n+1}{2}\right)^2}{2\sigma_n^2}\right]}}{\sum_{i=1}^{n}e^{-\left[\frac{\left(i-\frac{n+1}{2}\right)^2}{2\sigma_n^2}\right]}} = \frac{e^{-\left[\frac{(j-3)^2}{4}\right]}}{\sum_{i=1}^{n}e^{-\left[\frac{(i-3)^2}{4}\right]}} = \frac{e^{-\left[\frac{(j-3)^2}{4}\right]}}{3.2932}.$$

At, $j = 1, w_1 = \dfrac{e^{-\left[\frac{(j-3)^2}{4}\right]}}{3.2932} = \dfrac{e^{-\left[\frac{(1-3)^2}{4}\right]}}{3.2932} = \dfrac{e^{-1}}{3.2932} = 0.112.$

Likewise, for $j = 2, 3, 4, 5$, we get, $w = [0.112, 0.236, 0.304, 0.236, 0.112]^T.$

Now, introducing the weight vector in the aggregation definition, we get

$IFHA_{w,\omega}(p_1, p_2, p_3, ..., p_n)$

$$= \left[1 - \prod_{j=1}^{5} \left(1 - \mu_{\dot{p}_{\sigma(j)}} \right)^{w_j}, \sum_{j=1}^{5} \nu_{\dot{p}_{\sigma(j)}}^{w_j} \right]$$

$$= \left[\begin{array}{l} 1 - \left((1 - 0.5335)^{0.112} \times (1 - 0.497)^{0.236} \times (1 - 0.4)^{0.304} \right. \\ \times (1 - 0.2424)^{0.236} \times (1 - 0.0905)^{0.112} \right), 0.266^{0.112} \times 0.299^{0.236} \\ \times 0.3^{0.304} \times 0.5281^{0.236} \times 0.7254^{0.112} \end{array} \right]$$

$$= \left[\begin{array}{l} 1 - (0.9181 \times 0.8503 \times 0.8562 \times 0.9366 \times 0.9894), \\ 0.8622 \times 0.7521 \times 0.6935 \times 0.8601 \times 0.9647 \end{array} \right]$$

$$= [0.3806, 0.3731].$$

5.7.4 Intuitionistic Fuzzy Weighted Geometric Operator (IFWG)

Xui and Yager [30] introduced intuitionistic fuzzy geometric averaging operators. If $p_j = \left(\mu_{p_j}, \nu_{p_j} \right)$, with $j = 1, 2, 3, ..., n$ be a collection of intuitionistic fuzzy values, then

$$IFWG(p_1, p_2, p_3, ..., p_n) = p_1^{w_1} \otimes p_2^{w_2} \oplus p_3^{w_3} \oplus \cdots \oplus p_n^{w_n}$$

$w = (w_1, w_2, w_3, ..., w_n)^T$ is a weight vector of p_j and $\sum_{i=1}^{n} w_j = 1, j = 1, 2, 3, ... n$. From Eqs. (5.45, 5.46), we have

$$p_1^{w_1} = \left(\mu_{p_1}^{w_1}, 1 - \left(1 - \nu_{p_1} \right)^{w_1} \right) \quad \text{and} \quad p_2^{w_2} = \left(\mu_{p_2}^{w_2}, 1 - \left(1 - \nu_{p_2} \right)^{w_2} \right).$$

Now,

$$p_1^{w_1} \otimes p_2^{w_2} = \left[\mu_{p_1}^{w_1} \cdot \mu_{p_2}^{w_2}, 1 - \left(1 - \nu_{p_1} \right)^{w_1} + 1 - \left(1 - \nu_{p_1} \right)^{w_1} - \left(1 - \left(1 - \nu_{p_1} \right)^{w_1} \right) \cdot \left(1 - \left(1 - \nu_{p_2} \right)^{w_2} \right) \right]$$

$$= \left[\begin{array}{l} \mu_{p_1}^{w_1} \cdot \mu_{p_2}^{w_2}, 2 - \left(1 - \nu_{p_1} \right)^{w_1} - \left(1 - \nu_{p_2} \right)^{w_2} - 1 + \left(1 - \nu_{p_1} \right)^{w_1} + \left(1 - \nu_{p_2} \right)^{w_2} \\ - \left(1 - \nu_{p_1} \right)^{w_1} \cdot \left(1 - \nu_{p_2} \right)^{w_2} \end{array} \right]$$

$$= \left[\mu_{p_1}^{w_1} \cdot \mu_{p_2}^{w_2}, 1 - \left(1 - \nu_{p_1} \right)^{w_1} \cdot \left(1 - \nu_{p_2} \right)^{w_2} \right].$$

So,

$$IFWG(p_1,p_2,p_3,...,p_n) = \left[\prod_{j=1}^{n} \mu_{p_j}^{w_j}, 1 - \prod_{j=1}^{n} \left(1 - \nu_{p_j}\right)^{w_j} \right].$$ (5.56)

5.7.5 Intuitionistic Fuzzy Ordered Weighted Geometric Operator

Similar to IFWG averaging operator, intuitionistic fuzzy ordered weighted geometric averaging operator is defined as:

$$IFWG(p_1,p_2,p_3,...,p_n) = \left[\prod_{j=1}^{n} \mu_{p_{\sigma(j)}}^{w_j}, 1 - \prod_{j=1}^{n} \left(1 - \nu_{p_{\sigma(j)}}\right)^{w_j} \right],$$ (5.57)

where $p_{\sigma(j)}$ is the jth largest value in the set $(p_1,p_2,p_3, \dots ,p_n)$ such that $p_{\sigma(j)} \geq p_{\sigma(j-1)}$.

5.7.6 Induced Generalized Intuitionistic Fuzzy Ordered Averaging Operator

Similar to induced fuzzy ordered aggregation operator, induced generalized intuitionistic fuzzy ordered aggregation operator was suggested by Xu and Wang [24]. In this operator, as described in fuzzy case, reordering is carried out by another variable u_j where the ordered position of p_j depends on order-induced variable u_j. Let the operator of dimension n has an associated weighting vector of $w = (w_1,w_2,w_3, \dots ,w_n)^T$ of dimension n with $\sum_{j=1}^{n} w_j = 1$, a parameter $\lambda \in [-\infty,\infty]$, and an order-inducing variable u_j, then $IGOWA$ is written as:

$$I-GIFOWA(\langle u_1,a_1 \rangle, \langle u_2,a_2 \rangle, \langle u_3,a_3 \rangle, \langle u_4,a_4 \rangle)$$

$$= \left(\sum_{i=1}^{n} w_i p_{\sigma(i)}^{\lambda} \right)^{1/\lambda}$$

$$= \left(1 - \left(\prod_{j=1}^{n} \left(\mu_{p_{\sigma(j)}} \right)^{w_j} \right)^{1/\lambda}, 1 - \left(\prod_{j=1}^{n} \left(1 - \left(1 - \nu_{\sigma(j)} \right)^{\lambda} \right)^{w_j} \right)^{1/\lambda} \right),$$ (5.58)

where $p_{\sigma(j)} = \left(\mu_{p_{\sigma(j)}}, \nu_{p_{\sigma(j)}} \right)$ is p_j reordered in decreasing order of the value of u_j in $\langle u_j,a_j \rangle$, i.e. $(p_{\sigma(1)},p_{\sigma(2)},p_{\sigma(3)}, \dots ,p_{\sigma(n)})$ are similar to $(p_1,p_2,p_3, \dots ,p_n)$ but reordered in decreasing order of the values of u_j, i.e. from largest to smallest.

For all the above intuitionistic fuzzy aggregation operators, for simplicity, one may use $\sigma(j)$ as (j). So, $p_{\sigma(j)}$ will be written as $p_{(j)}$.

5.7.7 Intuitionistic Fuzzy Choquet Integral Operator

Tan and Chen [31] introduced intuitionistic fuzzy Choquet integral operator similar to induced fuzzy Choquet (IFC) integral operator, where the nonmembership function is considered.

If $p_j = \left(\mu_{p_j}, \nu_{p_j} \right)$, with $j = 1, 2, 3, \ldots, n$ be a collection of intuitionistic fuzzy values on X, then intuitionistic fuzzy Choquet integral of p_j with respect to fuzzy measure μ on space $P(X)$ is written as:

$$IFC_m(p_1, p_2, p_3, \ldots, p_n) = p_{(1)} \left(\mu\left(A_{(1)}\right) - \mu\left(A_{(0)}\right) \right) \oplus p_{(2)} \left(\mu\left(A_{(2)}\right) - \mu\left(A_{(1)}\right) \right) \oplus \cdots$$

$$\oplus p_{(n)} \left(\mu\left(A_{(n)}\right) - \mu\left(A_{(n-1)}\right) \right)$$

$$= \sum_{j=1}^{n} \oplus p_{(j)} \left(\mu\left(A_{(j)}\right) - \mu\left(A_{(j-1)}\right) \right),$$

where $((1), (2), (3), \ldots, (n))$ is the permutation of $(1, 2, 3, \ldots, n)$ such that $p_{(1)} \geq p_{(2)} \geq p_{(3)} \geq \cdots \geq p_{(n)}$. Said in another way, $p_{(j)}$ is the jth largest component of $(p_1, p_2, p_3, \ldots, p_n)$, i.e. $(p_{(1)}, p_{(2)}, p_{(3)}, \ldots, p_{(n)})$ are similar to $(p_1, p_2, p_3, \ldots, p_n)$ but reordered in decreasing order.

$$A_{(j)} = \left\{ x_{(1)}, x_{(2)}, x_{(3)}, \ldots, x_{(j)} \right\}, j = 1, 2, 3, \ldots, n \text{ and } A_{(0)} = \varnothing.$$

Using intuitionistic fuzzy operators in Eqs. (5.43–5.49), as explained in Section 5.6, their aggregation value using intuitionistic fuzzy Choquet operator will be

$$IFC_m(p_1, p_2, p_3, \ldots, p_n)$$

$$= \left(1 - \prod_{j=1}^{n} \left(1 - \mu_{p(j)} \right)^{\mu\left(A_{(j)}\right) - \mu\left(A_{(j-1)}\right)}, \prod_{j=1}^{n} \nu_{p(j)}^{\mu\left(A_{(j)}\right) - \mu\left(A_{(j-1)}\right)} \right). \tag{5.59}$$

5.7.8 Induced Intuitionistic Fuzzy Choquet Integral Operator

Just like IFC integral operator as discussed in Section 5.7.7, in this operator, input arguments are arranged using an order-inducing variable [31]. If $p_j = \left(\mu_{p_j}, \nu_{p_j} \right)$, with $j = 1, 2, 3, \ldots, n$, is a collection of intuitionistic fuzzy values on X, then an induced IFC integral operator of dimension n is defined to aggregate the set of second arguments of a list $(\langle u_1, p_1 \rangle, \langle u_2, p_2 \rangle, \ldots, \langle u_n, p_n \rangle)$ which is given as:

$$IFC_m(\langle u_1, p_1 \rangle, \langle u_2, p_2 \rangle, \ldots, \langle u_n, p_n \rangle) = \sum_{j=1}^{n} \oplus p_{(j)} \left(\mu\left(A_{(j)}\right) - \mu\left(A_{(j-1)}\right) \right),$$

(u_j, p_j) is a two-tuple *with* $u_{(j)}$ the jth largest value in $\{u_1, u_2, u_3, \dots, u_n\}$, i.e. $u_{(1)} \geq u_{(2)} \geq u_{(3)} \geq \cdots \geq u_{(n)}$.

$$A_{(j)} = \{x_{(1)}, x_{(2)}, x_{(3)}, \dots, x_{(j)}\}, \quad j = 1, 2, 3, \dots, n \text{ and } A_{(0)} = \varnothing$$

$$IFC_m(\langle u_1, p_1 \rangle, \langle u_2, p_2 \rangle, \dots, \langle u_n, p_n \rangle)$$

$$= \left(1 - \prod_{j=1}^{n} \left(1 - \mu_{p(j)} \right)^{\mu\left(A_{(j)}\right) - \mu\left(A_{(j-1)}\right)}, \prod_{j=1}^{n} v_{p(j)}^{\mu\left(A_{(j)}\right) - \mu\left(A_{(j-1)}\right)} \right), \tag{5.60}$$

where μ is fuzzy measure on $P(X)$.

Example 10   Let μ be a fuzzy measure on space $P(X) = \{x_1, x_2, x_3\}$ in which

$$\mu(0) = \varnothing, \quad \mu(x_1) = 0.2, \quad \mu(x_2) = 0.5, \quad \mu(x_3) = 0.6$$

$$\mu(x_1, x_2) = 0.6378, \mu(x_1, x_3) = 0.7254, \mu(x_2, x_3) = 0.9135, \mu(x_1, x_2, x_3) = 1$$

Consider three intuitionistic fuzzy argument values on $X = \{x_1, x_2, x_3\}$, respectively, as:

$a_1 = (0.3, 0.5)$, $a_2 = (0.6, 0.3)$, $a_3 = (0.5, 0.2)$ with order-inducing variable of the three pairs are $\langle 3, a_1 \rangle$, $\langle 7, a_2 \rangle$, $\langle 6, a_3 \rangle$ where the first component is the order-inducing variable.

Compute aggregation of intuitionistic fuzzy values using induced intuitionistic fuzzy Choquet integral operator.

Solution

On the basis of order-inducing variable, $\langle 3, a_1 \rangle$, $\langle 7, a_2 \rangle$, $\langle 6, a_3 \rangle$, the values are reordered and listed as:

$$\langle 7, a_2 \rangle = \langle 7, (0.6, 0.3) \rangle,$$
$$\langle 6, a_3 \rangle = \langle 6, (0.5, 0.2) \rangle,$$
$$\langle 3, a_1 \rangle = \langle 3, (0.3, 0.5) \rangle.$$

So, it is seen that the initial *u-index* is reordered.
$u\text{-}index(1) = 2$, $u\text{-}index(2) = 3$, $u\text{-}index(3) = 1$.

So the ordered vector $B = \begin{bmatrix} a_{u-index(1)} \\ a_{u-index(2)} \\ a_{u-index(3)} \end{bmatrix} = \begin{bmatrix} a_{(1)} \\ a_{(2)} \\ a_{(3)} \end{bmatrix} = \begin{bmatrix} a_2 \\ a_3 \\ a_1 \end{bmatrix} = \begin{bmatrix} (0.6, 0.3) \\ (0.5, 0.2) \\ (0.3, 0.5) \end{bmatrix}$,

$u - index(j)$ is taken as (j) for simplicity.

Next, we will compute $A_{(j)}$ values for finding the weights where we will use *u-index* values.

For $j = 1, 2, 3, \dots, n$, we know $A_{(j)} = \{x_{(1)}, x_{(2)}, x_{(3)}, \dots, x_{(j)}\}$ and $A_{(0)} = \varnothing$.

So, $A_{(1)} = \{x_{(1)}\} = \{x_2\}$,

$$A_{(2)} = \left\{x_{(1)}, x_{(2)}\right\} = \left\{x_2, x_3\right\}, \quad A_{(3)} = \left\{x_{(1)}, x_{(2)}, x_{(3)}\right\} = \left\{x_2, x_3, x_1\right\}.$$

From fuzzy measure, we have

$$\mu\left(A_{(1)}\right) = \mu(\{x_2\}) = 0.5, \quad \mu\left(A_{(2)}\right) = \mu\{x_2, x_3\} = 0.9135,$$
$$\mu\left(A_{(3)}\right) = \mu(x_2, x_3, x_1) = 1.$$

For different values of j, we compute $\mu(A_{(j)}) - \mu(A_{(j-1)})$:

$$\mu\left(A_{(1)}\right) = 0.5,$$

$$\mu\left(A_{(2)}\right) - \mu\left(A_{(1)}\right) = 0.9135 - 0.5 = 0.4135,$$

$$\mu\left(A_{(3)}\right) - \mu\left(A_{(2)}\right) = 1 - 0.9135 = 0.0865.$$

$$IFC_m(\langle u_1, p_1 \rangle, \langle u_2, p_2 \rangle, \langle u_n, p_3 \rangle)$$

$$= \left(1 - \prod_{j=1}^{3}\left(1 - \mu_{p(j)}\right)^{\mu\left(A_{(j)}\right) - \mu\left(A_{(j-1)}\right)}, \prod_{j=1}^{3}\nu_{p(j)}^{\mu\left(A_{(j)}\right) - \mu\left(A_{(j-1)}\right)}\right)$$

$$= \left(1 - \left(1 - \mu_{a(1)}\right)^{0.5} \times \left(1 - \mu_{a(2)}\right)^{0.4135} \times \left(1 - \mu_{a(3)}\right)^{0.0865}, \nu_{a(1)}^{0.5} \times \nu_{a(2)}^{0.4135} \times \nu_{a(3)}^{0.0865}\right)$$

If the values of arguments are $a_1 = (0.6, 0.3)$, $a_2 = (0.5, 0.2)$, $a_3 = (0.3, 0.5)$, then

$$IFC_m(\langle u_1, a_1 \rangle, \langle u_2, a_2 \rangle, \ldots, \langle u_n, a_4 \rangle)$$

$$= \left(1 - \left(0.4^{0.5} \times 0.5^{0.4135} \times 0.7^{0.0865}\right), 0.3^{0.5} \times 0.2^{0.4135} \times 0.5^{0.0865}\right)$$

$$= [0.5396, 0.2652].$$

5.8 Example on Decision-making Problems

Multi-criteria decision-making problem is a method for finding the best alternative among all the alternatives evaluated using a set of attributes/criteria. Alternatives are evaluated on the basis of criterion or attribute. Let $B = \{b_1, b_2, b_3, \ldots, b_m\}$ be a set of attributes and $C = \{c_1, c_2, c_3, \ldots, c_n\}$ be a set of alternatives. Partial evaluation of the alternatives, $c_i(i = 1, 2, 3, \ldots, n)$, is carried out with respect to the attributes or criteria, $b_j(j = 1, 2, 3, \ldots, m)$. Partial evaluation, c_{ij}, is expressed using intuitionistic fuzzy values, $c_{ij} = (\mu_{ij}, \nu_{ij})$, where μ_{ij} is the satisfaction degree which means that c_i satisfies the criterion b_j and ν_{ij} is the dissatisfaction degree which means that c_i does not satisfy the criterion b_j with the condition $0 \leq \mu_{ij} \leq 1$, $0 \leq \nu_{ij} \leq 1$. A multi-attribute decision-making problem is expressed in matrix form:

$$
D = \begin{bmatrix}
 & b_1 & b_2 & b_3 & \cdots & b_m \\
c_1 & (\mu_{11}, \nu_{11}) & (\mu_{12}, \nu_{12}) & (\mu_{13}, \nu_{13}) & \cdots & (\mu_{1n}, \nu_{1m}) \\
c_2 & (\mu_{21}, \nu_{21}) & (\mu_{22}, \nu_{22}) & (\mu_{23}, \nu_{23}) & \cdots & (\mu_{2n}, \nu_{2m}) \\
& & & \vdots & & \\
c_n & (\mu_{n1}, \nu_{n1}) & (\mu_{n2}, \nu_{n2}) & (\mu_{n3}, \nu_{n3}) & \cdots & (\mu_{nm}, \nu_{nm})
\end{bmatrix}
$$

Score function, $S(c_{ij})$, of the partial evaluation c_{ij} of the alternative c_i is evaluated to rank c_{ij}. If there is no difference between the two score functions, then accuracy function, $H(c_{ij})$, is used to rank c_{ij} based on the accuracy.

An example of decision-making problem to find an expert supplier based on supplier's competencies in making a machine is given. Suppose there are three suppliers and their ability is judged by the attributes of the machine such as (i) innovative level, (ii) longevity, and (iii) cost. We are to take a decision, which supplier is to be selected for ordering a machine. Attributes are denoted as (b_1, b_2, b_2) and the three suppliers, which are the alternatives, are denoted as (c_1, c_2, c_3). To evaluate the competencies of the experts, 10 candidates are invited. Suppose there are six candidates who judge the attribute b_1 of the expert c_1 as strong and other three candidates who judge the attribute b_1 of the expert c_1 as not strong and the remaining one candidate does not judge the candidate as strong or not strong. Then, evaluating value of the attribute b_1 of c_1 may be expressed using an intuitionistic fuzzy value $c_{ij} = (0.6, 0.3)$, where c_{ij} is the partial evaluation of any alternatives, c_i with respect to attributes, b_j. Likewise, the results of 10 candidates to the 3 experts according to the 3 criteria (attributes) together are performed and an intuitionistic fuzzy decision matrix of the experts is formed:

$$
\begin{array}{cccc}
 & b_1 & b_2 & b_3 \\
c_1 & (0.6, 0.3) & (0.5, 0.4) & (0.7, 0.2) \\
c_2 & (0.4, 0.3) & (0.6, 0.2) & (0.5, 0.2) \\
c_3 & (0.4, 0.3) & (0.3, 0.5) & (0.6, 0.3)
\end{array}
$$

Partial evaluation c_{ij} of candidate supplier c_i with respect to the attributes or criteria, b_j, is reordered such that $c_{i(j)} \leq c_{i(j+1)}$.
Reordering is done on the basis of score function [32, 33].

As has been said, the partial evaluation of the alternatives c_i with respect to the attributes b_j is made by intuitionistic fuzzy values $c_{ij} = (\mu_{ij}, \nu_{ij})$ and the decision matrix (D) is formed:

$$
D = \begin{bmatrix}
c_{11} & c_{12} & c_{13} \\
c_{21} & c_{22} & c_{23} \\
c_{31} & c_{32} & c_{33}
\end{bmatrix}
$$

Then, score on c_{ij} is given as:

$$S(c_{ij}) = \mu_{c(ij)} - \nu_{c(ij)} \text{ and } S(c) \in [-1,1].$$

It represents the difference of membership and non-membership values. If the score values are similar, the accuracy degree is evaluated.

$$H(c_{ij}) = \mu_{c_{ij}} + \nu_{c_{ij}} \text{ where } H(c) \in [0,1].$$

After reordering, which is done on the basis of score function $c_{i(j)} \leq c_{i(j+1)}$, we get:

$$c_{1(1)} = (0.5,0.4), c_{1(2)} = (0.6,0.3), c_{1(3)} = (0.7,0.2)$$

$$c_{2(1)} = (0.4,0.3), c_{2(2)} = (0.5,0.2), c_{2(3)} = (0.6,0.2)$$

$$c_{3(1)} = (0.3,0.5), c_{3(2)} = (0.4,0.3), c_{3(3)} = (0.6,0.3)$$

Let the fuzzy measure of the criterion b_1, b_2, b_3 or group of criteria i.e, the importance of each criterion, be given as:

$$\mu(b_1) = 0.4, \mu(b_2) = 0.2, \mu(b_3) = 0.3.$$

Using Eq. (4.4), we obtain the value of λ and we get,

$$\mu(b_1,b_2) = 0.63, \mu(b_2,b_3) = 0.52, \mu(b_1,b_3) = 0.7445, \mu(b_1,b_2,b_3) = 1,$$

where μ is a fuzzy measure.

Using intuitionistic fuzzy Choquet integral operator:

$$c_i = IFC_\mu(c_{i1},c_{i2},c_{i3},...,c_{in}) = \left(1 - \prod_{j=1}^{n}\left(1 - \mu_{c_{i(j)}}\right)^{\mu(A_j)-\mu(A_{j+1})}, \prod_{j=1}^{n}\nu_{c_{i(j)}}^{\mu(A_j)-\mu(A_{j+1})}\right),$$

where $A_{(j)} = \{b_{(j)}, ..., b_{(3)}\}$, $A_{(3+1)} = \emptyset$. We are to aggregate c_{ij} corresponding to the supplier $c_i(i = 1, 2, 3)$.

For computing c_1, we require to compute $\mu(A_{(j)})$.
We have $c_{1(1)} = (0.5,0.4)$, $c_{1(2)} = (0.6,0.3)$, $c_{1(3)} = (0.7,0.2)$.
So, $(1) = 2$, $(2) = 1$, $(3) = 3$.
Hence, we get

$$A_{(1)} = \{b_{(1)}, b_{(2)}, b_{(3)}\} = \{b_2, b_1, b_3\} \text{ and } \mu(A_{(1)}) = 1,$$

$$A_{(2)} = \{b_{(2)}, b_{(3)}\} = \{b_1, b_3\} \text{ and } \mu(A_{(2)}) = 0.7445,$$

$$A_{(3)} = \{b_{(3)}\} = \{b_3\} \text{ and } \mu(A_{(3)}) = 0.3.$$

$$c_1 = IFC_\mu(c_{11}, c_{12}, c_{13})$$

$$= \left(1 - \prod_{j=1}^{3} \left(1 - \mu_{c_{1(j)}} \right)^{\mu\left(A_{(j)}\right) - \mu\left(A_{(j+1)}\right)}, \prod_{j=1}^{3} \nu_{c_{1(j)}}^{\mu\left(A_{(j)}\right) - \mu\left(A_{(j+1)}\right)} \right)$$

$$\Rightarrow c_1 = \left[\left(1 - (1-0.5)^{1-0.7445} \times (1-0.6)^{0.7445-0.3} \times (1-0.7)^{0.3}, 0.4^{1-0.7445} \right. \right.$$

$$\left. \left. \times 0.3^{0.7445-0.33} \times 0.2^{0.3} \right) \right] = \left(1 - 0.5^{0.255} \times 0.4^{0.4445} \times 0.3^{0.3}, 0.4^{0.255} \right.$$

$$\left. \times 0.3^{0.4445} \times 0.2^{0.3} \right) = (0.6115, 0.2964).$$

Likewise, for computing c_2, we have

$$c_{2(1)} = (0.4, 0.3), c_{2(2)} = (0.5, 0.2), c_{2(3)} = (0.6, 0.2).$$

Proceeding as above, we obtain

$$A_{(1)} = \left\{ b_{(1)}, b_{(2)}, b_{(3)} \right\} = \{b_1, b_3, b_2\}, \quad \mu(A_1) = 1,$$

$$A_{(2)} = \left\{ b_{(2)}, b_{(3)} \right\} = \{b_3, b_2\}, \quad \mu(A_2) = 0.52,$$

$$A_{(3)} = \left\{ b_{(3)} \right\} = \{b_2\}, \quad \mu(A_3) = 0.2.$$

$$c_2 = \left(1 - \prod_{j=1}^{4} \left(1 - \mu_{c_{2(j)}} \right)^{\mu\left(A_{(j)}\right) - \mu\left(A_{(j+1)}\right)}, \prod_{j=1}^{4} \nu_{c_{2(j)}}^{\mu\left(A_{(j)}\right) - \mu\left(A_{(j+1)}\right)} \right)$$

$$= \left[\left(1 - 0.6^{1-0.52} \times 0.5^{0.52-0.2} \times 0.4^{0.2}, 0.3^{1-0.52} \times 0.2^{0.52-0.2} \times 0.2^{0.2} \right) \right]$$

$$= (0.4781, 0.2430)$$

Likewise, for computing c_3, we have

$$c_{3(1)} = (0.3, 0.5), c_{3(2)} = (0.4, 0.3), c_{3(3)} = (0.6, 0.3)$$

Thus, we get

$$A_{(1)} = \left\{ b_{(1)}, b_{(2)}, b_{(3)} \right\} = \{b_2, b_1, b_3\}, \quad \mu(A_1) = 1$$

$$A_{(2)} = \left\{ b_{(2)}, b_{(3)} \right\} = \{b_1, b_3\}, \quad \mu(A_2) = 0.7445$$

$$A_{(3)} = \left\{ b_{(3)} \right\} = \{b_3\}, \quad \mu(A_3) = 0.3$$

$$c_3 = \left(1 - \prod_{j=1}^{4} \left(1 - \mu_{c_{3(j)}} \right)^{\mu\left(A_{(j)}\right) - \mu\left(A_{(j+1)}\right)}, \prod_{j=1}^{4} \nu_{c_{3(j)}}^{\mu\left(A_{(j)}\right) - \mu\left(A_{(j+1)}\right)} \right)$$

$$= \left(1 - (1-0.3)^{1-0.7445} \times (1-0.4)^{0.7445-0.3} \times (1-0.6)^{0.3}, 0.5^{1-0.7445} \right.$$

$$\left. \times 0.3^{0.7445-0.3} \times 0.3^{0.3} \right) = 0.4474, 0.3418$$

So, $c_1 = (0.6115, 0.2964)$, $c_2 = (0.4781, 0.2430)$, $c_3 = (0.4474, 0.3418)$

Using the score function (difference between membership and nonmembership values), ranking is done as: $c_1 > c_2 > c_3$.

Thus, c_1 supplier is the best.

5.9 Summary

In this chapter, different types of fuzzy operators are discussed. Fuzzy operators include fuzzy union (T-norm), fuzzy intersection (T-conorm), algebraic product sum, algebraic difference, bounded sum, bounded difference, and complement. Different types of fuzzy complements suggested by different authors are also discussed. Fuzzy T operators, suggested by different authors are discussed that contain either max–min terms or algebraic terms, i.e. they do not contain the max–min term. Fuzzy aggregation operators are discussed where different objects are combined into a single object of a same set. These include weighted averaging operator, ordered weighting operator, hybrid averaging operator, fuzzy GOWA operator, fuzzy induced generalized aggregation operator, fuzzy quasi arithmetic weighted averaging operator, fuzzy induced generalized averaging operator, Choquet ordered aggregation operator, and induced Choquet ordered aggregation operator. These are explained along with examples. In the second phase, intuitionistic fuzzy operators are discussed. Different types of intuitionistic fuzzy aggregation operators similar to fuzzy aggregation operators are explained with examples. These are generalized intuitionistic fuzzy aggregation operator, generalized intuitionistic fuzzy ordered aggregation operator, generalized intuitionistic fuzzy hybrid operator, IFWG operator, Choquet ordered aggregation operator, and induced Choquet ordered aggregation operator. Example on multi-criteria decision making is also given.

References

1 Zadeh, L.A. (1975). Calculus of fuzzy restrictions. In: *Fuzzy Sets and Their Applications to Cognitive and Decision Processes* (ed. L.A. Zadeh, K. Sun Fu, K. Tanaka and M. Shimura). New York: Academic Press.

2 Sugeno, M. (1977). Fuzzy measures and fuzzy integrals: a survey. In: *Fuzzy Automata and Decision Process* (ed. M. Gupta, G.N. Saridis and B.R. Gaines), 82–110. Amsterdam, New York: Elsevier.

3 Yager, R.R. (1980). On the general class of fuzzy connectives. *Fuzzy Sets and Systems* 4: 235–242.

4 Roychowdhury, S. and Wang, B.H. (1994). Composite generalization of Dombi class and a new family of T operators using additive-product connective generator. *Fuzzy Sets and Systems* 66: 329–346.

5 Chaira, T. (2015). *Medical Image Processing: Advanced Fuzzy Set Theoretic Techniques*. Boca Raton, FL: CRC Press.

6 Klir, G.J. and Yuan, B. (1995). *Fuzzy Sets and Fuzzy Logic: Theory and Application*. New Mexico: Prentice Hall.

7 Alsina, C., Trillas, E., and Valverde, L. (1983). On some logical connectives for fuzzy set theory. *Journal of Mathematical Analysis and Application* 93: 15–26.

8 Bandler, W. and Kohout, L. (1980). Fuzzy power sets and fuzzy implification operators. *Fuzzy Sets and Systems* 4: 13–30.

9 Dombi, J. (1982). A general class of fuzzy operators A De Morgan's class of fuzzy operators and fuzziness induced by fuzzy operators. *Fuzzy Sets and Systems* 8: 149–163.

10 Frank, M.J. (1979). On simultaneous associativity of $F(x, y)$ and $x+y$ $-F(x, y)$. *Aequationes Math* 19: 194–226.

11 Giles, R. (1976). Lukasiewicz logic and fuzzy set theory. *International Journal of Man-Machine Studies* 8: 313–327.

12 Gupta, M.M. and Qi, J. (1991). Theory of T norms and fuzzy inference methods. *Fuzzy Sets and Systems* 40: 431–450.

13 Merigo, J.M. and Casanovas, M. (2010). Fuzzy generalized hybrid aggregation operators and its application in fuzzy decision making. *International Journal of Fuzzy Systems* 12 (1): 15–24.

14 Weber, S. (1983). A general concept of fuzzy connectives, negations and implications based on t Norms and t Co norms. *Fuzzy Sets and Systems* 11: 115–134.

15 Dubois, D. and Prade, H. (1980). New results about properties and semantics of fuzzy set theoretic operators. In: *Fuzzy Sets: Theory and Applications to Policy Analysis and Information Systems* (ed. P. Wang and S. Chang), 59–65. New York: Plenum Press.

16 Hamacher, H. (1978). *Über Logische Aggregationen nicht-binar explizierter Entscheidungs-Kriterion*. Frankfurt am Main: R.G. Fisher Verlag.

17 Yager, R. (2003). Induced aggregation operator. *Fuzzy Sets and Systems* 137: 59–69.

18 Fodor, J. and Roubens, M. (1994). *Fuzzy Preference Modelling and Multicriteria Decision Support*. Dordrecht: Kluwer Academic Publishers.

19 Yager, R.R. (2004). On some new classes of implication operators and their role in approximate reasoning. *Information Sciences* 167: 193–216.

20 Yager, R. (1988). On ordered weighted averaging aggregation operators in multi criteria decision making. *IEEE Transaction on Systems, Man and Cybernetics* 18 (1): 183–190.

21 Yager, R. (1994). Generalized OWA aggregation operators. *Fuzzy Optimization and Decision Making* 3: 93–107.

22 Yang, W. and Chen, Z. (2012). The quasi-arithmetic intuitionistic fuzzy OWA operators. *Knowledge based Systems* 27: 219–233.

23 Merigo, J.M. and Casanovas, M. (2011). Uncertain induced Quasi –Arithmetic OWA operator. *International Journal of Intelligent Systems* 26: 1–24.

24 Xu, Y. and Wang, H. (2012). The induced generalized aggregation operators for intuitionistic fuzzy sets and their application in group decision making. *Applied Soft Computing* 12: 1168–1179.

25 Yager, R. (1999). Induced ordered weighted averaging operators. *IEEE Transaction on Systems, Man and Cybernetics: Part B* 29 (2): 183–190.

26 Chen, S.M. and Tan, J.M. (1994). Handling multicriteria fuzzy decision making problems based on vague set theory. *Fuzzy Sets and Systems* 67: 163–172.

27 Xu, Z. (2007). Intuitionistic fuzzy aggregation operators. *IEEE Transaction on Fuzzy Systems* 15 (6): 1179–1187.

28 Zhao, H., Xu, Z., Ni, M., and Liu, S. (2010). Generalized aggregation operators for intuitionistic fuzzy sets. *Journal of Intelligent and Fuzzy Systems* 25: 1–30.

29 Xu, Z. (2005). An overview method for determining OWA weights. *International Journal of Intelligent Systems* 20: 843–865.

30 Xui, Z.Z. and Yager, R.R. (2006). Some geometric aggregation operators based on intuitionistic fuzzy sets. *International Journal of General Systems* 35 (4): 417–433.

31 Tan, C. and Chen, X. (2011). Induced intuitionistic fuzzy Choquet integral operator for multicriteria decision making. *International Journal of Intelligent Systems* 26: 659–686.

32 Hong, D.H. and Choi, C.H. (2003). Multicriteria fuzzy decision-making problems based on vague set theory. *Fuzzy Sets and Systems* 114: 103–113.

33 Tan, C. and Chen, X. (2010). Intuitionistic fuzzy Choquet integral operator for multicriteria decision making. *Experts Systems and Applications* 37: 149–157.

6

Fuzzy Linear Equations

6.1 Introduction

Many real-time engineering systems are too complex to be defined in precise terms and imprecision is often involved. Linear system of equations with uncertainty parameters plays a significant role in the areas of economics, finance, engineering, control system, and so on. To analyze such situation, fuzzy information is required. Fuzzifying either parameters or variables or both in these systems has been one of the research areas, since these kinds of systems are encountered in many applications.

This chapter deals in solving fuzzy system of linear equations. There are many non-fuzzy classical methods to solve linear equations but in this chapter, solving fuzzy linear equation is discussed. A general model for solving a fuzzy linear system of equations where coefficient matrix is crisp and the right-hand-side column is a fuzzy vector was first proposed by Friedman et al. [1], and later with his colleagues, they replaced the original fuzzy linear system by a crisp linear system and solved it.

Then, Dehgan et al. [2] considered all parameters in a fuzzy linear system as fuzzy numbers and is called fully fuzzy linear system. The method is used in computing inverse of a matrix in fuzzy case that employs a linear equation system and identity matrix.

A matrix equation is represented as $\tilde{A} \otimes \tilde{x} = \tilde{b}$, where $\tilde{A} = \tilde{a}_{ij}$ is a fuzzy matrix $\tilde{A}$ of size $(n \times n)$ and $\tilde{b}$ and $\tilde{x}$ are fuzzy vectors of size $(n \times 1)$. If the vector or matrix elements are in intervals, then this system is called interval linear system.

We describe fuzzy matrix whose entries are fuzzy numbers and these may be used to model uncertain and imprecise aspects of real-world problems. When entries in the matrices are numbers, and if we want to find the inverse of the fuzzy matrix, then the best computing method is to write in the form of linear

Fuzzy Set and Its Extension: The Intuitionistic Fuzzy Set, First Edition. Tamalika Chaira.
© 2019 John Wiley & Sons, Inc. Published 2019 by John Wiley & Sons, Inc.

equations composing of product of $n \times n$ fuzzy matrices. Then, like ordinary matrix computation, this system is equated with identity matrix and is solved.

A matrix $\tilde{A} = \left(\tilde{a}_{ij}\right)$ is called a fuzzy matrix if each element of $\tilde{A}$ is a fuzzy number [3, 4]. $\tilde{A}$ is positive if each element of $\tilde{A} > 0$ and is negative if each element of $\tilde{A} < 0$.

Let $\tilde{A}$ $\left(\tilde{A} = \left(\tilde{a}_{ij}\right)\right)$ and $\tilde{B}$ $\left(\tilde{B} = \left(\tilde{b}_{ij}\right)\right)$ be two matrices of $p \times q$ and $q \times r$. The size of the product of fuzzy matrices is $p \times r$. The product is written as:

$$\tilde{A} \otimes B = \tilde{C} \left(\tilde{c}_{ij}\right),$$

where $\otimes$ is approximated as multiplication. $\tilde{c}_{ij} = \sum_{k=1,2,3,\ldots n} \tilde{a}_{ik} \otimes \tilde{b}_{kj}$

A fuzzy matrix $\tilde{A}$, just like a fuzzy number, consists of center, left spread, and right spread and is represented as $\tilde{A} = (A, L, R)$ where A, L, R are crisp matrices and these are denoted as center, left spread, and right spread, respectively.

A fuzzy number P is called LR-type fuzzy number if the membership function μ is of the following form [3]:

$$\mu_P = \begin{cases} L\left(\dfrac{p-x}{\alpha}\right), & \text{for } x \le p, \alpha > 0 \\[2mm] R\left(\dfrac{x-p}{\beta}\right), & \text{for } x \ge p, \beta > 0 \end{cases}$$

where L and R are continuous decreasing function in the interval $[0, \infty +)$. It fulfills the condition $L(0) = R(0) = 1$. p is the mean value of the fuzzy number P which is denoted as $P = (p, \alpha, \beta)$, where α and β are left and right spread, respectively, and these are positive real numbers.

6.2 Fuzzy Linear Equation

Solving the system of fuzzy linear equation has been studied by many authors. In this section, the problem is solved using approximated fuzzy arithmetic [2].

Consider a $n \times n$ linear system of linear equations:

$$(\tilde{a}_{11} \otimes \tilde{x}_1) \oplus (\tilde{a}_{12} \otimes \tilde{x}_1) \oplus \cdots \oplus (\tilde{a}_{1n} \otimes \tilde{x}_n) = \tilde{b}_1$$

$$(\tilde{a}_{21} \otimes \tilde{x}_1) \oplus (\tilde{a}_{22} \otimes \tilde{x}_1) \oplus \cdots \oplus (\tilde{a}_{2n} \otimes \tilde{x}_n) = \tilde{b}_1$$

$$\vdots \tag{6.1}$$

$$(\tilde{a}_{n1} \otimes \tilde{x}_1) \oplus (\tilde{a}_{n2} \otimes \tilde{x}_1) \oplus \cdots \oplus (\tilde{a}_{nn} \otimes \tilde{x}_n) = \tilde{b}_n$$

The matrix form of above equation is written as:

$$\tilde{A} \otimes \tilde{x} = \tilde{b} \text{ or } \tilde{A}\tilde{x} = \tilde{b}$$

$\tilde{x}$ is a fuzzy approximate solution of $\tilde{A} \otimes \tilde{x} = \tilde{b}$ with left and right shape functions as $L(.)$ and $R(.)$ that are used to represent $\tilde{A}$ and $\tilde{b}$, where $\tilde{A} = \left[\tilde{a}_{ij}\right]$ is a coefficient matrix of $n \times n$, and the vector $\tilde{a}_{ij} = \left(a_{ij}, \alpha_{ij}, \beta_{ij}\right)_{LR}$ and $\tilde{b} = (b, m, n)$.

Let the unknown vector $\tilde{x}$ be represented as $\tilde{x} = (x, y, z)$.

Considering $x \geq 0$, then we may write $\tilde{A} \otimes \tilde{x} = \tilde{b}$ as $(A, P, Q) \otimes (x, y, z) = (b, m, n)$, where $\alpha_{ij} = P$, $\beta_{ij} = Q$ is assumed.

Using the theory of multiplication of two fuzzy numbers, we know from Chapter 2 using Eq. (2.22):

For two positive fuzzy numbers $P = (p, \alpha, \beta)_{LR}$ and $Q = (q, \gamma, \delta)_{LR}$,

Multiplication: $(p, \alpha, \beta)_{LR} \otimes (q, \gamma, \delta)_{LR} = (p \cdot q, p\gamma + q\alpha, p\,\delta + q\beta)_{LR}$, $P > 0$, $Q > 0$

Then, $(A, P, Q) \otimes (x, y, z) = (Ax, Ay + Px, Az + Qx)$.

$$\text{Thus,} (Ax, Ay + Px, Az + Qx) = (b, m, n). \tag{6.2}$$

6.2.1 Problem of Finding an Unknown Number

Definition 1 [2]

Let $\tilde{x}$ is a fuzzy approximate solution of $\tilde{A} \otimes \tilde{x} = \tilde{b}$ with left and right shape functions similar to that of $L(.)$ and $R(.)$ functions which are used in $\tilde{A}$ and $\tilde{b}$ with approximate operator $\tilde{x} = (x, y, z) \geq 0$, then $\tilde{x}$ is said to be a fuzzy solution of $(A, P, Q) \otimes \tilde{x} = (b, m, n)$ if and only if:

$Ax = b$, $Ay + Px = m$, $Az + Qx = n$ (from Eq. (6.2))

The membership function of each element, x, $\mu_x > 0$ is defined from L and R functions that are used in $\tilde{A}$ and $\tilde{b}$.

Assuming A to be a crisp matrix, then we can write

$$(Ax, Ay + Px, Az + Qx) = (b, m, n)$$

Thus, we have three equations

$$\text{i)}\, Ax = b \tag{6.3}$$

$$\text{ii)}\, Ay + Px = m$$
$$\Rightarrow Ay = m - Px \tag{6.4}$$

$$\text{iii)}\, Az + Qx = n$$
$$\Rightarrow Az = n - Qx \tag{6.5}$$

So, we get

$$x = A^{-1}b$$

$$y = A^{-1}m - A^{-1}Px$$

$$z = A^{-1}n - A^{-1}Qx$$

Let us take an example to find the unknown vector $\tilde{x}$ of a fuzzy linear system, where the unknown vector is LR-type fuzzy numbers with left and right spread.

Example 1 Consider a fuzzy linear system

$$5\tilde{x}_1 + 6\tilde{x}_2 = 50$$

$$7\tilde{x}_1 + 4\tilde{x}_2 = 48$$

where x is a fuzzy approximate solution with y, z as the left and right spread functions. Compute $\tilde{x} = (x,y,z)$.

Solution

Let the system in fuzzy form with left and right spread be written as:

$$(5,1,2) \otimes (x_1, y_1, z_1) \oplus (6,1,1) \otimes (x_2, y_2, z_2) = (50,12,16)$$

$$(7,1,1) \otimes (x_1, y_1, z_1) \oplus (4,0,1) \otimes (x_2, y_2, z_2) = (48,7,11)$$

From the two equations, we write the values of A, P, Q as

$$A = \begin{bmatrix} 5 & 6 \\ 7 & 4 \end{bmatrix}, \quad P = \begin{bmatrix} 1 & 1 \\ 1 & 0 \end{bmatrix}, \quad Q = \begin{bmatrix} 2 & 1 \\ 1 & 1 \end{bmatrix}$$

In matrix form using Eq. (6.3), i.e. $Ax = b$, we can write:

$$\begin{bmatrix} 5 & 6 \\ 7 & 4 \end{bmatrix} \begin{bmatrix} x_1 \\ x_2 \end{bmatrix} = \begin{bmatrix} 50 \\ 48 \end{bmatrix}$$

$$\Rightarrow \begin{bmatrix} x_1 \\ x_2 \end{bmatrix} = \begin{bmatrix} 5 & 6 \\ 7 & 4 \end{bmatrix}^{-1} \begin{bmatrix} 50 \\ 48 \end{bmatrix}$$

On solving, we get

$$\begin{bmatrix} x_1 \\ x_2 \end{bmatrix} = \frac{1}{-22} \begin{bmatrix} 4 & -6 \\ -7 & 5 \end{bmatrix} \begin{bmatrix} 50 \\ 48 \end{bmatrix} = \frac{1}{-22} \begin{bmatrix} 200 - 288 \\ -350 + 240 \end{bmatrix} = \frac{1}{-22} \begin{bmatrix} -88 \\ -110 \end{bmatrix} = \begin{bmatrix} 4 \\ 5 \end{bmatrix}$$

For obtaining $\begin{bmatrix} y_1 \\ y_2 \end{bmatrix}$, we use Eq. (6.3), $Ay = m - Px$.

We get $\begin{bmatrix} 5 & 6 \\ 7 & 4 \end{bmatrix} \begin{bmatrix} y_1 \\ y_2 \end{bmatrix} = \begin{bmatrix} 12 \\ 7 \end{bmatrix} - \begin{bmatrix} 1 & 1 \\ 1 & 0 \end{bmatrix} \begin{bmatrix} 4 \\ 5 \end{bmatrix}$, where

$$\begin{bmatrix} 5 & 6 \\ 7 & 4 \end{bmatrix} \begin{bmatrix} y_1 \\ y_2 \end{bmatrix} = \begin{bmatrix} 12 \\ 7 \end{bmatrix} - \begin{bmatrix} 9 \\ 4 \end{bmatrix} = \begin{bmatrix} 3 \\ 3 \end{bmatrix} \Rightarrow \begin{bmatrix} y_1 \\ y_2 \end{bmatrix} = \begin{bmatrix} 5 & 6 \\ 7 & 4 \end{bmatrix}^{-1} \begin{bmatrix} 3 \\ 3 \end{bmatrix}$$

$$\begin{bmatrix} y_1 \\ y_2 \end{bmatrix} = \frac{1}{-22} \begin{bmatrix} 4 & -6 \\ -7 & 5 \end{bmatrix} \begin{bmatrix} 3 \\ 3 \end{bmatrix}$$

$$= \frac{1}{-22} \begin{bmatrix} -6 \\ -6 \end{bmatrix}$$

$$\Rightarrow \begin{bmatrix} y_1 \\ y_2 \end{bmatrix} = \begin{bmatrix} \dfrac{3}{11} \\ \dfrac{3}{11} \end{bmatrix}$$

Likewise, for finding $\begin{bmatrix} z_1 \\ z_2 \end{bmatrix}$, we use Eq. (6.4), $Az = n - Qx$

$$\begin{bmatrix} 5 & 6 \\ 7 & 4 \end{bmatrix} \begin{bmatrix} z_1 \\ z_2 \end{bmatrix} = \begin{bmatrix} 16 \\ 11 \end{bmatrix} - \begin{bmatrix} 2 & 1 \\ 1 & 1 \end{bmatrix} \begin{bmatrix} 4 \\ 5 \end{bmatrix}$$

$$\begin{bmatrix} 5 & 6 \\ 7 & 4 \end{bmatrix} \begin{bmatrix} z_1 \\ z_2 \end{bmatrix} = \begin{bmatrix} 3 \\ 2 \end{bmatrix} \Rightarrow \begin{bmatrix} z_1 \\ z_2 \end{bmatrix} = \begin{bmatrix} 5 & 6 \\ 7 & 4 \end{bmatrix}^{-1} \begin{bmatrix} 3 \\ 2 \end{bmatrix} = \frac{1}{-22} \begin{bmatrix} 4 & -6 \\ -7 & 5 \end{bmatrix} \begin{bmatrix} 3 \\ 2 \end{bmatrix}$$

$$\Rightarrow \begin{bmatrix} z_1 \\ z_2 \end{bmatrix} = \frac{1}{-22} \begin{bmatrix} 12-12 \\ -21+10 \end{bmatrix} = \begin{bmatrix} 0 \\ \dfrac{1}{2} \end{bmatrix}$$

Thus, the solution of the unknown vector $\tilde{x}$ is:

$$\tilde{x} = \begin{bmatrix} \left(4, \dfrac{3}{11}, 0\right) \\ \left(5, \dfrac{3}{11}, \dfrac{1}{2}\right) \end{bmatrix}$$

Example 2 Consider another fuzzy linear system

$$2\tilde{x}_1 + \tilde{x}_2 = 16$$
$$3\tilde{x}_1 + 5\tilde{x}_2 = 31$$

Compute $\tilde{x} = (x, y, z)$ where $\tilde{x}$ is fuzzy approximate solution with y, z as the left and right spread functions, respectively.

Solution

Let the system in fuzzy form with left and right spread be written as:

$$(2,1,0) \otimes (x_1, y_1, z_1) \oplus (1,0,2) \otimes (x_2, y_2, z_2) = (16, 8, 5)$$
$$(3,1,1) \otimes (x_1, y_1, z_1) \oplus (5,1,1) \otimes (x_2, y_2, z_2) = (31, 11, 13)$$

From the two equations, we write the values of A, P, Q as

$$A = \begin{bmatrix} 2 & 1 \\ 3 & 5 \end{bmatrix}, \quad P = \begin{bmatrix} 1 & 0 \\ 1 & 1 \end{bmatrix}, \quad Q = \begin{bmatrix} 0 & 2 \\ 1 & 1 \end{bmatrix}$$

In matrix form, equation $Ax = b$ may be written as:

$$\begin{bmatrix} 2 & 1 \\ 3 & 5 \end{bmatrix} \begin{bmatrix} x_1 \\ x_2 \end{bmatrix} = \begin{bmatrix} 16 \\ 31 \end{bmatrix}$$

$$\Rightarrow \begin{bmatrix} x_1 \\ x_2 \end{bmatrix} = \begin{bmatrix} 2 & 1 \\ 3 & 5 \end{bmatrix}^{-1} \begin{bmatrix} 16 \\ 31 \end{bmatrix}$$

On solving we get $\begin{bmatrix} x_1 \\ x_2 \end{bmatrix} = \dfrac{1}{7} \begin{bmatrix} 5 & -1 \\ -3 & 2 \end{bmatrix} \begin{bmatrix} 16 \\ 31 \end{bmatrix} = \dfrac{1}{7} \begin{bmatrix} 80-31 \\ -48+62 \end{bmatrix} = \dfrac{1}{7} \begin{bmatrix} 49 \\ 14 \end{bmatrix} = \begin{bmatrix} 7 \\ 2 \end{bmatrix}$

For obtaining $\begin{bmatrix} y_1 \\ y_2 \end{bmatrix}$, we use Eq. (6.3), $Ay = m - Px$

$$\begin{bmatrix} 2 & 1 \\ 3 & 5 \end{bmatrix} \begin{bmatrix} y_1 \\ y_2 \end{bmatrix} = \begin{bmatrix} 8 \\ 11 \end{bmatrix} - \begin{bmatrix} 1 & 0 \\ 1 & 1 \end{bmatrix} \begin{bmatrix} 7 \\ 2 \end{bmatrix} = \begin{bmatrix} 8 \\ 11 \end{bmatrix} - \begin{bmatrix} 7 \\ 9 \end{bmatrix}$$

or $\begin{bmatrix} 2 & 1 \\ 3 & 5 \end{bmatrix} \begin{bmatrix} y_1 \\ y_2 \end{bmatrix} = \begin{bmatrix} 2 & 1 \\ 3 & 5 \end{bmatrix}^{-1} \begin{bmatrix} 1 \\ 2 \end{bmatrix}$

or $\begin{bmatrix} y_1 \\ y_2 \end{bmatrix} = \dfrac{1}{7} \begin{bmatrix} 5 & -1 \\ -3 & 2 \end{bmatrix} \begin{bmatrix} 1 \\ 2 \end{bmatrix} = \dfrac{1}{7} \begin{bmatrix} 5-2 \\ -3+4 \end{bmatrix} = \dfrac{1}{7} \begin{bmatrix} 3 \\ 1 \end{bmatrix} = \begin{bmatrix} \frac{3}{7} \\ \frac{1}{7} \end{bmatrix}$

$$\text{or } \begin{bmatrix} y_1 \\ y_2 \end{bmatrix} = \begin{bmatrix} \dfrac{3}{7} \\ \dfrac{1}{7} \end{bmatrix}$$

Likewise, for $\begin{bmatrix} z_1 \\ z_2 \end{bmatrix}$, using Eq. (6.4) $Az = n - Qx$ we get

$$\begin{bmatrix} 2 & 1 \\ 3 & 5 \end{bmatrix} \begin{bmatrix} z_1 \\ z_2 \end{bmatrix} = \begin{bmatrix} 5 \\ 13 \end{bmatrix} - \begin{bmatrix} 0 & 2 \\ 1 & 1 \end{bmatrix} \begin{bmatrix} 7 \\ 2 \end{bmatrix}$$

$$\begin{bmatrix} 2 & 1 \\ 3 & 5 \end{bmatrix} \begin{bmatrix} z_1 \\ z_2 \end{bmatrix} = \begin{bmatrix} 1 \\ 4 \end{bmatrix}$$

$$\begin{bmatrix} z_1 \\ z_2 \end{bmatrix} = \frac{1}{7} \begin{bmatrix} 5 & -1 \\ -3 & 2 \end{bmatrix} \begin{bmatrix} 1 \\ 4 \end{bmatrix} = \frac{1}{7} \begin{bmatrix} 5-4 \\ -3+8 \end{bmatrix} = \begin{bmatrix} \dfrac{1}{7} \\ \dfrac{5}{7} \end{bmatrix}$$

Thus, the solution is:

$$\tilde{x} = \begin{bmatrix} \left(7, \dfrac{1}{7}, \dfrac{1}{7}\right) \\ \left(2, \dfrac{1}{7}, \dfrac{5}{7}\right) \end{bmatrix}$$

6.3 Solving Linear Equation Using Cramer's Rule

The above problem may be solved using Cramer's rule [2, 5].

Cramer's rule is a formula for the solution of a system of linear equations with as many equations as unknowns. The solution is expressed in terms of determinants of a square coefficient matrix and of matrices obtained from it by replacing one column of the coefficient matrix by the column vector of right-hand side of the equations.

Considering Eqs. (6.3–6.5), we can write

$$x_i = \frac{\det\left(A_1^{(i)}\right)}{\det(A)}, \ i = 1, 2, 3, \ldots, n$$

where $A_1^{(i)}$ is obtained by replacing the ith column of A by b.

Then, using the value of x from Eqs. (6.4, 6.5), $Ay = m - Px$ and $Az = n - Qx$ are computed and the following values are obtained as:

$$y_i = \frac{\det\left(A_2^{(i)}\right)}{\det(A)}, \; i = 1, 2, 3, \ldots, n$$

where $A_2^{(i)}$ is obtained by replacing ith column of A by $m - Px$.

$$z_i = \frac{\det\left(A_3^{(i)}\right)}{\det(A)}, \; i = 1, 2, 3, \ldots, n$$

where $A_3^{(i)}$ is obtained by replacing ith column of A by $n - Qx$.

An example will explain the procedure clearly.

Example 3   Consider a fuzzy linear system

$$4\tilde{x}_1 + 5\tilde{x}_2 + 3\tilde{x}_3 = 71$$
$$7\tilde{x}_1 + 10\tilde{x}_2 + 2\tilde{x}_3 = 118$$
$$6\tilde{x}_1 + 7\tilde{x}_2 + 15\tilde{x}_3 = 155$$

Compute the vector $\tilde{x}$, $\tilde{x} = (x, y, z)$ with y, z denoting the left and right spreads, respectively.

Solution

Considering $\tilde{x} = (x, y, z)$ to be a fuzzy approximate solution with left and right spread-shape functions, the fuzzy linear system may be written as:

$$(4,1,0)\otimes(x_1,y_1,z_1)\oplus(5,3,2)\otimes(x_2,y_2,z_2)\oplus(3,0,3)\otimes(x_3,y_3,z_3) = (71,54,76)$$
$$(7,4,3)\otimes(x_1,y_1,z_1)\oplus(10,6,5)\otimes(x_2,y_2,z_2)\oplus(2,1,1)\otimes(x_3,y_3,z_3) = (118,113,129)$$
$$(6,2,4)\otimes(x_1,y_1,z_1)\oplus(7,1,2)\otimes(x_2,y_2,z_2)\oplus(15,5,4)\otimes(x_3,y_3,z_3) = (155,89,151)$$

In matrix form, the above equations may be written as:

$$\begin{bmatrix} (4,1,0) & (5,3,2) & (3,0,3) \\ (7,4,3) & (10,6,5) & (2,1,1) \\ (6,2,4) & (7,1,2) & (15,5,4) \end{bmatrix} \begin{bmatrix} \tilde{x}_1 \\ \tilde{x}_2 \\ \tilde{x}_3 \end{bmatrix} = \begin{bmatrix} (71,54,76) \\ (118,113,129) \\ (155,89,151) \end{bmatrix}$$

Here, $A = \begin{bmatrix} 4 & 5 & 3 \\ 7 & 10 & 2 \\ 6 & 7 & 15 \end{bmatrix}$, $P = \begin{bmatrix} 1 & 3 & 0 \\ 4 & 6 & 1 \\ 2 & 1 & 5 \end{bmatrix}$, $Q = \begin{bmatrix} 0 & 2 & 3 \\ 3 & 5 & 1 \\ 4 & 2 & 4 \end{bmatrix}$

First, we will solve for x using the equation: $Ax = b$ from Eq. (6.3).

$$A = \begin{bmatrix} 4 & 5 & 3 \\ 7 & 10 & 2 \\ 6 & 7 & 15 \end{bmatrix}$$

$$\det A = 4 \begin{vmatrix} 10 & 2 \\ 7 & 15 \end{vmatrix} - 5 \begin{vmatrix} 7 & 2 \\ 6 & 15 \end{vmatrix} + 3 \begin{vmatrix} 7 & 10 \\ 6 & 7 \end{vmatrix}$$

$$= 4(150 - 14) - 5(105 - 12) + 3(49 - 60)$$

$$= 544 - 465 - 33 = 46.$$

$A_1^{(1)}$ is obtained by replacing the first column of matrix A by b
So,

$$A_1^{(1)} = \begin{bmatrix} 71 & 5 & 3 \\ 118 & 10 & 2 \\ 155 & 7 & 15 \end{bmatrix}$$

$$\det A_1^{(1)} = 71 \begin{vmatrix} 10 & 2 \\ 7 & 15 \end{vmatrix} - 5 \begin{vmatrix} 118 & 2 \\ 155 & 15 \end{vmatrix} + 3 \begin{vmatrix} 118 & 10 \\ 155 & 7 \end{vmatrix}$$

$$= 71(150 - 14) - 5(1770 - 310) + 3(826 - 1550)$$

$$= 9656 - 7300 - 2172 = 184.$$

$\det A_1^{(2)}$ is obtained by replacing the second column of matrix A by b
So,

$$A_1^{(2)} = \begin{bmatrix} 4 & 71 & 3 \\ 7 & 118 & 2 \\ 6 & 155 & 15 \end{bmatrix}$$

$$\det A_1^{(2)} = 4 \begin{vmatrix} 118 & 2 \\ 155 & 15 \end{vmatrix} - 71 \begin{vmatrix} 7 & 2 \\ 6 & 15 \end{vmatrix} + 3 \begin{vmatrix} 7 & 118 \\ 6 & 155 \end{vmatrix}$$

$$= 5840 - 6603 + 1131 = 368.$$

$\det A_1^{(3)}$ is obtained by replacing the third column of matrix A by b
So,

$$A_1^{(3)} = \begin{bmatrix} 4 & 5 & 71 \\ 7 & 10 & 118 \\ 6 & 7 & 155 \end{bmatrix}$$

$$\det A_1^{(3)} = 4 \begin{vmatrix} 10 & 118 \\ 7 & 155 \end{vmatrix} - 5 \begin{vmatrix} 7 & 118 \\ 6 & 155 \end{vmatrix} + 71 \begin{vmatrix} 7 & 10 \\ 6 & 7 \end{vmatrix}$$

$$= 2896 - 1885 - 781 = 230.$$

We get

$$x_1 = \frac{\det A_1^{(1)}}{\det A} = \frac{184}{46} = 4, \qquad x_2 = \frac{\det A_1^{(2)}}{\det A} = \frac{368}{46} = 8, \qquad x_3 = \frac{\det A_1^{(3)}}{\det A} = \frac{230}{46} = 5.$$

So,

$$\begin{bmatrix} x_1 \\ x_2 \\ x_3 \end{bmatrix} = \begin{bmatrix} 4 \\ 8 \\ 5 \end{bmatrix}$$

Likewise, we compute for y and z. From Eq. (6.4), we have

$$Ay = m - Px$$

$$= \begin{bmatrix} 54 \\ 113 \\ 89 \end{bmatrix} - \begin{bmatrix} 1 & 3 & 0 \\ 4 & 6 & 1 \\ 2 & 1 & 5 \end{bmatrix} \begin{bmatrix} 4 \\ 8 \\ 5 \end{bmatrix}$$

$$= \begin{bmatrix} 54 \\ 113 \\ 89 \end{bmatrix} - \begin{bmatrix} 28 \\ 69 \\ 41 \end{bmatrix} = \begin{bmatrix} 26 \\ 44 \\ 48 \end{bmatrix}$$

$$y = \begin{bmatrix} 4 & 5 & 3 \\ 7 & 10 & 2 \\ 6 & 7 & 15 \end{bmatrix}^{-1} \begin{bmatrix} 26 \\ 44 \\ 48 \end{bmatrix}$$

Proceeding a similar manner as above, we obtain $A_2^{(1)}, A_2^{(2)}, A_2^{(3)}$ by replacing first, second, and third column of matrix A by $(m - Px)$, respectively, we get

$$\det A_2^{(1)} = \begin{bmatrix} 26 & 5 & 3 \\ 44 & 10 & 2 \\ 48 & 7 & 15 \end{bmatrix} = 26(150 - 14) - 5(660 - 96) + 3(308 - 480) = 200.$$

$$\det A_2^{(2)} = \begin{bmatrix} 4 & 26 & 3 \\ 7 & 44 & 2 \\ 6 & 48 & 15 \end{bmatrix} = 4(660 - 96) - 26(105 - 12) + 3(336 - 264) = 54.$$

$$\det A_2^{(3)} = \begin{bmatrix} 4 & 5 & 26 \\ 7 & 10 & 44 \\ 6 & 7 & 48 \end{bmatrix} = 4(480 - 308) - 5(336 - 264) + 26(49 - 60) = 42.$$

We get $y_1 = \dfrac{\det A_2^{(1)}}{\det A} = \dfrac{200}{46} = 100/23,\quad y_2 = \dfrac{\det A_2^{(2)}}{\det A} = \dfrac{54}{46} = 27/23,$

$y_3 = \dfrac{\det A_2^{(3)}}{\det A} = \dfrac{42}{46} = 21/23.$

So, $\begin{bmatrix} y_1 \\ y_2 \\ y_3 \end{bmatrix} = \begin{bmatrix} \dfrac{100}{23} \\ \dfrac{27}{23} \\ \dfrac{21}{23} \end{bmatrix}$

Likewise, for z, using Eq. (6.5) $Az = n - Qx$

$$Az = \begin{bmatrix} 76 \\ 129 \\ 151 \end{bmatrix} - \begin{bmatrix} 0 & 2 & 3 \\ 3 & 5 & 1 \\ 4 & 2 & 4 \end{bmatrix} \begin{bmatrix} 4 \\ 8 \\ 5 \end{bmatrix}$$

$$= \begin{bmatrix} 76 \\ 129 \\ 151 \end{bmatrix} - \begin{bmatrix} 31 \\ 57 \\ 52 \end{bmatrix} = \begin{bmatrix} 45 \\ 72 \\ 99 \end{bmatrix}$$

$$z = \begin{bmatrix} 4 & 5 & 3 \\ 7 & 10 & 2 \\ 6 & 7 & 15 \end{bmatrix}^{-1} \begin{bmatrix} 45 \\ 72 \\ 99 \end{bmatrix}$$

Proceeding in a similar manner as above we obtain $A_3^{(1)}, A_3^{(2)}, A_3^{(3)}$ by replacing the first, second and third columns of matrix A by $(n - Qx)$, respectively, we get

$$\det A_3^{(1)} = \begin{bmatrix} 45 & 5 & 3 \\ 72 & 10 & 2 \\ 99 & 7 & 15 \end{bmatrix} = 45(150 - 14) - 5(1080 - 198) + 3(504 - 990) = 252.$$

$$\det A_3^{(2)} = \begin{bmatrix} 4 & 45 & 3 \\ 7 & 72 & 2 \\ 6 & 99 & 15 \end{bmatrix} = 4(1080 - 198) - 45(105 - 12) + 3(693 - 432) = 126.$$

$$\det A_3^{(3)} = \begin{bmatrix} 4 & 5 & 45 \\ 7 & 10 & 72 \\ 6 & 7 & 99 \end{bmatrix} = 4(990 - 504) - 5(693 - 432) + 45(49 - 60) = 144.$$

We get

$$z_1 = \frac{\det A_3^{(1)}}{\det A} = \frac{252}{46} = \frac{126}{23}, \ z_2 = \frac{\det A_3^{(2)}}{\det A} = \frac{126}{46} = \frac{63}{23}, \ z_3 = \frac{\det A_3^{(3)}}{\det A} = \frac{144}{46} = \frac{72}{23}.$$

So,
$$\begin{bmatrix} z_1 \\ z_2 \\ z_3 \end{bmatrix} = \begin{bmatrix} \dfrac{126}{23} \\ \dfrac{63}{23} \\ \dfrac{72}{23} \end{bmatrix}.$$

So, the fuzzy solution of $\tilde{x}$ is:

$$\tilde{x} = \begin{bmatrix} \left(4, \dfrac{100}{23}, \dfrac{126}{23}\right) \\ \left(8, \dfrac{27}{23}, \dfrac{63}{23}\right) \\ \left(5, \dfrac{21}{23}, \dfrac{72}{23}\right) \end{bmatrix}.$$

6.4 Inverse of a Fuzzy Matrix

When we want to find the inverse of square and non-singular matrices, whose values are real numbers, then a linear equation is formed that is composed of product of two matrices – $n \times n$ coefficients matrix and $n \times n$ unknown matrix. This system is equated with identity matrix and is solved. But this method is not applicable for a fuzzy matrix whose entries are LR fuzzy numbers as there is no such fuzzy identity matrix. Basaram [6] suggested a method for computing the inverse of a fuzzy matrix using LR-type fuzzy numbers and the identity matrix mimics the identity matrix in real case. Fuzzy identity matrix and fuzzy zero number are defined as follows [6]:

Fuzzy one number – For a fuzzy number with center value 1 and left and right spreads α and β, respectively, where $\alpha > 0$ and $\beta < 1$, then this fuzzy number is called fuzzy one number denoted as $\tilde{1}(1, \alpha, \beta)$. The spread values take the value between 0 and 1. When $\alpha = \beta$, then fuzzy one number is a symmetric number.

Fuzzy zero number – In a similar fashion, for a fuzzy zero number, if the center value is 0 and the left and right spread values are ρ and σ with $\rho > 0$ and $\sigma < 1$, then this fuzzy number is called fuzzy zero number which is denoted as $\tilde{0}(0, \rho, \sigma)$. When $\rho = \sigma$, then fuzzy zero number is a symmetric number.

Identity matrix – If the diagonal elements of a fuzzy matrix are fuzzy one number and the off-diagonal elements are fuzzy zero number, then this matrix is called fuzzy identity matrix. A 3×3 fuzzy identity matrix is shown.

$$\tilde{I} = \begin{bmatrix} \tilde{1} & \tilde{0} & \tilde{0} \\ \tilde{0} & \tilde{1} & \tilde{0} \\ \tilde{0} & \tilde{0} & \tilde{1} \end{bmatrix}.$$

For computing fuzzy inverse matrix, fuzzy unknown vector $\tilde{x}$ and fuzzy right-hand-side vector $\tilde{b}$ in Eq. (6.1) are replaced by fuzzy unknown matrix and fuzzy identity matrix, respectively, which is written as:

$$\begin{aligned} (\tilde{a}_{11} \otimes \tilde{x}_{11}) \oplus (\tilde{a}_{12} \otimes \tilde{x}_{21}) \oplus \cdots \oplus (\tilde{a}_{1n} \otimes \tilde{x}_{n1}) &= \tilde{1} \\ (\tilde{a}_{21} \otimes \tilde{x}_{11}) \oplus (\tilde{a}_{21} \otimes \tilde{x}_{21}) \oplus \cdots \oplus (\tilde{a}_{21} \otimes \tilde{x}_{n1}) &= \tilde{0} \\ &\vdots \\ (\tilde{a}_{n1} \otimes \tilde{x}_{1n}) \oplus (\tilde{a}_{n2} \otimes \tilde{x}_{2n}) \oplus \cdots \oplus (\tilde{a}_{nn} \otimes \tilde{x}_{nn}) &= \tilde{1}. \end{aligned} \qquad (6.6)$$

Let us consider a linear fuzzy matrix LR type $\tilde{A}$ of size 2×2, which is given as:

$$\tilde{A} = \begin{bmatrix} \tilde{a}_{11} & \tilde{a}_{12} \\ \tilde{a}_{21} & \tilde{a}_{22} \end{bmatrix}$$

and the inverse of the matrix $\tilde{A}$ that is required to find out is:

$$\tilde{X} = \begin{bmatrix} \tilde{x}_{11} & \tilde{x}_{12} \\ \tilde{x}_{21} & \tilde{x}_{22} \end{bmatrix},$$

where $\tilde{a}_{ij}$ and $\tilde{x}_{ij}$ for $i, j = 1, 2$ are triangular symmetric fuzzy numbers.

Then the system of equation in Eq. (6.6) may be written as:

$$\tilde{A} \otimes \tilde{X} = \tilde{I}, \text{where } \tilde{I} \text{ is fuzzy identity matrix.} \qquad (6.7)$$

This expression may be written as:

$$\begin{bmatrix} \tilde{a}_{11} & \tilde{a}_{12} \\ \tilde{a}_{21} & \tilde{a}_{22} \end{bmatrix} \otimes \begin{bmatrix} \tilde{x}_{11} & \tilde{x}_{12} \\ \tilde{x}_{21} & \tilde{x}_{22} \end{bmatrix} = \begin{bmatrix} \tilde{1} & \tilde{0} \\ \tilde{0} & \tilde{1} \end{bmatrix},$$

where $\tilde{1} = (1, \alpha)$ and $\tilde{0} = (0, \rho)$, $\tilde{a}_{ij} = \left(a_{ij}, b_{ij}\right)$ and $\tilde{x}_{ij} = \left(x_{ij}, z_{ij}\right)$, $j = 1, 2$ are symmetric triangular fuzzy numbers.

Example 4 Let us consider a 2×2 matrix consists of symmetric fuzzy numbers

$$\tilde{A} = \begin{bmatrix} (12,6) & (9,4) \\ (6,2) & (5,3) \end{bmatrix}$$

Compute the inverse of the matrix, considering fuzzy number to be symmetrical triangular fuzzy numbers.

Solution

For computing the inverse of matrix $\tilde{A}$, we follow the procedure as described in Section 6.3. We write the fuzzy equation in the form $\tilde{A} \otimes \tilde{X} = \tilde{I}$ as:

$$\begin{bmatrix} (12,6) & (9,4) \\ (6,2) & (5,3) \end{bmatrix} \otimes \begin{bmatrix} \tilde{x}_{11} & \tilde{x}_{12} \\ \tilde{x}_{21} & \tilde{x}_{22} \end{bmatrix} = \begin{bmatrix} \tilde{1} & \tilde{0} \\ \tilde{0} & \tilde{1} \end{bmatrix}$$

Then, we perform matrix multiplication which is in the form:

$$(12,6)\tilde{x}_{11} \oplus (9,4)\tilde{x}_{21} = (1,\alpha)$$
$$(12,6)\tilde{x}_{12} \oplus (9,4)\tilde{x}_{22} = (0,\rho)$$
$$(6,2)\tilde{x}_{11} \oplus (5,3)\tilde{x}_{21} = (0,\rho) \qquad (6.8)$$
$$(6,2)\tilde{x}_{12} \oplus (5,3)\tilde{x}_{22} = (1,\alpha)$$

When both α and ρ approach 1, then fuzziness in the fuzzy identity matrix is considered to be the largest and when both α and ρ approach 0, then fuzziness in the fuzzy identity matrix is considered to be the least.

Substituting the values of $\tilde{x}_{ij} = \left(x_{ij}, z_{ij}\right)$ in Eqs. (6.8), the above system of equations may be written as:

$$(12,6)(x_{11}, z_{11}) \oplus (9,4)(x_{21}, z_{21}) = (1,\alpha)$$
$$(12,6)(x_{12}, z_{12}) \oplus (9,4)(x_{22}, z_{22}) = (0,\rho)$$

$(6,2)(x_{11},z_{11}) \oplus (5,3)(x_{21},z_{21}) = (0,\rho)$

$(6,2)(x_{12},z_{12}) \oplus (5,3)(x_{22},z_{22}) = (1,\alpha)$

From the properties of multiplication of two LR fuzzy numbers, we have

$$(p,\alpha,\beta)_{LR} \otimes (q,\gamma,\delta)_{LR} = (p{\cdot}q, p\gamma + q\alpha, p\delta + q\beta)_{LR}, M > 0, N > 0. \qquad (6.9)$$

If $\alpha = \beta$ and $\gamma = \delta$, then it becomes a symmetrical triangular fuzzy numbers. Considering these properties, the center part is written as:

$12x_{11} + 9x_{21} = 1$

$12x_{12} + 9x_{22} = 0$

$6x_{11} + 5x_{21} = 0$

$6x_{12} + 5x_{22} = 1$

On solving two equations: $12x_{11} + 9x_{21} = 1$ and $6x_{11} + 5x_{21} = 0$, we get

$$x_{11} = \frac{5}{6} \text{ and } x_{21} = -1.$$

On solving other two equations: $12x_{12} + 9x_{22} = 0$ and $6x_{12} + 5x_{22} = 1$,

we get $x_{12} = -\dfrac{3}{2} = -0.75$ and $x_{22} = 2$.

The spread part of the equation is written as (using multiplication of fuzzy numbers in Eq. (6.8)):

$12z_{11} + 6x_{11} + 9z_{21} + 4x_{21} = \alpha$

$12z_{12} + 6x_{12} + 9z_{22} + 4x_{22} = \rho$

$6z_{11} + 2x_{11} + 5z_{21} + 3x_{21} = \rho$

$6z_{12} + 2x_{12} + 5z_{22} + 3x_{22} = \alpha$

On solving, we get

$12z_{11} + 9z_{21} = \alpha - 1$

$12z_{12} + 9z_{22} = \rho + 1$

$6z_{11} + 5z_{21} = \rho + \dfrac{4}{3}$

$6z_{12} + 5z_{22} = \alpha - 3$

On solving we get,

$$z_{11} = \frac{5\alpha}{6} - \frac{3\rho}{2} - \frac{17}{6}$$

$$z_{21} = 2\rho - \alpha + \frac{11}{3}$$

$$z_{12} = -\frac{3\alpha}{2} + \frac{5\rho}{6} + \frac{16}{3}$$

$$z_{22} = 2\alpha - \rho - 7$$

If the values of x_{ij} are known, the center values z_{ij} can be computed

$$\tilde{A}^{-1} = \begin{bmatrix} \left(\frac{5}{6}, \left|\frac{5\alpha}{6} - \frac{3\rho}{2} - \frac{17}{6}\right|\right) & \left(-\frac{3}{2}, \left|-\frac{3\alpha}{2} + \frac{5\rho}{6} + \frac{16}{3}\right|\right) \\ \left(-1, \left|2\rho - \alpha + \frac{11}{3}\right|\right) & \left(2, \left|2\alpha - \rho - 7\right|\right) \end{bmatrix}.$$

z_{ij} is obtained on the parameters α and ρ. On different values of α and ρ, different values of z_{ij} are obtained. The spread values of LR fuzzy numbers are not negative and if the spread values α and ρ are negative, then absolute values are taken.

If $\alpha = 0.5$, $\rho = 0.5$, then

$$\tilde{A}^{-1} = \begin{bmatrix} (0.833, 3.16) & (-1.5, 5) \\ (-1, 4.17) & (2, 6.5) \end{bmatrix}.$$

But Mosleh and Otadi [7] confirmed that in the above problem, negative values of x_{12}, x_{21} are not considered. They modified the solution as follows:

On solving the center part, it is seen that x_{11}, x_{22} are positive and the values of x_{12}, x_{21} are negative.

The spread part of the equation is written as:

$$12z_{11} + 6x_{11} + 9z_{21} - 4x_{21} = \alpha$$

$$12z_{12} - 6x_{12} + 9z_{22} + 4x_{22} = \rho$$

$$6z_{11} + 2x_{11} + 5z_{21} - 3x_{21} = \rho$$

$$6z_{12} - 2x_{12} + 5z_{22} + 3x_{22} = \alpha$$

On solving we get,

$$z_{11} = \frac{5\alpha}{6} - \frac{3\rho}{2} - \frac{1}{2}$$

$$z_{21} = 2\rho - \alpha - \frac{1}{3}$$

$$z_{12} = -\frac{3\alpha}{2} + \frac{5\rho}{6} - \frac{2}{3}$$

$$z_{22} = 2\alpha - \rho - 1$$

If the values of x_{ij} are known, the center values z_{ij} can be computed

$$\tilde{A}^{-1} = \begin{bmatrix} \left(\dfrac{5}{6}, \left|\dfrac{5\alpha}{6} - \dfrac{3\rho}{2} - \dfrac{1}{2}\right|\right) & \left(-\dfrac{3}{2}, \left|-\dfrac{3\alpha}{2} + \dfrac{5\rho}{6} - \dfrac{2}{3}\right|\right) \\ \left(-1, \left|2\rho - \alpha - \dfrac{1}{3}\right|\right) & (2, |2\alpha - \rho - 1|) \end{bmatrix}$$

As the spread parameters are positive due to the definition of LR-type fuzzy numbers, absolute values of the spreads are taken.
If $\alpha = 0.5$, $\rho = 0.5$, then

$$\tilde{A}^{-1} = \begin{bmatrix} (0.833, 0.833) & (-1.5, 1) \\ (2, 0.167) & (-1, 0.5) \end{bmatrix}$$

Let us take another example where the fuzzy matrix is negative. Here, the method Mosleh and Otadi is used.

Example 5   Consider a negative fuzzy 2×2 matrix with symmetrical triangular fuzzy number

$$\tilde{A} = \begin{bmatrix} (-8, 4) & (-6, 3) \\ (-5, 1) & (-3, 1) \end{bmatrix} \quad \text{and compute the inverse of the matrix.}$$

Solution

For computing the inverse of matrix $\tilde{A}$, we follow the procedure as described above.

$$\begin{bmatrix} (-8, 4) & (-6, 3) \\ (-5, 1) & (-3, 1) \end{bmatrix} \otimes \begin{bmatrix} \tilde{x}_{11} & \tilde{x}_{12} \\ \tilde{x}_{21} & \tilde{x}_{22} \end{bmatrix} = \begin{bmatrix} \tilde{1} & \tilde{0} \\ \tilde{0} & \tilde{1} \end{bmatrix}$$

Then, we perform matrix multiplication which is in the form:

$$(-8, 4)\tilde{x}_{11} \oplus (-6, 3)\tilde{x}_{21} = (1, \alpha)$$
$$(-8, 4)\tilde{x}_{12} \oplus (-6, 3)\tilde{x}_{22} = (0, \rho)$$
$$(-5, 1)\tilde{x}_{11} \oplus (-3, 1)\tilde{x}_{21} = (0, \rho)$$
$$(-5, 1)\tilde{x}_{12} \oplus (-3, 1)\tilde{x}_{22} = (1, \alpha)$$

On substituting the values of $\tilde{x}_{ij} = (x_{ij}, z_{ij})$, the above system of equations may be written as:

$$(-8, 4)(x_{11}, z_{11}) \oplus (-6, 3)(x_{21}, z_{21}) = (1, \alpha)$$
$$(-8, 4)(x_{12}, z_{12}) \oplus (-6, 1)(x_{22}, z_{22}) = (0, \rho)$$

$$(-5,1)(x_{11},z_{11}) \oplus (-3,1)(x_{21},z_{21}) = (0,\rho)$$
$$(-5,1)(x_{12},z_{12}) \oplus (-3,1)(x_{22},z_{22}) = (1,\alpha)$$

The center part is written as:

$$-8x_{11} - 6x_{21} = 1$$
$$-8x_{12} - 6x_{22} = 0$$
$$-5x_{11} - 3x_{21} = 0$$
$$-5x_{12} - 3x_{22} = 1$$

On solving two equations: $-8x_{11} - 6x_{21} = 1$ and $-5_{11} - 3x_{21} = 0$, we get

$$x_{11} = \frac{1}{2} = 0.5 \text{ and } x_{21} = \frac{-5}{6} = -0.833.$$

On solving other two equations: $-8x_{12} - 6x_{22} = 0$ and $-5x_{12} - 3x_{22} = 1$, we get $x_{12} = -1$ and $x_{22} = \frac{4}{3} = 1.33$.

According to Mosleh and Otadi [7], the negative values of x_{12}, x_{21} are considered. Then the spread part of the equation becomes:

$$-8z_{11} + 4x_{11} - 6z_{21} - 3x_{21} = \alpha$$
$$-8z_{12} - 4x_{12} - 6z_{22} + x_{22} = \rho$$
$$-5z_{11} + x_{11} - 3z_{21} - x_{21} = \rho$$
$$-5z_{12} - x_{12} - 3z_{22} + x_{22} = \alpha.$$

On solving, we get

$$-8z_{11} - 6z_{21} = \alpha - \frac{9}{2}$$
$$-8z_{12} - 6z_{22} = \rho - \frac{16}{3}$$
$$-5z_{11} - 3z_{21} = \rho - \frac{4}{3}$$
$$-5z_{12} - 3z_{22} = \alpha - \frac{7}{3}$$

which implies

$$z_{11} = -\rho + \frac{\alpha}{2} - \frac{11}{12}$$
$$z_{21} = \frac{4\rho}{3} - \frac{5\alpha}{6} + \frac{71}{36}$$
$$z_{12} = -\alpha + \frac{\rho}{2} - \frac{1}{3}$$

$$z_{22} = \frac{4\alpha}{3} - \frac{5\rho}{6} + \frac{4}{3}.$$

If the values of x_{ij} are known, the center values z_{ij} can be computed

$$\tilde{A}^{-1} = \begin{bmatrix} \left(\frac{1}{2}, \left|-\rho + \frac{\alpha}{2} - \frac{11}{12}\right|\right) & \left(-1, \left|-\alpha + \frac{\rho}{2} - \frac{1}{3}\right|\right) \\ \left(-0.833, \left|\frac{4\rho}{3} - \frac{5\alpha}{6} + \frac{71}{36}\right|\right) & \left(1.33, \left|\frac{4\alpha}{3} - \frac{5\rho}{6} + \frac{4}{3}\right|\right) \end{bmatrix}.$$

As the spread parameters are positive due to the definition of LR-type fuzzy numbers, absolute values of the spreads are taken. If $\alpha = 0.5$, $\rho = 0.5$, then

$$\tilde{A}^{-1} = \begin{bmatrix} (0.5, 1.167) & (-1, 0.58) \\ (-0.833, 2.22) & (1.33, 1.58) \end{bmatrix}.$$

6.5 Summary

In this chapter, we have presented a method for fuzzy linear systems with fuzzy coefficients that involves fuzzy variables. Fuzzy linear equation has been solved using both direct method and Cramer's rule method. It is useful in analysis of physical systems and other topics in real engineering problems, where uncertainty aspects are present. For example, in finite elements for heat-transfer problems or finite-element formulation of equilibrium and steady-state problems in other areas, solving a set of simultaneous algebraic linear equations is required. Computing inverse of a fuzzy matrix that consists of LR fuzzy numbers is also discussed. It requires solving $n \times n$ by $n \times n$ equation system in fuzzy sense where all the unknown coefficients and the right-hand side of the equation are fuzzy numbers. Explanation of fuzzy identity matrix, fuzzy one number, and fuzzy zero number is also given as these are required in computing the inverse of fuzzy matrix with LR-type fuzzy numbers.

References

1 Friedman, M., Ming, M., and Kandel, A. (2000). Duality in fuzzy linear systems. *Fuzzy Sets Systems* 109: 55–58.
2 Dehgan, M., Hashemi, B., and Ghatee, M. (2006). Computational methods for solving fully fuzzy linear systems. *Applied Mathematics and Computation* 179: 328–343.

3 Dubois, D. and Prade, H. (1980). *Fuzzy Sets and Systems: Theory and Applications.* Boston, MA: Academic Press.

4 Dubois, D. and Prade, H. (1980). Systems of linear fuzzy constraints. *Fuzzy Sets and Systems* 3: 37–48.

5 Watkins, D.S. (2002). *Fundamentals of Matrix Computation.* New York: Wiley.

6 Basaram, M.A. (2012). Calculating fuzzy inverse matrix using fuzzy linear equation system. *Applied Soft Computing* 12: 1810–1813.

7 Mosleh, M. and Otadi, M. (2015). A discussion on "calculating fuzzy inverse matrix using fuzzy linear equation system". *Applied Soft Computing* 28: 511–513.

7

Fuzzy Matrices and Determinants

7.1 Basic Matrix Theory

In this section we give some basic notions of ordinary matrix theory that is essential to make the book self-contained one. A matrix is a set of numbers with finite rows and columns. Since data is written in tabular form, it is very easy to consider the table as matrix. Only by removing the lines between the data, the table can become a matrix. The horizontal entries of the table become the rows of the matrix and the vertical entries of the table become the columns of the matrix. An example of four rows and five columns is shown,

$$A = \begin{bmatrix} 1 & 5 & 6 & 8 & 9 \\ 4 & 2 & 6 & 5 & 0 \\ 3 & 0 & 1 & 4 & 9 \\ 2 & 1 & 5 & 6 & 7 \end{bmatrix}.$$

From the number of rows and columns, one can find the order of a matrix. In this case, the order of matrix is 4×5.

We can also interchange the rows and columns of the matrix. If we interchange the second row with the fourth row, we get an interchanged matrix,

$$A' = \begin{bmatrix} 1 & 5 & 6 & 8 & 9 \\ 2 & 1 & 5 & 6 & 7 \\ 3 & 0 & 1 & 4 & 9 \\ 4 & 2 & 6 & 5 & 0 \end{bmatrix}.$$

Similarly, one can interchange the columns.

Fuzzy Set and Its Extension: The Intuitionistic Fuzzy Set, First Edition. Tamalika Chaira.
© 2019 John Wiley & Sons, Inc. Published 2019 by John Wiley & Sons, Inc.

If a matrix has only columns then it is called a column matrix and if a matrix has only one row, it is called row matrix. If a matrix has equal number of rows and columns, it is called a square matrix.

A matrix can be represented in a general way with m rows and n columns. Each element is represented by a_{ij} which means the element a is in ith row and jth column of the matrix where $i = 1, 2, 3, ..., m, j = 1, 2, 3, ..., n$. A $n \times m$ matrix is written as:

$$A = \begin{bmatrix} a_{11} & a_{12} & \cdots & a_{1m} \\ a_{21} & a_{22} & \cdots & a_{2m} \\ \vdots & \vdots & \vdots & \vdots \\ a_{n1} & a_{n2} & \cdots & a_{nm} \end{bmatrix}.$$

A matrix is called a null matrix if all the elements of the matrix is 0.

A matrix is called an identity matrix if all the diagonal terms are 1 which is shown as

$$I = \begin{bmatrix} 1 & 0 & 0 & 0 \\ 0 & 1 & 0 & 0 \\ 0 & 0 & 1 & 0 \\ 0 & 0 & 0 & 1 \end{bmatrix}.$$

Some basic algebra of matrices:

7.1.1 Matrix Addition

For matrix addition, the order of the matrix should be equal. Examples are given to explain clearly.

Example 1

$$A = \begin{bmatrix} 1 & 2 & 0 \\ 4 & 6 & 1 \\ 2 & 3 & 0 \end{bmatrix}, \quad B = \begin{bmatrix} 1 & 0 & 4 \\ 2 & 5 & 2 \\ 1 & 3 & 0 \end{bmatrix},$$

Then,

$$A + B = \begin{bmatrix} 1+1 & 2+0 & 0+4 \\ 4+2 & 6+5 & 1+2 \\ 2+1 & 3+3 & 0+0 \end{bmatrix} = \begin{bmatrix} 2 & 2 & 4 \\ 6 & 11 & 3 \\ 3 & 6 & 0 \end{bmatrix}.$$

Likewise, one can perform matrix subtraction.

7.1.2 Matrix Multiplication

For matrix multiplication, two matrices are such that the number of columns of the matrix A is equal to the number of rows of matrix B. Suppose, a matrix A is of the order 3×1 and a matrix B is of the order 1×3, then the matrix AB is of the order 3×3.

Example 2 Consider a matrix

$$A = \begin{bmatrix} 1 \\ 5 \\ 6 \end{bmatrix}, \quad B = \begin{bmatrix} 6 & 5 & 7 \end{bmatrix},$$

Then,

$$AB = \begin{bmatrix} 1 \\ 5 \\ 6 \end{bmatrix} \begin{bmatrix} 6 & 5 & 7 \end{bmatrix} = \begin{bmatrix} 1 \times 6 & 1 \times 5 & 1 \times 7 \\ 5 \times 6 & 5 \times 5 & 5 \times 7 \\ 6 \times 6 & 6 \times 5 & 6 \times 7 \end{bmatrix} = \begin{bmatrix} 6 & 5 & 7 \\ 30 & 25 & 35 \\ 36 & 30 & 42 \end{bmatrix}.$$

If we calculate BA, then BA will be of the order 1×1.

$$BA = \begin{bmatrix} 6 & 5 & 7 \end{bmatrix} \begin{bmatrix} 1 \\ 5 \\ 6 \end{bmatrix} = \begin{bmatrix} 6 \times 1 + 5 \times 5 + 7 \times 6 \end{bmatrix} = \begin{bmatrix} 73 \end{bmatrix}.$$

So, $AB \neq BA$ implies matrix multiplication is not commutative.

7.1.3 Transpose of a Matrix

Transpose of a matrix is defined as $90°$ rotation, denoted by A^T.

Example 3 Consider a matrix

$$A = \begin{bmatrix} 1 & 2 & 0 \\ 4 & 6 & 1 \\ 2 & 3 & 0 \end{bmatrix}, \text{ then transpose of the matrix, } A^T, \text{ is}$$

$$A^T = \begin{bmatrix} 1 & 4 & 2 \\ 2 & 6 & 3 \\ 0 & 1 & 0 \end{bmatrix}.$$

If we compute AA^T, we get

$$AA^T = \begin{bmatrix} 1 & 2 & 0 \\ 4 & 6 & 1 \\ 2 & 3 & 0 \end{bmatrix} \begin{bmatrix} 1 & 4 & 2 \\ 2 & 6 & 3 \\ 0 & 1 & 0 \end{bmatrix} = \begin{bmatrix} 5 & 16 & 8 \\ 16 & 53 & 26 \\ 8 & 26 & 13 \end{bmatrix}.$$

A symmetric matrix is one in which a square matrix is equal to its transpose, i.e. $A = A^T$.

Till now we have worked on addition and multiplication.

Apart from this, we will see few works on operations such as max or min operator.

Example 4 Let $A = \begin{bmatrix} 1 & 2 & 0 \\ 4 & 6 & 1 \\ 2 & 3 & 0 \end{bmatrix}$, $B = \begin{bmatrix} 1 & 0 & 4 \\ 2 & 5 & 2 \\ 1 & 3 & 0 \end{bmatrix}$ be two matrices.

Maximum of two matrices is:

$$\max(A \cdot B) = \begin{bmatrix} \max(1,1) & \max(2,0) & \max(0,4) \\ \max(4,2) & \max(6,5) & \max(1,2) \\ \max(2,1) & \max(3,3) & \max(0,0) \end{bmatrix} = \begin{bmatrix} 1 & 2 & 4 \\ 4 & 6 & 2 \\ 2 & 3 & 0 \end{bmatrix}.$$

Maximum operator compares corresponding elements and chooses the larger.

Likewise, we can also find the minimum of the two matrices as:

$$\max(A \cdot B) = \begin{bmatrix} \min(1,1) & \min(2,0) & \min(0,4) \\ \min(4,2) & \min(6,5) & \min(1,2) \\ \min(2,1) & \min(3,3) & \min(0,0) \end{bmatrix} = \begin{bmatrix} 1 & 0 & 0 \\ 2 & 6 & 1 \\ 1 & 3 & 0 \end{bmatrix}.$$

7.2 Fuzzy Matrices

Here, we describe few basic properties and concepts of fuzzy matrices. As we are dealing with fuzzy set, so throughout the section we will consider [0,1] as unit interval and this interval is known as fuzzy interval. Some operators on fuzzy matrices are defined in this section [1, 2]. For all $x, y \in [0,1]$, we have the following operators:

$$x \vee y = \max(x,y),$$

$$x \wedge y = \min(x,y),$$

$$x^\alpha (\text{upper } \alpha \text{ cut}) = \begin{cases} 1 & \text{if } x \geq \alpha \\ 0 & \text{if } x < \alpha \end{cases},$$

$$x_\alpha (\text{lower } \alpha \text{ cut}) = \begin{cases} X & \text{if } x \geq \alpha \\ 0 & \text{if } x < \alpha \end{cases},$$

$x^c = 1 - x$ (complement of x),

$x \oplus y = x + y - x \cdot y$,

$x \odot y = x \cdot y$.

For any two fuzzy matrices, same properties follow as fuzzy vectors described in Chapter 5.

$A \oplus B = \left[a_{ij} + b_{ij} - a_{ij} \cdot b_{ij} \right]$,

$A \odot B = \left[a_{ij} \cdot b_{ij} \right]$,

$A \vee B = \left[a_{ij} \vee b_{ij} \right]$,

$A \wedge B = \left[a_{ij} \wedge b_{ij} \right]$,

$A^c = \left[1 - a_{ij} \right]$.

For any two matrices, $A < B$, implies $a_{ij} < b_{ij}$ for all i, j.

Every fuzzy matrix may be visualized as a three-dimensional figure [2]. Three axes are represented as row, column, and membership value.

Consider two fuzzy matrices A and B shown in Figures 7.1 and 7.2. We will compute $A \vee B$ and show this geometrically in Figure 7.3.

$$A = \begin{bmatrix} 0.5 & 0.2 & 0.8 & 1.0 \\ 0.0 & 0.7 & 0.5 & 0.7 \\ 0.4 & 0.3 & 0.0 & 1.0 \\ 1.0 & 0.2 & 0.8 & 0.0 \end{bmatrix}$$

$$B = \begin{bmatrix} 0.1 & 0.2 & 0.6 & 1 \\ 0.4 & 0.7 & 0.5 & 0.0 \\ 0.7 & 0.1 & 0.5 & 0.3 \\ 1 & 0.2 & 0.2 & 0.9 \end{bmatrix}$$

Figure 7.1 Geometric representation fuzzy matrix A [Adapted from [2]].

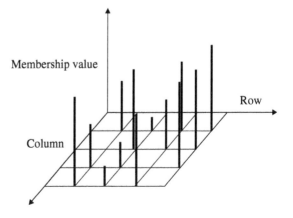

Membership value

Row

Column

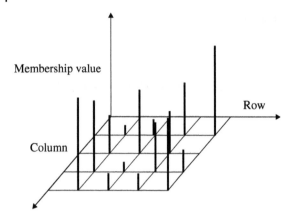

Figure 7.2 Geometric representation of fuzzy matrix *B* [Adapted from [2]].

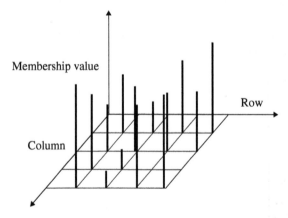

Figure 7.3 Geometric representation of fuzzy matrix *A* ∨ *B*.

Then,

$$
A \vee B =
\begin{bmatrix}
0.5 & 0.2 & 0.8 & 1 \\
0.4 & 0.7 & 0.5 & 0.7 \\
0.7 & 0.3 & 0.5 & 1 \\
1 & 0.2 & 0.8 & 0.9
\end{bmatrix}
$$

For any two fuzzy matrices, A and B, the following conditions hold [2]:
(i) $A \odot B \leq A$, (ii) $A \oplus B \geq A$, (iii) $A \oplus B \geq A \odot B$, (iv) $A \oplus A \geq A$, and (v) $A \odot A \leq A$. Here, the conditions are verified considering (i, j)th element in each case.

i) If $A = [a_{ij}]$, $B = [b_{ij}]$, then $A \odot B = [a_{ij} \cdot b_{ij}]$.
We know $0 \le b_{ij} \le 1$, so it is clear that $a_{ij} \cdot b_{ij} \le a_{ij}$.

ii) (i, j)th element of $A \oplus B = a_{ij} + b_{ij} - a_{ij} \cdot b_{ij} = a_{ij} + b_{ij}(1 - a_{ij}) \ge a_{ij}$.

iii) Let us assume $A \oplus B \ge A \odot B$, then

$$a_{ij} + b_{ij} - a_{ij} \cdot b_{ij} \ge a_{ij} \cdot b_{ij} \Rightarrow a_{ij} + b_{ij} - a_{ij} \cdot b_{ij} - a_{ij} \cdot b_{ij} \ge 0$$
$$\Rightarrow a_{ij}(1 - b_{ij}) + b_{ij}(1 - a_{ij}) \ge 0.$$

Now, as A and B are fuzzy, we know

$$0 \le a_{ij} \le 1, \quad 0 \le b_{ij} \le 1,$$

so $a_{ij}(1 - b_{ij}) + b_{ij}(1 - a_{ij}) \ge 0$ is true.

iv) (i, j)th element of $A \oplus A = \left(a_{ij} + a_{ij} - a_{ij} \cdot a_{ij}\right) = 2a_{ij} - a_{ij}^2 = a_{ij} + a_{ij}(1 - a_{ij}) \ge a_{ij}$. Hence, $A \oplus A \ge A$.

v) (i, j)th element of $A \odot A = a_{ij} \cdot a_{ij}$. It is obvious $a_{ij} \cdot a_{ij} \le a_{ij}$. Hence, $A \odot A \le A$.

7.2.1 Matrix Addition, Multiplication, Max, Min Operations

In fuzzy case, if we simply add like ordinary matrix multiplication, we will get the values more than 1. If we consider two fuzzy matrices:

$$A = \begin{bmatrix} 0.5 & 0.2 & 0.8 & 1.0 \\ 0.0 & 0.7 & 0.5 & 0.7 \\ 0.4 & 0.3 & 0.0 & 1.0 \\ 1.0 & 0.2 & 0.8 & 0.0 \end{bmatrix}, \quad B = \begin{bmatrix} 0.1 & 0.2 & 0.6 & 1.0 \\ 0.4 & 0.7 & 0.5 & 0.0 \\ 0.7 & 0.1 & 0.5 & 0.3 \\ 1.0 & 0.2 & 0.2 & 0.9 \end{bmatrix}$$

and simply add the two matrices, i.e. $A + B$ like ordinary matrix addition, we will get

$$A + B = \begin{bmatrix} 0.6 & 0.4 & 1.4 & 2.0 \\ 0.4 & 1.4 & 1.0 & 0.7 \\ 1.1 & 0.4 & 0.5 & 1.3 \\ 2.0 & 0.4 & 1.0 & 0.9 \end{bmatrix}.$$

We see that all entries in $A + B$ do not lie in the interval $[0,1]$, rather the values are more than 1. So, only in fuzzy cases, max or min operations are defined.

Fuzzy max operation is similar to fuzzy addition. An example will explain clearly.

Example 5

$$A \vee B = \begin{bmatrix} \max(0.5,0.1) & \max(0.2,0.2) & \max(0.8,0.6) & \max(1.0,1.0) \\ \max(0.0,0.4) & \max(0.7,0.7) & \max(0.5,0.5) & \max(0.7,0.0) \\ \max(0.4,0.7) & \max(0.3,0.1) & \max(0.0,0.5) & \max(1.0,0.3) \\ \max(1.0,1.0) & \max(0.2,0.2) & \max(0.8,0.2) & \max(0.0,0.9) \end{bmatrix}$$

$$= \begin{bmatrix} 0.5 & 0.2 & 0.8 & 1.0 \\ 0.4 & 0.7 & 0.5 & 0.7 \\ 0.7 & 0.3 & 0.5 & 1.0 \\ 1.0 & 0.2 & 0.8 & 0.9 \end{bmatrix}.$$

It is seen that all the entries lie in an interval [0,1]. This is a fuzzy matrix.

Example 6 Fuzzy min operation

$$A \wedge B = \begin{bmatrix} \min(0.5,0.1) & \min(0.2,0.2) & \min(0.8,0.6) & \min(1.0,1.0) \\ \min(0.0,0.4) & \min(0.7,0.7) & \min(0.5,0.5) & \min(0.7,0.0) \\ \min(0.4,0.7) & \min(0.3,0.1) & \min(0.0,0.5) & \min(1.0,0.3) \\ \min(1.0,1.0) & \min(0.2,0.2) & \min(0.8,0.2) & \min(0.0,0.9) \end{bmatrix}$$

$$= \begin{bmatrix} 0.1 & 0.2 & 0.6 & 1.0 \\ 0.0 & 0.7 & 0.5 & 0.0 \\ 0.4 & 0.1 & 0.0 & 0.3 \\ 1.0 & 0.2 & 0.2 & 0.0 \end{bmatrix}.$$

It is observed that $\min(A,B) \neq \max(A,B)$.

For fuzzy multiplication, similar to multiplication of ordinary matrices, if we compute for fuzzy multiplication, we obtain:

$$A = \begin{bmatrix} 0.5 & 0.2 & 0.8 & 1.0 \\ 0.0 & 0.7 & 0.5 & 0.7 \\ 0.4 & 0.3 & 0.0 & 1.0 \\ 1.0 & 0.2 & 0.8 & 0.0 \end{bmatrix}, \quad B = \begin{bmatrix} 0.1 & 0.2 & 0.6 & 1.0 \\ 0.4 & 0.7 & 0.5 & 0.0 \\ 0.7 & 0.1 & 0.5 & 0.3 \\ 1.0 & 0.2 & 0.2 & 0.9 \end{bmatrix} \quad \text{then}$$

$$AB = \begin{bmatrix} 0.5 & 0.2 & 0.8 & 1.0 \\ 0.0 & 0.7 & 0.5 & 0.7 \\ 0.4 & 0.3 & 0.0 & 1.0 \\ 1.0 & 0.2 & 0.8 & 0.0 \end{bmatrix} \begin{bmatrix} 0.1 & 0.2 & 0.6 & 1.0 \\ 0.4 & 0.7 & 0.5 & 0.0 \\ 0.7 & 0.1 & 0.5 & 0.3 \\ 1.0 & 0.2 & 0.2 & 0.9 \end{bmatrix} = \begin{bmatrix} 1.69 & 0.52 & 1.0 & 1.64 \\ 1.33 & 0.68 & 0.5 & 0.74 \\ 1.16 & 0.49 & 0.59 & 1.3 \\ 0.74 & 0.42 & 1.1 & 1.24 \end{bmatrix}.$$

Still the entries in the matrix are outside the interval range [0,1]. This implies that AB is not a fuzzy matrix. So, we need to define an operation that is analogous to matrix multiplication and the product should be a fuzzy matrix, i.e. the elements of the matrix lie in a unit interval [0,1]. In the new operation, the number of rows in one matrix should always be similar to the number of columns of the other matrix as in the case of ordinary matrix multiplication. The operator that is used for fuzzy matrix multiplication is max–min operator or min–max operators.

Fuzzy multiplication using max–min operator is explained with an example.

Example 7   Consider two fuzzy matrices

$$A = \begin{bmatrix} 0.5 & 0.2 & 0.8 & 1.0 \\ 0.0 & 0.7 & 0.5 & 0.7 \\ 0.4 & 0.3 & 0.0 & 1.0 \\ 1.0 & 0.2 & 0.8 & 0.0 \end{bmatrix}, \quad B = \begin{bmatrix} 0.1 & 0.2 & 0.6 & 1.0 \\ 0.4 & 0.7 & 0.5 & 0.0 \\ 0.7 & 0.1 & 0.5 & 0.3 \\ 1.0 & 0.2 & 0.2 & 0.9 \end{bmatrix}.$$

Then,

$$AB = \begin{bmatrix} C_{11} & C_{12} & C_{13} & C_{14} \\ C_{21} & C_{22} & C_{23} & C_{24} \\ C_{31} & C_{32} & C_{33} & C_{34} \\ C_{41} & C_{42} & C_{43} & C_{44} \end{bmatrix} = \begin{bmatrix} 0.5 & 0.2 & 0.8 & 1.0 \\ 0.0 & 0.7 & 0.5 & 0.7 \\ 0.4 & 0.3 & 0.0 & 1.0 \\ 1.0 & 0.2 & 0.8 & 0.0 \end{bmatrix} \begin{bmatrix} 0.1 & 0.2 & 0.6 & 1.0 \\ 0.4 & 0.7 & 0.5 & 0.0 \\ 0.7 & 0.1 & 0.5 & 0.3 \\ 1.0 & 0.2 & 0.2 & 0.9 \end{bmatrix}.$$

$$C_{11} = \max[\min(0.5, 0.1), \min(0.2, 0.4), \min(0.8, 0.7), \min(1, 1)]$$

$$= \max(0.1, 0.2, 0.7, 1) = 1,$$

$$C_{12} = \max[\min(0.5, 0.2), \min(0.2, 0.7), \min(0.8, 0.1), \min(1, 0.2)]$$

$$= \max(0.2, 0.2, 0.1, 0.2) = 0.2,$$

$$C_{13} = \max[\min(0.5, 0.6), \min(0.2, 0.5), \min(0.8, 0.5), \min(1, 0.2)]$$

$$= \max(0.5, 0.2, 0.5, 0.2) = 0.5,$$

$$C_{14} = \max[\min(0.5, 1), \min(0.2, 0), \min(0.8, 0.3), \min(1, 0.9)]$$

$$= \max(0.5, 0, 0.3, 0.9) = 0.9,$$

$$C_{21} = \max[\min(0, 0.1), \min(0.7, 0.4), \min(0.5, 0.7), \min(0.7, 1)]$$

$$= \max(0, 0.4, 0.5, 0.7) = 0.7,$$

$$C_{22} = \max[\min(0, 0.2), \min(0.7, 0.7), \min(0.5, 0.1), \min(0.7, 0.2)]$$

$$= \max(0.1, 0.7, 0.1, 0.2) = 0.7,$$

$$C_{23} = \max[\min(0,0.6),\min(0.7,0.5),\min(0.5,0.5),\min(0.7,0.2)]$$
$$= \max(0,0.5,0.5,0.2) = 0.5,$$

$$C_{24} = \max[\min(0,1),\min(0.7,0),\min(0.5,0.3),\min(0.7,0.9)]$$
$$= \max(0,0,0.3,0.7) = 0.7,$$

$$C_{31} = \max[\min(0.4,0.1),\min(0.3,0.4),\min(0,0.7),\min(1,1)]$$
$$= \max(0.1,0.3,0,1) = 1,$$

$$C_{32} = \max[\min(0.4,0.2),\min(0.3,0.7),\min(0,0.1),\min(1,0.2)]$$
$$= \max(0.2,0.3,0,0.2) = 0.3,$$

$$C_{33} = \max[\min(0.4,0.6),\min(0.3,0.5),\min(0,0.5),\min(1,0.2)]$$
$$= \max(0.4,0.3,0,0.2) = 0.4,$$

$$C_{34} = \max[\min(0.4,1),\min(0.3,0.0),\min(0,0.3),\min(1,0.9)]$$
$$= \max(0.4,0.0,0,0.9) = 0.9,$$

$$C_{41} = \max[\min(1,0.1),\min(0.2,0.4),\min(0.8,0.7),\min(0,1)]$$
$$= \max(0.1,0.2,0.7,0) = 0.7,$$

$$C_{42} = \max[\min(1,0.2),\min(0.2,0.7),\min(0.8,0.1),\min(0,0.2)]$$
$$= \max(0.2,0.2,0.1,0) = 0.2,$$

$$C_{43} = \max[\min(1,0.6),\min(0.2,0.5),\min(0.8,0.5),\min(0,0.2)]$$
$$= \max(0.6,0.2,0.5,0) = 0.6,$$

$$C_{44} = \max[\min(1,1),\min(0.2,0.0),\min(0.8,0.3),\min(0,0.9)]$$
$$= \max(1,0.0,0.3,0) = 1.$$

Thus, we have

$$AB = \begin{bmatrix} 1.0 & 0.2 & 0.5 & 0.9 \\ 0.7 & 0.7 & 0.5 & 0.7 \\ 0.1 & 0.3 & 0.4 & 0.9 \\ 0.7 & 0.2 & 0.6 & 1.0 \end{bmatrix}.$$

Fuzzy multiplication can also be done using min–max operation.

Example 8 Let $AB = \begin{bmatrix} B_{11} & B_{12} & B_{13} & B_{14} \\ B_{21} & B_{22} & B_{23} & B_{24} \\ B_{31} & B_{32} & B_{33} & B_{34} \\ B_{41} & B_{42} & B_{43} & B_{44} \end{bmatrix}$

Let
For two fuzzy matrices:

$$A = \begin{bmatrix} 0.5 & 0.2 & 0.8 & 1.0 \\ 0.0 & 0.7 & 0.5 & 0.7 \\ 0.4 & 0.3 & 0.0 & 1.0 \\ 1.0 & 0.2 & 0.8 & 0.0 \end{bmatrix}, \quad B = \begin{bmatrix} 0.1 & 0.2 & 0.6 & 1.0 \\ 0.4 & 0.7 & 0.5 & 0.0 \\ 0.7 & 0.1 & 0.5 & 0.3 \\ 1.0 & 0.2 & 0.2 & 0.9 \end{bmatrix},$$

we compute min-max operation like max–min operation as:

$B_{11} = \min[\max(0.5,0.1), \max(0.2,0.4), \max(0.8,0.7), \max(1,1)]$
$\quad = \min(0.5,0.4,0.8,1) = 0.4,$

$B_{12} = \min[\max(0.5,0.2), \max(0.2,0.7), \max(0.8,0.1), \max(1,0.2)]$
$\quad = \min(0.5,0.7,0.8,1) = 0.5,$

$B_{13} = \min[\max(0.5,0.6), \max(0.2,0.5), \max(0.8,0.5), \max(1,0.2)]$
$\quad = \min(0.6,0.5,0.8,1) = 0.5,$

$B_{14} = \min[\max(0.5,1), \max(0.2,0), \max(0.8,0.3), \max(1,0.9)]$
$\quad = \min(1,0.2,0.8,1) = 0.2,$

$B_{21} = \min[\max(0,0.1), \max(0.7,0.4), \max(0.5,0.7), \max(0.7,1)]$
$\quad = \min(0.1,0.7,0.7,1) = 0.1,$

$B_{22} = \min[\max(0,0.2), \max(0.7,0.7), \max(0.5,0.1), \max(0.7,0.2)]$
$\quad = \min(0.2,0.7,0.5,0.7) = 0.2,$

$B_{23} = \min[\max(0,0.6), \max(0.7,0.5), \max(0.5,0.5), \max(0.7,0.2)]$
$\quad = \min(0.6,0.7,0.5,0.7) = 0.5,$

$B_{24} = \min[\max(0,1), \max(0.7,0), \max(0.5,0.3), \max(0.7,0.9)]$
$\quad = \min(1,0.7,0.5,0.9) = 0.5,$

$B_{31} = \min[\max(0.4,0.1), \max(0.3,0.4), \max(0,0.7), \max(1,1)]$
$\quad = \min(0.4,0.4,0.7,1) = 0.4,$

$B_{32} = \min[\max(0.4,0.2), \max(0.3,0.7), \max(0,0.1), \max(1,0.2)]$
$\quad = \min(0.4,0.7,0.1,1) = 0.1,$

$B_{33} = \min[\max(0.4,0.6), \max(0.3,0.5), \max(0,0.5), \max(1,0.2)]$
$\quad = \min(0.6,0.5,0.5,1) = 0.5,$

$B_{34} = \min[\max(0.4,1), \max(0.3,0.0), \max(0,0.3), \max(1,0.9)]$
$\quad = \min(1,0.3,0.3,1) = 0.3,$

$B_{41} = \min[\max(1,0.1), \max(0.2,0.4), \max(0.8,0.7), \max(0,1)]$
$\quad = \min(1,0.4,0.8,1) = 0.4,$

$$B_{42} = \min[\max(1,0.2),\max(0.2,0.7),\max(0.8,0.1),\max(0,0.2)]$$
$$= \min(1,0.7,0.8,0.2) = 0.2,$$

$$B_{43} = \min[\max(1,0.6),\max(0.2,0.5),\max(0.8,0.5),\max(0,0.2)]$$
$$= \min(1,0.5,0.8,0.2) = 0.2,$$

$$B_{44} = \min[\max(1,1),\max(0.2,0.0),\max(0.8,0.3),\max(0,0.9)]$$
$$= \min(1,0.2,0.8,0.9) = 0.2.$$

Thus, we have

$$AB = \begin{bmatrix} 0.4 & 0.5 & 0.5 & 0.2 \\ 0.1 & 0.2 & 0.5 & 0.5 \\ 0.4 & 0.1 & 0.5 & 0.3 \\ 0.4 & 0.2 & 0.2 & 0.2 \end{bmatrix}.$$

So, we see that max–min operation on two matrices is not equal to min–max operation. Experts may wish to work either using max– min operator or min–max operator in their application.

Like an ordinary matrix multiplication, fuzzy matrix multiplication does not follow multiplicative commutative property: $AB \neq BA$.

7.2.2 Identity Matrix

If the diagonal elements of a fuzzy matrix are fuzzy one numbers and the off-diagonal elements are fuzzy zero numbers, then the fuzzy matrix is called fuzzy identity matrix and is denoted by I. Fuzzy one number and zero number are explained in Chapter 6.

7.3 Determinant of a Square Fuzzy Matrix

For any fuzzy matrices, $A=[a_{ij}]_{m \times m}$, $B = [b_{ij}]_{m \times p}$, $C = [c_{ij}]_{m \times p}$, the following operations holds [3, 4]:

i) $[B + C] = [b_{ij} + c_{ij}]$ where $b_{ij} + c_{ij} = \max(b_{ij},c_{ij})$,

ii) $AB = \sum_{k=1}^{n} [a_{ik} \cdot b_{kj}]$ where $a_{ik} \cdot b_{kj} = \min(a_{ik},b_{kj})$,

$B' = [b_{ji}]$,

iii) $B \leq C$ iff $b_{ij} \leq c_{ij}$,

$A^{r+1} = A^r \cdot A, r = 0, 1, 2,..., n.$

An $m \times n$ fuzzy matrix $A = [a_{ij}]$ is said to be constant if $a_{ik} = a_{jk}$ for all i, j, k. Matrix A is said to be symmetric if $A = A^T$. It is said to be idempotent if $A^2 = A$. A square $n \times n$ matrix $A = [a_{ij}]$ is said to be lower triangular matrix if $a_{ij} = 0$ for $i \leq j$ and is called upper triangular matrix if $a_{ij} = 0$ for $i \geq j$.

A determinant $|A|$ of a $n \times n$ fuzzy matrix A is defined as follows [3, 4]:

$$|A| = \sum_{\sigma \in S_n} a_{1\sigma(1)} a_{2\sigma(2)} \cdots a_{n\sigma(n)},$$

where S_n denotes the symmetric group of all permutations of the indices $(1,2,3, \ldots, n)$, and the summation is taken over all σ of S_n.

An example is given that can explain the determinant of a matrix.

Example 9   Consider a 2×2 matrix $A = \begin{bmatrix} a_{11} & a_{12} \\ a_{21} & a_{22} \end{bmatrix}$.

Here, S_n is the group of all permutations of the indices $(1,2)$. The permutations are $(1,2)$ and $(2,1)$.

So, $|A| = \sum_{\sigma \in S_n} a_{1\sigma(1)} a_{2\sigma(2)} = a_{11}a_{22} + a_{12}a_{21}$.

Likewise, for a 3×3 matrix $A = \begin{bmatrix} a_{11} & a_{12} & a_{13} \\ a_{21} & a_{22} & a_{23} \\ a_{31} & a_{32} & a_{33} \end{bmatrix}$.

Here, S_n is the group of all permutation of the indices $(1,2,3)$. So the permutations will be $(1,2,3), (1,3,2), (2,1,3), (2,3,1), (3,1,2), (3,2,1)$.

Then, $|A| = \sum_{\sigma \in S_n} a_{1\sigma(1)} a_{2\sigma(2)} a_{3\sigma(3)} = a_{11}a_{22}a_{33} + a_{11}a_{23}a_{32} + a_{12}a_{21}a_{33} + a_{12}a_{23}a_{31} + a_{13}a_{21}a_{32} + a_{13}a_{22}a_{31}$.

If we compute the determinant directly, then we get

$$A = \begin{bmatrix} a_{11} & a_{12} & a_{13} \\ a_{21} & a_{22} & a_{23} \\ a_{31} & a_{32} & a_{33} \end{bmatrix}$$

$$|A| = a_{11} \begin{vmatrix} a_{22} & a_{23} \\ a_{32} & a_{33} \end{vmatrix} + a_{12} \begin{vmatrix} a_{21} & a_{23} \\ a_{31} & a_{33} \end{vmatrix} + a_{13} \begin{vmatrix} a_{21} & a_{22} \\ a_{31} & a_{32} \end{vmatrix}$$

$$= a_{11}(a_{22}a_{33} + a_{23}a_{32}) + a_{12}(a_{21}a_{33} + a_{23}a_{31}) + a_{13}(a_{21}a_{32} + a_{22}a_{31})$$

$$= a_{11}a_{22}a_{33} + a_{11}a_{23}a_{32} + a_{12}a_{21}a_{33} + a_{12}a_{23}a_{31} + a_{13}a_{21}a_{32} + a_{13}a_{22}a_{31}.$$

7.3.1   Examples of Fuzzy Determinants

Example 10   Find the determinant of a fuzzy matrix:

$$A = \begin{bmatrix} 0.2 & 0.4 & 0.5 \\ 0.6 & 0.2 & 0.8 \\ 0 & 0.3 & 0.2 \end{bmatrix}.$$

Solution

Determinant of A det(A) is $|A|$:

$$|A| = 0.2 \begin{vmatrix} 0.2 & 0.8 \\ 0.3 & 0.2 \end{vmatrix} + 0.4 \begin{vmatrix} 0.6 & 0.8 \\ 0 & 0.2 \end{vmatrix} + 0.5 \begin{vmatrix} 0.6 & 0.2 \\ 0 & 0.3 \end{vmatrix}$$

$$= 0.2(\min(0.2,0.2) + \min(0.8,0.3)) + 0.4(\min(0.6,0.2) + \min(0.8,0))$$

$$+ 0.5(\min(0.6,0.3) + \min(0.2,0))$$

$$= 0.2(0.2 + 0.3) + 0.4(0.2 + 0) + 0.5(0.3 + 0)$$

$$= 0.2(0.3) + 0.4(0.2) + 0.5(0.3) = 0.2 + 0.2 + 0.3 = 0.3.$$

Example 11 If A is a fuzzy matrix [4], $A = \begin{vmatrix} a & b \\ c & d \end{vmatrix}$, then, show $\begin{vmatrix} a & b \\ a & b \end{vmatrix} \begin{vmatrix} c & d \\ c & d \end{vmatrix} \leq \begin{vmatrix} a & b \\ c & d \end{vmatrix}$.

Solution

$\begin{vmatrix} a & b \\ a & b \end{vmatrix} = ab$, $\begin{vmatrix} c & d \\ c & d \end{vmatrix} = cd$, $\begin{vmatrix} a & b \\ a & b \end{vmatrix} \begin{vmatrix} c & d \\ c & d \end{vmatrix} = abcd$, and det $A = \begin{vmatrix} a & b \\ c & d \end{vmatrix} = ad + bc$.

Hence, $abcd \leq ad + bc = $ det(A).

Example 12 Show $|AB| \geq |A||B|$ [4, 5].

Solution

Consider two fuzzy matrices A and B

$$A = \begin{bmatrix} a_{11} & a_{12} \\ a_{21} & a_{22} \end{bmatrix} \text{ and } B = \begin{bmatrix} b_{11} & b_{12} \\ b_{21} & b_{22} \end{bmatrix}.$$

We compute the following:

$$AB = \begin{bmatrix} a_{11}b_{11} + a_{12}b_{21} & a_{11}b_{12} + a_{12}b_{22} \\ a_{21}b_{11} + a_{22}b_{21} & a_{21}b_{12} + a_{22}b_{22} \end{bmatrix},$$

$$|A| = a_{11}a_{22} + a_{12}a_{21},$$

$$|B| = b_{11}b_{22} + b_{12}b_{21},$$

$$|A||B| = a_{11}a_{22}b_{11}b_{22} + a_{11}a_{22}b_{12}b_{21} + a_{12}a_{21}b_{11}b_{22} + a_{12}a_{21}b_{12}b_{21},$$

$$|AB| = (a_{11}b_{11} + a_{12}b_{21})(a_{21}b_{12} + a_{22}b_{22})$$

$$+ (a_{11}b_{12} + a_{12}b_{22})(a_{21}b_{11} + a_{22}b_{21})$$

$$= |A||B| + a_{11}a_{21}b_{11}b_{12} + a_{12}a_{22}b_{21}b_{22}$$

$$+ a_{11}b_{12}a_{21}b_{11} + a_{12}b_{22}a_{22}b_{21}.$$

Hence, $|AB| \geq |A||B|$.

Example 13   Consider two fuzzy matrices

$$A = \begin{bmatrix} 0.9 & 0.1 \\ 0.12 & 0.5 \end{bmatrix} \text{ and } B = \begin{bmatrix} 0.8 & 0.2 \\ 0.1 & 0.1 \end{bmatrix}, \text{ show } |AB| \geq |A||B|.$$

Solution

First we compute $AB = \begin{bmatrix} c_{11} & c_{12} \\ c_{13} & c_{14} \end{bmatrix}$, where

$$c_{11} = \max(\min(0.9, 0.8), \min(0.1, 0.1)) = \max(0.8, 0.1) = 0.8,$$

$$c_{12} = \max(\min(0.9, 0.2), \min(0.1, 0.1)) = \max(0.2, 0.1) = 0.2,$$

$$c_{21} = \max(\min(0.12, 0.8), \min(0.5, 0.1)) = \max(0.12, 0.1) = 0.12,$$

$$c_{22} = \max(\min(0.12, 0.2), \min(0.5, 0.1)) = \max(0.12, 0.1) = 0.12.$$

So, $AB = C = \begin{bmatrix} 0.8 & 0.2 \\ 0.12 & 0.12 \end{bmatrix}$.

Now, we compute $|AB|$, $|A|$, $|A||B|$, which are as follows:

$$|AB| = \min(0.8, 0.12) + \min(0.2, 0.12) = 0.12 + 0.12 = 0.12,$$

$$|A| = \min(0.9, 0.5) + \min(0.1, 0.12) = 0.5 + 0.1 = 0.5,$$

$$|B| = \min(0.8, 0.1) + \min(0.2, 0.1) = 0.1 + 0.1 = 0.1,$$

$$|A||B| = (0.5)(0.1) = \min(0.5, 0.1) = 0.1.$$

Hence, $|AB| > |A||B|$.

Example 14   Consider two fuzzy matrices

$$A = \begin{bmatrix} 0.2 & 0.1 \\ 0.5 & 0.3 \end{bmatrix} \text{ and } B = \begin{bmatrix} 0.4 & 0.2 \\ 0.1 & 0.7 \end{bmatrix}, \text{ show } |AB| \geq |A||B|.$$

Solution

We compute $|A| = 0.2$, $|B| = 0.4$, $|A||B| = 0.2$,

$$AB = \begin{bmatrix} 0.2 & 0.1 \\ 0.5 & 0.3 \end{bmatrix} \begin{bmatrix} 0.4 & 0.2 \\ 0.1 & 0.7 \end{bmatrix} = \begin{bmatrix} 0.2 & 0.2 \\ 0.4 & 0.3 \end{bmatrix},$$

$|AB| = 0.2.$

So, $|AB| = |A||B|$.
So, from Examples 13 and 14, we see that $|AB| \geq |A||B|$.

7.4 Adjoint of a Square Fuzzy Matrix

The adjoint of a matrix is defined only for square matrix. For an ordinary matrix, adjoint of a matrix is defined as the transpose of the matrix obtained by replacing the elements of A by their respective cofactors in A.

Let A be a fuzzy matrix, $A = [a_{ij}]$. Let us create a matrix B by deleting the ith row and jth column. The adjoint matrix of a square matrix of $m \times m$ fuzzy matrix A is denoted by $B = \mathrm{adj}(A)$ where b_{ij} is defined as [6]:

$$b_{ij} = |A_{ji}|,$$

and $|A_{ji}|$ is the determinant of the $(m-1) \times (m-1)$ fuzzy matrix formed by deleting jth row and ith column from A.

If A be a fuzzy matrix, then

$$B = \mathrm{adj}\, A = \begin{bmatrix} |A_{11}| & |A_{21}| & \cdots & |A_{n1}| \\ |A_{12}| & |A_{22}| & \cdots & |A_{n2}| \\ \vdots & \vdots & \vdots & \vdots \\ |A_{1n}| & |A_{21}| & \cdots & |A_{nn}| \end{bmatrix}.$$

Element b_{ij} can also be written as:

$$b_{ij} = \sum_{\rho \in S_{n_j n_i}} \prod_{t \in n_j} a_{t\rho(t)},$$

where $n_j = \{1,2,3,\dots,n\}\setminus\{j\}$, and $S_{n_j n_i}$ is the set of all permutation of set n_j over set n_i.

7.4.1 Few Proposition of Adjoint of Fuzzy Matrices

i) For two $n \times n$ fuzzy matrices, A and B, if

$A \leq B$, then $\mathrm{adj}(A) \leq \mathrm{adj}(B)$.
Let $P = \mathrm{adj}(A)$ and $Q = \mathrm{adj}(B)$, then

$$p_{ij} = \sum_{\rho \in S_{n_j n_i}} \prod_{t \in n_j} a_{t\rho(t)}, \quad q_{ij} = \sum_{\rho \in S_{n_j n_i}} \prod_{t \in n_j} b_{t\rho(t)},$$

Now, as $A \leq B$ so, $a_{t\rho(t)} \leq b_{t\rho(t)}$ implies $p_{ij} \leq q_{ij}$.

ii) For a fuzzy matrix A, prove $|A| = |\mathrm{adj}(A)|$ [6].

Proof: We know

$$\mathrm{adj}\, A = \begin{bmatrix} |A_{11}| & |A_{21}| & \cdots & |A_{n1}| \\ |A_{12}| & |A_{22}| & \cdots & |A_{n2}| \\ \vdots & \vdots & \vdots & \vdots \\ |A_{1n}| & |A_{21}| & \cdots & |A_{nn}| \end{bmatrix}.$$

Then,

$$|\mathrm{adj}(A)| = \sum_{\rho \in S_n} |A_{1\rho(1)}| |A_{2\rho(2)}| |A_{3\rho(3)}| \cdots |A_{n\rho(n)}|$$

$$= \sum_{\rho \in S_n} \prod_{i=1}^{n} |A_{i\rho(i)}|$$

$$= \sum_{\rho \in S_n} \prod_{i=1}^{n} \left[\sum_{\tau \in S_{n_i n_{\rho(i)}}} \prod_{t \in n_i} a_{t\tau(t)} \right],$$

from the theory of determinant

$$= \sum_{\rho \in S_n} \left[\left(\sum_{\tau \in S_{n_1 n_{\rho(1)}}} \prod_{t \in n_1} a_{t\tau(t)} \right) \left(\sum_{\tau \in S_{n_2 n_{\rho(2)}}} \prod_{t \in n_2} a_{t\tau(t)} \right) \left(\sum_{\tau \in S_{n_3 n_{\rho(3)}}} \prod_{t \in n_3} a_{t\tau(t)} \right) \right.$$
$$\left. \cdots \left(\sum_{\tau \in S_{n_n n_{\rho(n)}}} \prod_{t \in n_n} a_{t\tau(t)} \right) \right]$$

Considering $\tau_1 = S_{n_1 n_{\rho(1)}}, \tau_2 = S_{n_2 n_{\rho(2)}}$, and so on, we get

$$= \sum_{\rho \in S_n} \left[\left(\prod_{t \in n_1} a_{t\tau_{1(t)}} \right) \left(\prod_{t \in n_2} a_{t\tau_{2(t)}} \right) \cdots \left(\prod_{t \in n_n} a_{t\tau_{n(t)}} \right) \right]$$

$$= \sum_{\rho \in S_n} \left[\left(a_{2\tau_{1(2)}} a_{3\tau_{1(3)}} \cdots a_{n\tau_{1(n)}} \right) \left(a_{1\tau_{2(1)}} a_{3\tau_{2(3)}} \cdots a_{n\tau_{2(n)}} \right) \left(a_{1\tau_{3(1)}} a_{3\tau_{3(2)}} \cdots a_{n\tau_{3(n)}} \right) \right.$$
$$\left. \cdots \left(a_{1\tau_{n(1)}} a_{2\tau_{n(2)}} \cdots a_{n-1\tau_{n(n-1)}} \right) \right]$$

$$= \sum_{\rho \in S_n} \left[\begin{array}{c} \left(a_{1\tau_{2(1)}} a_{1\tau_{3(1)}} a_{1\tau_{4(1)}} \cdots a_{1\tau_{n(1)}} \right) \left(a_{2\tau_{1(2)}} a_{2\tau_{3(2)}} a_{2\tau_{4(2)}} \cdots a_{2\tau_{n(2)}} \right) \\ \left(a_{3\tau_{1(3)}} a_{3\tau_{2(3)}} a_{3\tau_{4(3)}} \cdots a_{3\tau_{n(3)}} \right) \cdots \left(a_{n\tau_{1(n)}} a_{n\tau_{2(n)}} \cdots a_{n\tau_{n-1(n)}} \right) \end{array} \right].$$

Assuming $g_h = \{1,2,3,\ldots,n\} \backslash \{h\}$, $h = 1, 2, 3, \ldots, n$, then

$$|adj(A)| = \sum_{\rho \in S_n} \left(a_{1\tau_{g_1(1)}} a_{2\tau_{g_2(2)}} a_{3\tau_{g_3(3)}} \cdots a_{n\tau_{g_n(n)}} \right).$$

We have $a_{h\tau_{g_h(h)}} = a_{h\rho(h)}$, then

$$|adj(A)| = \sum_{\rho \in S_n} \left(a_{1\rho(1)} a_{2\rho(2)} a_{3\rho(3)} \cdots a_{1\rho(n)} \right),$$

and we know $|A| = \sum_{\rho \in S_n} \left(a_{1\rho(1)} a_{2\rho(2)} a_{3\rho(3)} \cdots a_{1\rho(n)} \right) = |adj(A)|$.

Example 15   Let us take a square matrix:

$$A = \begin{bmatrix} 0.2 & 0.4 & 0.5 \\ 0.6 & 0.2 & 0.8 \\ 0.1 & 0.3 & 0.2 \end{bmatrix}$$

Show $|adj\ A| = |A|$.

Solution

First, we will compute adjoint of the matrix A.
The cofactors of a_{ij} are

Cofactor of a_{11}, $|A_{11}| = \begin{vmatrix} 0.2 & 0.8 \\ 0.3 & 0.2 \end{vmatrix} = 0.2 + 0.3 = 0.3$,

Cofactor of a_{12}, $|A_{12}| = \begin{vmatrix} 0.6 & 0.8 \\ 0.1 & 0.2 \end{vmatrix} = 0.2 + 0.1 = 0.2$,

Cofactor of a_{13}, $|A_{13}| = \begin{vmatrix} 0.6 & 0.2 \\ 0.1 & 0.3 \end{vmatrix} = 0.3 + 0.1 = 0.3$,

Cofactor of a_{21}, $|A_{21}| = \begin{vmatrix} 0.4 & 0.5 \\ 0.3 & 0.2 \end{vmatrix} = 0.2 + 0.3 = 0.3$,

Cofactor of a_{22}, $|A_{22}| = \begin{vmatrix} 0.2 & 0.5 \\ 0.1 & 0.2 \end{vmatrix} = 0.2 + 0.1 = 0.2$,

Cofactor of a_{23}, $|A_{23}| = \begin{vmatrix} 0.2 & 0.4 \\ 0.1 & 0.3 \end{vmatrix} = 0.2 + 0.1 = 0.2$,

Cofactor of a_{31}, $|A_{31}| = \begin{vmatrix} 0.4 & 0.5 \\ 0.2 & 0.8 \end{vmatrix} = 0.4 + 0.2 = 0.4,$

Cofactor of a_{32}, $|A_{32}| = \begin{vmatrix} 0.2 & 0.5 \\ 0.6 & 0.8 \end{vmatrix} = 0.2 + 0.5 = 0.5,$

Cofactor of a_{33}, $|A_{33}| = \begin{vmatrix} 0.2 & 0.4 \\ 0.6 & 0.2 \end{vmatrix} = 0.2 + 0.4 = 0.4.$

So, adj $A = \begin{bmatrix} |A_{11}| & |A_{21}| & |A_{31}| \\ |A_{12}| & |A_{22}| & |A_{32}| \\ |A_{13}| & |A_{23}| & |A_{33}| \end{bmatrix} = \begin{bmatrix} 0.3 & 0.3 & 0.4 \\ 0.2 & 0.2 & 0.5 \\ 0.3 & 0.2 & 0.4 \end{bmatrix}.$

Now, let us compute determinant of the adjoint matrix adj A:

$$|adj\ A| = 0.3 \begin{vmatrix} 0.2 & 0.5 \\ 0.2 & 0.4 \end{vmatrix} + 0.3 \begin{vmatrix} 0.2 & 0.5 \\ 0.3 & 0.4 \end{vmatrix} + 0.4 \begin{vmatrix} 0.2 & 0.2 \\ 0.3 & 0.2 \end{vmatrix}$$

$$= 0.3\,(0.2 + 0.2) + 0.3(0.2 + 0.3) + 0.4(0.2 + 0.2)$$

$$= 0.3\,(0.2) + 0.3(0.3) + 0.4(0.2)$$

$$= 0.2 + 0.3 + 0.2 = 0.3.$$

We will now compute determinant of the matrix A:

$$|A| = 0.2 \begin{vmatrix} 0.2 & 0.8 \\ 0.3 & 0.2 \end{vmatrix} + 0.4 \begin{vmatrix} 0.6 & 0.8 \\ 0.1 & 0.2 \end{vmatrix} + 0.5 \begin{vmatrix} 0.6 & 0.2 \\ 0.1 & 0.3 \end{vmatrix}$$

$$= 0.2\,(0.2 + 0.3) + 0.4(0.2 + 0.1) + 0.5(0.3 + 0.1)$$

$$= 0.2 + 0.2 + 0.3 = 0.3.$$

We see that $|adj\ A| = |A|$.

Example 16 We will show another example to compute the adjoint and determinant of the fuzzy matrix A to prove $|adj\ A| = |A|$, where

$$A = \begin{bmatrix} 0.3 & 0.2 & 0.1 \\ 0.4 & 0.2 & 0.4 \\ 0.2 & 0.6 & 0.2 \end{bmatrix}.$$

Solution

We compute

$$|A_{11}| = \begin{vmatrix} 0.2 & 0.4 \\ 0.6 & 0.2 \end{vmatrix} = 0.2 + 0.4 = 0.4,$$

$$|A_{12}| = \begin{vmatrix} 0.4 & 0.4 \\ 0.2 & 0.2 \end{vmatrix} = 0.2 + 0.2 = 0.2,$$

$$|A_{13}| = \begin{vmatrix} 0.4 & 0.2 \\ 0.2 & 0.6 \end{vmatrix} = 0.4 + 0.2 = 0.4,$$

$$|A_{21}| = \begin{vmatrix} 0.2 & 0.1 \\ 0.6 & 0.2 \end{vmatrix} = 0.2 + 0.1 = 0.2,$$

$$|A_{22}| = \begin{vmatrix} 0.3 & 0.1 \\ 0.2 & 0.2 \end{vmatrix} = 0.2 + 0.1 = 0.2,$$

$$|A_{23}| = \begin{vmatrix} 0.3 & 0.2 \\ 0.2 & 0.6 \end{vmatrix} = 0.3 + 0.2 = 0.3,$$

$$|A_{31}| = \begin{vmatrix} 0.2 & 0.1 \\ 0.2 & 0.4 \end{vmatrix} = 0.2 + 0.1 = 0.2,$$

$$|A_{32}| = \begin{vmatrix} 0.3 & 0.1 \\ 0.4 & 0.4 \end{vmatrix} = 0.3 + 0.1 = 0.3,$$

$$|A_{33}| = \begin{vmatrix} 0.3 & 0.2 \\ 0.4 & 0.2 \end{vmatrix} = 0.2 + 0.2 = 0.2.$$

$$\text{So, adj } A = \begin{bmatrix} |A_{11}| & |A_{21}| & |A_{31}| \\ |A_{12}| & |A_{22}| & |A_{32}| \\ |A_{13}| & |A_{23}| & |A_{33}| \end{bmatrix} = \begin{bmatrix} 0.4 & 0.2 & 0.2 \\ 0.2 & 0.2 & 0.3 \\ 0.4 & 0.3 & 0.2 \end{bmatrix}.$$

Determinant of the adjoint matrix adj A is

$$|\text{adj } A| = 0.4 \begin{vmatrix} 0.2 & 0.3 \\ 0.3 & 0.2 \end{vmatrix} + 0.2 \begin{vmatrix} 0.2 & 0.3 \\ 0.4 & 0.2 \end{vmatrix} + 0.2 \begin{vmatrix} 0.2 & 0.2 \\ 0.4 & 0.3 \end{vmatrix}$$

$$= 0.3 (0.2 + 0.3) + 0.3(0.2 + 0.3) + 0.4(0.2 + 0.2)$$

$$= 0.3(0.3) + 0.3(0.3) + 0.4(0.2)$$

$$= 0.3 + 0.3 + 0.2 = 0.3.$$

Determinant of the matrix A is

$$|A| = 0.3 \begin{vmatrix} 0.2 & 0.4 \\ 0.6 & 0.2 \end{vmatrix} + 0.2 \begin{vmatrix} 0.4 & 0.4 \\ 0.2 & 0.2 \end{vmatrix} + 0.1 \begin{vmatrix} 0.4 & 0.2 \\ 0.2 & 0.6 \end{vmatrix}$$

$$= 0.3 (0.2 + 0.4) + 0.2(0.2 + 0.2) + 0.1(0.4 + 0.2)$$

$$= 0.3 (0.4) + 0.2(0.2) + 0.1(0.4) = 0.3 + 0.2 + 0.1 = 0.3.$$

We see that $|\text{adj } A| = |A|$.

iii) Consider two fuzzy $n \times n$ matrices A and B such that $A < B$, then adj A + adj $B \le \text{adj}(A + B)$ [6].

Proof: As $A < B$, then $A \le A + B$ and $B \le A + B$.
From this, it is obvious that $\text{adj}(A) \le \text{adj}(A + B)$ and $\text{adj}(B) \le \text{adj}(A + B)$. This implies $\text{adj}(A) + \text{adj}(B) \le \text{adj}(A + B)$.
An example will explain the above property clearly.
Let us consider two matrices $A = [a_{ij}]$ and $B = [b_{ij}]$.

Example 17 Consider two fuzzy matrices, A, B and $A \le B$

$$A = \begin{bmatrix} 0.2 & 0.3 & 0.2 \\ 0.4 & 0.3 & 0.3 \\ 0.1 & 0.1 & 0.2 \end{bmatrix}, \quad B = \begin{bmatrix} 0.3 & 0.4 & 0.3 \\ 0.6 & 0.5 & 0.6 \\ 0.2 & 0.2 & 0.3 \end{bmatrix},$$

$$A + B = \begin{bmatrix} 0.3 & 0.4 & 0.3 \\ 0.6 & 0.5 & 0.6 \\ 0.2 & 0.2 & 0.3 \end{bmatrix} \quad \text{(we take the maximum of the fuzzy values)},$$

$$\text{adj}(A + B) = \begin{bmatrix} 0.3 & 0.3 & 0.4 \\ 0.3 & 0.3 & 0.3 \\ 0.2 & 0.2 & 0.4 \end{bmatrix},$$

$$\text{adj}(A) = \begin{bmatrix} 0.2 & 0.2 & 0.3 \\ 0.2 & 0.2 & 0.2 \\ 0.1 & 0.1 & 0.3 \end{bmatrix},$$

$$\text{adj}(B) = \begin{bmatrix} 0.3 & 0.3 & 0.4 \\ 0.3 & 0.3 & 0.3 \\ 0.2 & 0.2 & 0.4 \end{bmatrix},$$

$$\text{adj}(A) + \text{adj}(B) = \begin{bmatrix} 0.3 & 0.3 & 0.4 \\ 0.3 & 0.3 & 0.3 \\ 0.2 & 0.2 & 0.4 \end{bmatrix} \leq \text{adj}(A+B) = \begin{bmatrix} 0.3 & 0.3 & 0.4 \\ 0.6 & 0.3 & 0.3 \\ 0.2 & 0.2 & 0.4 \end{bmatrix}.$$

Here, we define some special types of fuzzy matrices.

For a $n \times n$ fuzzy matrix, $A = [a_{ij}]$, the following properties hold [7]:

i) A is transitive if $A^2 \leq A$,

ii) A is reflexive if and only if $a_{ii} = 1$, $i = 1, 2, 3, \dots, n$,

iii) A is irreflexive if and only if $a_{ii} = 0$, $i = 1, 2, 3, \dots, n$,

iv) A is called idempotent matrix if and only if $A^2 = A$,

v) A is called symmetric matrix if $A' = A$.

Definition.

A $n \times n$ fuzzy matrix, $A = [a_{ij}]$, is called constant if $a_{ik} = a_{jk}$ for all $i, j \in \{1,2,3,4,\dots,n\}$ and $k \in \{1,2,3,4,\dots,n\}$.

7.5   Properties of Reflexive Matrices

For any reflexive square $n \times n$ fuzzy matrix A, i.e. $a_{ii} = 1$, the following properties hold [6, 8]:

i) $\text{adj}(A) = A^p$, where A^p is an idempotent matrix and $p \leq n - 1$. Idempotent matrix is a matrix which when multiplied by itself, we get the same matrix.

ii) For any two fuzzy matrices, A and B, and a fuzzy matrix C, if $A \leq B$, then $AC \leq BC$,

iii) $\text{adj}(A^2) = (\text{adj } A)^2 = \text{adj}(A)$,

Proof: We know $\text{adj}(A) = A^p$.

As A is reflexive, A^2 is also reflexive.

Then, $\text{adj}(A^2) = (A^2)^p = (A^p)^2 = (\text{adj } A)^2$

Thus, $\text{adj}(A^2) = (\text{adj } A)^2$.

Again, as A^p is idempotent, so $(\text{adj } A)^2 = \text{adj}(A)$.

Hence, $\text{adj}(A^2) = (\text{adj } A)^2 = \text{adj}(A)$.

iv) $\text{adj } A \geq A$.

Proof: Let $P = \text{adj}(A)$. Let us compute $\text{adj}(A)$ for a 4×4 matrix. Considering $i = 2$, $j = 1$, we get

$$p_{21} = |A_{12}| = \begin{vmatrix} a_{21} & a_{23} & a_{24} \\ a_{31} & a_{33} & a_{34} \\ a_{41} & a_{43} & a_{44} \end{vmatrix}$$

$$= a_{21}a_{33}a_{44} + a_{21}a_{34}a_{43} + a_{23}a_{31}a_{44} + a_{23}a_{34}a_{41} + a_{24}a_{31}a_{43} + a_{24}a_{33}a_{41},$$

$$\Rightarrow p_{21} \geq a_{21}a_{33}a_{44} = a_{21}, \quad \text{as } A \text{ is reflexive.}$$

Likewise, for $n = 5$ and using the formula, $p_{ij} = \sum_{\rho \in S_{n_j n_i}} \prod_{t \in n_j} a_{t\rho(t)}$, we get, say at $i = 2$, $j = 5$

$$p_{25} = \sum_{\rho \in S_{n_j n_i}} a_{1\rho(1)} a_{2\rho(2)} a_{3\rho(3)} a_{4\rho(4)} \text{ here } n_j = n_5 = \{1,2,3,4,5\} \backslash \{5\}$$

$$\implies p_{25} \geq a_{11} a_{25} a_{33} a_{44} = a_{25}.$$

Likewise, at $i = 4$, $j = 5$, $p_{45} \geq a_{11} a_{22} a_{33} a_{45} = a_{45}$

So, we see $p_{ij} \geq a_{ij}$.

We can write it in a general form by considering the permutation $\rho(t) = t$, $\rho(i) = j$, $t \neq i$, i.e. the permutations:

$$\begin{pmatrix} 1\,2\,3...i...(j-1)...(j+1)...n \\ 1\,2\,3...j...(j-1)...(j+1)...n \end{pmatrix},$$

then, $p_{ij} \geq a_{11} a_{22} a_{33} ... a_{ij} a_{(j-1)(j-1)} a_{(j+1)(j+1)} ... a_{nn} = a_{ij}$, as A is reflexive implies $p_{ij} \geq a_{ij}$.

Hence, adj$(A) \geq A$.

v) adj A is reflexive,

Proof: Let $B = \text{adj}(A)$ and $b_{ij} = \sum_{\rho \in S_{n_j n_i}} \prod_{t \in n_j} a_{t\rho(t)}$.

As we are concerned about reflexivity, substituting $j = i$, we get

$$b_{ii} = \sum_{\rho \in S_{n_j n_i}} a_{1\rho(1)} a_{2\rho(2)} a_{3\rho(3)} ... a_{(i-1)\rho(i-1)} a_{(i+1)\rho(i+1)} ... a_{n\rho(n)},$$

$$n_i = \{1,2,3,...,n\} \backslash \{i\}.$$

Considering identity permutation, i.e. $\rho(t) = t$, we get $b_{ii} \geq a_{11} a_{22} a_{33} a_{44} ... a_{(i-1)(j-1)} a_{(i+1)(j+1)} ... a_{nn} = 1$, as A is reflexive.

Hence, adj A is reflexive.

vi) $A(\text{adj } A) = (\text{adj } A)A = \text{adj}(A)$.

Proof: Let us consider a 3×3 matrix. Let $P = A(\text{adj } A)$, $Q = (\text{adj } A)A$

$$P = A(\text{adj } A) = \begin{bmatrix} a_{11} & a_{12} & a_{13} \\ a_{21} & a_{22} & a_{23} \\ a_{31} & a_{32} & a_{33} \end{bmatrix} \begin{bmatrix} |A_{11}| & |A_{21}| & |A_{31}| \\ |A_{12}| & |A_{22}| & |A_{32}| \\ |A_{13}| & |A_{23}| & |A_{33}| \end{bmatrix},$$

$$p_{ij} = \sum_{k=1}^{3} a_{ik} |A_{jk}|$$

Consider at $i = 2$, $j = 3$,

$$p_{23} = \sum_{k=1}^{3} a_{2k} |A_{3k}| = a_{21} |A_{31}| + a_{22} |A_{32}| + a_{23} |A_{33}|,$$

which implies $p_{23} \geq a_{22}|A_{32}| = |A_{32}|$, as A is reflexive.

So, we can write in a general form $p_{ij} \geq a_{ii}|A_{ji}| = |A_{ji}|$.

$$Q = (\text{adj}\,A)A = \begin{bmatrix} |A_{11}| & |A_{21}| & |A_{31}| \\ |A_{12}| & |A_{22}| & |A_{32}| \\ |A_{13}| & |A_{23}| & |A_{33}| \end{bmatrix} \begin{bmatrix} a_{11} & a_{12} & a_{13} \\ a_{21} & a_{22} & a_{23} \\ a_{31} & a_{32} & a_{33} \end{bmatrix}.$$

Say, at $i = 2$, $j = 3$,

$$q_{23} = \sum_{k=1}^{3} |A_{k2}|a_{k3} = |A_{12}|a_{13} + |A_{22}|a_{23} + |A_{32}|a_{33},$$

this implies $q_{23} \geq |A_{32}|a_{33} = |A_{32}|$, as A s reflexive.

So, we can write in a general form $q_{ij} \geq |A_{ji}|a_{jj} = |A_{ji}|$.

Thus, $p_{ij} \geq |A_{ji}|$ and $q_{ij} \geq |A_{ji}|$.

This implies $A(\text{adj}\,A) \geq \text{adj}\,A$ and $(\text{adj}\,A)A \geq \text{adj}\,A$.

Now, from the properties, $(\text{adj}\,A)^2 = \text{adj}(A)$ and $\text{adj}(A) \geq A$ from propositions (iii–iv), we get

$$\text{adj}(A) = \text{adj}(A)\text{adj}(A) \geq A\,\text{adj}(A).$$

Thus, we have $A(\text{adj}\,A) \geq \text{adj}\,A$ and $\text{adj}(A) \geq A\,\text{adj}(A)$ and from these two equations we get

$$A(\text{adj}\,A) = \text{adj}(A).$$

Likewise, $\text{adj}(A) = \text{adj}(A)\text{adj}(A) \geq \text{adj}(A)A$.

Hence, $\text{adj}(A) = (\text{adj}\,A)A$.

Example 18 Consider a reflexive fuzzy matrix $A = \begin{bmatrix} 1.0 & 0.4 & 0.5 \\ 0.6 & 1.0 & 0.8 \\ 0.1 & 0.3 & 1.0 \end{bmatrix}$

Prove $A(\text{adj}\,A) = (\text{adj}\,A)A$.

Cofactor of $a_{11} = \begin{vmatrix} 1.0 & 0.8 \\ 0.3 & 1.0 \end{vmatrix} = 1.0$, Cofactor of $a_{12} = \begin{vmatrix} 6.0 & 0.8 \\ 1.0 & 1.0 \end{vmatrix} = 0.6$,

Cofactor of $a_{13} = \begin{vmatrix} 0.6 & 1.0 \\ 0.1 & 0.3 \end{vmatrix} = 0.3$, Cofactor of $a_{21} = \begin{vmatrix} 0.4 & 0.5 \\ 0.3 & 1.0 \end{vmatrix} = 0.4$,

Cofactor of $a_{22} = \begin{vmatrix} 1.0 & 0.5 \\ 0.1 & 1.0 \end{vmatrix} = 1.0$, Cofactor of $a_{23} = \begin{vmatrix} 1.0 & 0.4 \\ 0.1 & 0.3 \end{vmatrix} = 0.3$,

Cofactor of $a_{31} = \begin{vmatrix} 0.4 & 0.5 \\ 1.0 & 0.8 \end{vmatrix} = 0.5$, Cofactor of $a_{32} = \begin{vmatrix} 1.0 & 0.5 \\ 0.6 & 0.8 \end{vmatrix} = 0.8$,

Cofactor of $a_{33} = \begin{vmatrix} 1.0 & 0.4 \\ 0.6 & 1.0 \end{vmatrix} = 1.0.$

So, we get adj $A = \begin{bmatrix} 1.0 & 0.4 & 0.5 \\ 0.6 & 1.0 & 0.8 \\ 0.3 & 0.3 & 1.0 \end{bmatrix} \geq A.$

Thus, we see that adj A is also reflexive and adj $A \geq A$.
Now, let us compute

$$A\,(\text{adj } A) = \begin{bmatrix} 1.0 & 0.4 & 0.5 \\ 0.6 & 1.0 & 0.8 \\ 0.1 & 0.3 & 1.0 \end{bmatrix} \begin{bmatrix} 1.0 & 0.4 & 0.5 \\ 0.6 & 1.0 & 0.8 \\ 0.3 & 0.3 & 1.0 \end{bmatrix} = \begin{bmatrix} 1.0 & 0.4 & 0.5 \\ 0.6 & 1.0 & 0.8 \\ 0.3 & 0.3 & 1.0 \end{bmatrix},$$

$$(\text{adj } A)A = \begin{bmatrix} 1.0 & 0.4 & 0.5 \\ 0.6 & 1.0 & 0.8 \\ 0.3 & 0.3 & 1.0 \end{bmatrix} \begin{bmatrix} 1.0 & 0.4 & 0.5 \\ 0.6 & 1.0 & 0.8 \\ 0.1 & 0.3 & 1.0 \end{bmatrix} = \begin{bmatrix} 1.0 & 0.4 & 0.5 \\ 0.6 & 1.0 & 0.8 \\ 0.3 & 0.3 & 1.0 \end{bmatrix}.$$

Thus, $A(\text{adj } A) = (\text{adj } A)A$.

7.6 Generalized Inverse of a Fuzzy Matrix

Kim and Roush [9] introduced generalized fuzzy matrix. A matrix A is called regular if and only if there exists a matrix B such that $ABA = A$. Such a matrix is called a generalized inverse, or g-inverse of A.

Example 19 Consider matrix $A = \begin{bmatrix} 0.4 & 1 \\ 0 & 1 \end{bmatrix}, B = \begin{bmatrix} 1 & 0.4 \\ 0 & 1 \end{bmatrix}.$

Then,

$$ABA = \begin{bmatrix} 0.4 & 1 \\ 0 & 1 \end{bmatrix} \begin{bmatrix} 1 & 0.4 \\ 0 & 1 \end{bmatrix} \begin{bmatrix} 0.4 & 1 \\ 0 & 1 \end{bmatrix} = \begin{bmatrix} 0.4 & 1 \\ 0 & 1 \end{bmatrix} \begin{bmatrix} 0.4 & 1 \\ 0 & 1 \end{bmatrix} = \begin{bmatrix} 0.4 & 1 \\ 0 & 1 \end{bmatrix} = A.$$

Thus, the matrix A is regular and matrix B is a generalized inverse of matrix A. Again, we see that

$$A^2 = \begin{bmatrix} 0.4 & 1 \\ 0 & 1 \end{bmatrix} \begin{bmatrix} 0.4 & 1 \\ 0 & 1 \end{bmatrix} = \begin{bmatrix} 0.4 & 1 \\ 0 & 1 \end{bmatrix} = A.$$

So, matrix A is idempotent.

7.7 Intuitionistic Fuzzy Matrix

Like fuzzy matrix, intuitionistic fuzzy matrix is a matrix $A = [a_{ij}, (\mu_{ij}, \nu_{ij})]$ where μ_{ij}, ν_{ij} are the membership and nonmembership values of a set. Addition and multiplication operations on two intuitionistic fuzzy matrices are given.

For two intuitionistic fuzzy matrices, $A = [a_{ij}, (\mu_{ij}, \nu_{ij})]_{m \times n}, B = [b_{ij}, (\mu_{ij}, \nu_{ij})]_{m \times n}, C = [c_{ij}, (\mu_{ij}, \nu_{ij})]_{n \times p}$, then

$$[A + B] = [a_{ij} + b_{ij}],$$

where

$$a_{ij} + b_{ij} = \left\{ \max\left(\mu_{aij}, \mu_{bij}\right), \min\left(\nu_{aij}, \nu_{bij}\right) \right\}.$$

For multiplication of two intuitionistic fuzzy matrices, B, C we need to compute for membership and nonmembership part.

Multiplication of the membership part of $D = BC$ is similar to fuzzy multiplication, which is computed as: $\mu_{dij} = \max_k \left\{ \min\left(\mu_{bik}, \mu_{ckj}\right) \right\}$.

Multiplication of the nonmembership part of BC is computed as: $\nu_{dij} = \min_k \left\{ \max\left(\nu_{bik}, \nu_{ckj}\right) \right\}$.

Following is an example that will explain the operations clearly.

Consider two intuitionistic fuzzy matrices,

$$A = \begin{bmatrix} (0.5, 0.4) & (0.6, 0.2) & (0.4, 0.4) \\ (0.7, 0.2) & (0.6, 0.3) & (0.9, 0.0) \\ (0.6, 0.3) & (0.8, 0.1) & (0.3, 0.4) \end{bmatrix},$$

$$B = \begin{bmatrix} (0.6, 0.3) & (0.7, 0.2) & (0.4, 0.3) \\ (0.5, 0.2) & (0.7, 0.2) & (0.6, 0.2) \\ (0.4, 0.3) & (0.9, 0.1) & (0.5, 0.4) \end{bmatrix},$$

$$A + B = \left\{ \begin{bmatrix} 0.6 & 0.7 & 0.4 \\ 0.7 & 0.7 & 0.9 \\ 0.6 & 0.9 & 0.5 \end{bmatrix}, \begin{bmatrix} 0.3 & 0.2 & 0.3 \\ 0.2 & 0.2 & 0.0 \\ 0.3 & 0.1 & 0.4 \end{bmatrix} \right\}.$$

Like fuzzy matrix multiplication, intuitionistic fuzzy matrix multiplication is a max–min composition for membership values and min–max composition for nonmembership values.

Considering the membership part (using max–min composition)

$$A = \begin{bmatrix} 0.5 & 0.6 & 0.4 \\ 0.7 & 0.6 & 0.9 \\ 0.6 & 0.8 & 0.3 \end{bmatrix}, \quad B = \begin{bmatrix} 0.6 & 0.7 & 0.4 \\ 0.5 & 0.7 & 0.6 \\ 0.3 & 0.9 & 0.5 \end{bmatrix},$$

$$AB = \begin{bmatrix} \max(0.5,0.5,0.3) & \max(0.5,0.6,0.4) & \max(0.4,0.6,0.4) \\ \max(0.6,0.5,0.3) & \max(0.7,0.6,0.9) & \max(0.4,0.6,0.5) \\ \max(0.6,0.5,0.3) & \max(0.6,0.7,0.3) & \max(0.4,0.6,0.3) \end{bmatrix}$$

$$AB = \begin{bmatrix} 0.5 & 0.6 & 0.6 \\ 0.6 & 0.9 & 0.6 \\ 0.6 & 0.7 & 0.6 \end{bmatrix}.$$

Considering the nonmembership part, we get (using min–max composition)

$$A = \begin{bmatrix} 0.4 & 0.2 & 0.4 \\ 0.2 & 0.3 & 0.0 \\ 0.3 & 0.1 & 0.4 \end{bmatrix}, \quad B = \begin{bmatrix} 0.3 & 0.2 & 0.3 \\ 0.2 & 0.2 & 0.2 \\ 0.3 & 0.1 & 0.4 \end{bmatrix}$$

$$AB = \begin{bmatrix} \min(0.4,0.2,0.4) & \min(0.4,0.2,0.4) & \min(0.4,0.2,0.4) \\ \min(0.3,0.3,0.3) & \min(0.2,0.3,0.1) & \min(0.3,0.3,0.4) \\ \min(0.3,0.2,0.4) & \min(0.3,0.2,0.4) & \min(0.3,0.2,0.4) \end{bmatrix}$$

$$AB = \begin{bmatrix} 0.2 & 0.2 & 0.2 \\ 0.3 & 0.1 & 0.3 \\ 0.2 & 0.2 & 0.2 \end{bmatrix}.$$

7.7.1 Identity Matrix

An intuitionistic fuzzy identity matrix of order $n \times n$ is denoted by I which is defined by $(\delta_{\mu ij}, \delta_{\nu ij})$ and is denoted as:

$$\delta_{\mu ij} = \begin{bmatrix} 1 & 0 & 0 \\ 0 & 1 & 0 \\ 0 & 0 & 1 \end{bmatrix}, \delta_{\nu ij} = \begin{bmatrix} 0 & 1 & 1 \\ 1 & 0 & 1 \\ 1 & 1 & 0 \end{bmatrix}$$

or it can be written as:

$$I = \begin{bmatrix} (1,0) & (0,1) & (0,1) \\ (0,1) & (1,0) & (0,1) \\ (0,1) & (0,1) & (1,0) \end{bmatrix}.$$

7.7.2 Null Matrix

A matrix where all elements are $(0,0)$ and is denoted by $I_{(0,0)}$.

Like fuzzy matrix, an intuitionistic fuzzy matrix A is called idempotent matrix if and only if $A^2 = A$.

An intuitionistic fuzzy matrix A is called symmetric matrix if $A' = A$.

7.7.3 Generalized Inverse of Intuitionistic Fuzzy Matrix

An intuitionistic fuzzy matrix A is said to be regular if there exists another intuitionistic fuzzy matrix X such that $AXA = A$, where X is called a generalized inverse (g-inverse) of A. The g-inverse of an intuitionistic fuzzy matrix is not unique, that is the matrix has many g-inverses.

7.8 Summary

This chapter presents a detailed description of fuzzy matrices and determinants along with their properties. Adjoint and inverse of a matrix is also explained along with properties and examples. Examples help the readers to understand the subject more clearly.

References

1 Jian-Xin, L. (1999). Controllable fuzzy matrices. *Fuzzy Sets and Systems* 45: 313–319.

2 Shyamal, A.K. and Pal, M. (2004). Two new operators on fuzzy matrices. *Journal of Applied Mathematics and Computing* 15 (1–2): 91–100.

3 Hemasinha, R., Pal, N.R., and Bezdek, J.C. (1997). Determinant of a fuzzy matrix with respect to t norms and co-t norms. *Fuzzy Sets and Systems* 87: 297–306.

4 Kim, J., Baartamans, A., and Shahadin, N.S. (1989). Determinant theory of fuzzy matrices. *Fuzzy Sets and Systems* 29: 349–356.

5 Pradhan, R. and Pal, M. (2014). The Generalized Inverse of Atanassov's Intuitionistic Fuzzy Matrices. *International Journal of Computational Intelligence Systems* 7 (6): 1083–1095.

6 Raghab, M.Z. and Emam, E.G. (1994). Determinant and adjoint of a square fuzzy matrix. *Fuzzy Sets and Systems* 61 (3): 297–307.

7 Raghab, M.Z. and Emam, E.G. (1994). On the min-max composition of fuzzy matrices. *Fuzzy Sets and Systems* 75: 83–92.

8 Thomason, M.G. (1977). Convergence of powers of a fuzzy matrix. *Journal of Mathematical Analysis and Applications* 77: 476–480.

9 Kim, K.H. and Roush, F.W. (1980). Generalized fuzzy matrices. *Fuzzy Sets and Systems* 4: 293–315.

8

Fuzzy Subgroups

8.1 Introduction

In crisp logic, a group is an algebraic structure that is equipped with mathematical operations where two elements a and b are combined to form a third element and satisfies four conditions – closure, identity, associativity, and invertibility. The most familiar example of a group is a set of integers under the operation "addition."

i) If two integers a, b are added, we get a sum $(a + b)$, which is an integer – closure property.
ii) If 0 is added to an integer, we get the same integer – identity.
iii) For any three integers, three elements follow the property $(a + b) + c = a + (b + c)$ – associativity.
iv) For each integer, a, another number exists b, such that $a + b = b + a = 0$ implies that the integer b is an inverse element of a which is – a. "$-a$" is also an integer.

In fuzzy set theory, fuzzy subgroup was first defined by Rosenfeld [1]. Then the definition was generalized by Negoite and Ralescu [2] and Anthony and Sherwood [3]. Here, we give some elementary theory of groups and groupoids.
We know that if X be a set and fuzzy subset, A of X is a function $A : X \rightarrow [0,1]$. The definitions of fuzzy subgroup are as follows [1,4]:

Definition 1(a) [1]: Let G be a group. A fuzzy set A in G is said to be a fuzzy subgroup of G if

i) $A(xy) \geq \min[A(x), A(y)]$
ii) $A\left(x^{-1}\right) \geq A(x)$

Fuzzy Set and Its Extension: The Intuitionistic Fuzzy Set, First Edition. Tamalika Chaira.
© 2019 John Wiley & Sons, Inc. Published 2019 by John Wiley & Sons, Inc.

Definition 1(b) Let G be a group. A fuzzy set A in G is said to be a fuzzy subgroup of G if for all x, y in G [4]

i) $A(xy) \geq \min[A(x), A(y)]$,

ii) $A(x^{-1}) = A(x)$,

iii) $A(e) = 1, e$ is the identity in G.

where the product of x and y is denoted as xy and the inverse of x is x^{-1}.

The identity element of any group will always be denoted by "e."

Let G be a groupoid, i.e. a set closed under binary composition (denoted multiplicatively). Then, A is called fuzzy subgroupoid of G if and only if [1]

$$A(xy) \geq \min[A(x), A(y)], \tag{8.1}$$

A is called fuzzy left ideal if $A(xy) \geq A(y)$,

A is called fuzzy right ideal if $A(xy) \geq A(x)$.

It is called left and right ideal if $A(xy) \geq \max[A(x), A(y)]$.

Fuzzy (left, right) ideal is a fuzzy subgroupoid.

For any fuzzy subgroupoid in G, $A(x^n) \geq A(x), \forall x \in G$

where x^n is any composite of x.

If G is a group and A is fuzzy subgroupoid, then A is a fuzzy subgroup of G defined by [3], iff

$$A(x^{-1}) = A(x). \tag{8.2}$$

According to [1], if A is a fuzzy subgroup, then $A(x^{-1}) = A(x)$ and $A(e)$ $A(xx^{-1}) \geq \min(A(x), A(x^{-1})) = \min(A(x), A(x)) = A(x)$, e is the identity in G.

Thus, $A(e) \geq A(x)$.

8.2 Theorems of Fuzzy Subgroup

Few theorems/lemmas and propositions of fuzzy subgroups given by several authors are given:

Theorem 1 [4]: If G is a group and A is a fuzzy subgroup of G, then $A(xy) = \min[A(x) \cdot A(y)]$ for each x, $y \in G$ with $A(x) \neq A(y)$.

Proof: Let us assume $A(x) > A(y)$.

We can write $A(y) = A(x^{-1}xy) \geq \min[A(x^{-1}), A(xy)]$ from Eq. (8.1).

As A is a fuzzy subgroup, $A(x^{-1}) = A(x)$, so

$$A(y) \geq \min[A(x), A(xy)] = A(xy), \text{as } A(x) > A(y). \tag{8.3}$$

Again we find $A(xy) \geq \min[A(x), A(y)] = A(y)$.

So, we get $A(xy) \geq A(y)$ and $A(y) \geq A(xy)$.

Hence, $A(xy) = A(y) = \min[A(x), A(y)]$.

Theorem 2 [4]: If G is a group, A is a fuzzy subgroup of G, and if $A(xyx^{-1}) = A(y)$, then $A(xy) = A(yx)$ $\forall x,y \in G$.
Proof: We can write $xy = x(yx)x^{-1}$.

Then, $A(xy) = A\left[x(yx)x^{-1}\right] = A(yx)$. $\qquad$ (8.4)

Theorem 3 [4]: If G is a group, A is a fuzzy subgroup of G, and $A(yx) = A(yx)$, then $A(xyx^{-1}) = A(y)$.
Proof:

$$A\left(xyx^{-1}\right) = A\left[(xy)x^{-1}\right] = A\left[x^{-1}(xy)\right] = A\left[x^{-1}xy\right] = A(y). \qquad (8.5)$$

Theorem 4 [1]: If A is a fuzzy subgroup of G, then $A(x) \le A(e)$, $\forall x \in G$.
Proof:

$$A(e) = A\left(xx^{-1}\right) \ge \min\left[A(x), A\left(x^{-1}\right)\right]$$
$$\ge \min[A(x), A(x)] \text{ as } A \text{ is a fuzzy subgroup, then } A\left(x^{-1}\right) = A(x)$$
$$\ge A(x).$$
Thus, $A(x) \le A(e)$. $\qquad$ (8.6)

Theorem 5 [5]: Let G be a group and A be the fuzzy subgroup of G. If $x,y \in G$ and $A(x) < A(y)$, then $A(xy) = A(x) = A(yx)$.
Proof: We know $A(xy) \ge \min[A(x), A(y)]$ $\forall x,y \in G$

As $A(x) < A(y)$, then $A(xy) \ge A(x)$ $\qquad$ (i)

Again, we can write

$$A(x) = A\left(xyy^{-1}\right) \ge \min\left[A(xy), A\left(y^{-1}\right)\right] = \min[A(xy), A(y)]. \qquad (8.7)$$
But $A(x) < A(y)$ so $A(x) \ge A(xy)$ $\qquad$ (ii)

Thus from (i) and (ii), we get $A(xy) = A(x)$.
In a similar manner we can prove $A(yx) = A(x)$.
We know $A(yx) \ge \min[A(y), A(x)]$ $\forall x,y \in G$.
As $A(x) < A(y)$, then $A(yx) \ge A(x)$.
Again, $A(x) = A(y^{-1}yx) \ge \min[A(y^{-1}), A(yx)] = \min[A(y), A(yx)]$.
But $A(x) < A(y)$, so $A(x) \ge A(yx)$.
Thus, with $A(yx) \ge A(x)$ and $A(yx) \le A(x)$, we get $A(yx) = A(x)$.

So, we arrive $A(xy) = A(x) = A(yx)$. $\qquad$ (8.8)

Theorem 6 [6]: If G is a group and A is a fuzzy subgroup that satisfies the group conditions $A(xy) \ge \min(A(x), A(y))$, $A(x^{-1}) = A(x)$, and $A(e) > 0$, then the function γ defined for each x in G through $\gamma(x) = \dfrac{A(x)}{A(e)}$ is a fuzzy subgroup of G with respect to "min" such that $\gamma(e) = 1$.

Proof: We are to prove three conditions for γ to be a fuzzy subgroup:

a) $\gamma(e) = 1$, b) $\gamma(xy) \geq \min[\gamma(x),\gamma(y)]$, c) $\gamma(x^{-1}) = \gamma(x)$

For any x in G, we may write

$$\gamma(e) = \frac{A(e)}{A(e)} = \frac{A(xx^{-1})}{A(e)}$$

$$\geq \frac{\min[A(x),A(x^{-1})]}{A(e)}$$

$$= \frac{\min[A(x),A(x)]}{A(e)}$$

$$= \frac{A(x)}{A(e)}, \text{ as } A(x^{-1}) = A(x)$$

$$= \gamma(x).$$

Thus, we see that we started at $\gamma(e) = \dfrac{A(e)}{A(e)}$ and arrived at $\gamma(x) = \dfrac{A(x)}{A(e)}$.

Thus, the condition $\gamma(e) = 1$ is verified.

From $\gamma(x) = \dfrac{A(x)}{A(e)}$, we may write $\gamma(xy) = \dfrac{A(xy)}{A(e)} \geq \dfrac{\min[A(x),A(y)]}{A(e)}$

$$= \min\left[\frac{A(x)}{A(e)}, \frac{A(y)}{A(e)}\right] = \min[\gamma(x),\gamma(y)]$$

$$= \min[\gamma(x),\gamma(y)]$$

Thus. the second condition is verified

Third condition: $\gamma(x^{-1}) = \dfrac{A(x^{-1})}{A(e)}$

$$= \frac{A(x)}{A(e)} = \gamma(x).$$

Thus, γ is a fuzzy subgroup of G with respect to "min."

Theorem 7 [7]: If A is a fuzzy subgroup of a group G and $x \in G$, then $A(xy) = A(y), \forall y \in G \iff A(x) = A(e)$.

Proof: First, we assume that $A(xy) = A(y)$. Then, choosing $y = e$, we get $A(x) = A(e)$.

Second, we assume that $A(x) = A(e)$ and we are to prove $A(xy) = A(y)$.

As y is in a group G, $A(y) \leq A(e), \forall y \in G$ from Eq. (8.6).

So, $A(y) \leq A(x)$.

From the definition of group, we know

$A(xy) \geq \min[A(x), A(y)] = A(y)$

So, $A(xy) \geq A(y)$ (i)

Again, we can write,

$A(y) = A\left(x^{-1}xy\right) \geq \min\left[A\left(x^{-1}\right), A(xy)\right] = \min[A(x), A(xy)]$

$\Longrightarrow A(y) \geq \min[A(x), A(xy)] = A(xy)$, as $A(y) \leq A(x)$.

Thus, $A(y) \geq A(xy)$ (ii)

Hence, from (i) and (ii), we get $A(xy) = A(y)$.

Theorem 8   [7]: Let $x \in G$. If $A(x) = A(e)$, then $A(xy) = A(yx), \forall\, y \in G$.
Proof: From Theorem 7, we see that if $A(x) = A(e)$, then we get $A(xy) = A(y)$.
As y is in a group G, from Eq. (8.6) we have

$A(y) \leq A(e), \forall y \in G \Rightarrow A(y) \leq A(x)$ as $A(x) = A(e)$.

We know $A(yx) \geq \min[A(y), A(x)]$

$\Rightarrow A(yx) \geq A(y)$, as $A(y) \leq A(x)$

Again, we can write $A(y) = A(yxx^{-1}) \geq \min[A(yx), A(x^{-1})] = \min[A(yx), A(x)]$
implies $A(y) \geq A(yx)$
So, we have $A(yx) \geq A(y)$ and $A(y) \geq A(yx)$ implies

$A(y) = A(yx)$.

Hence, $A(xy) = A(yx)$.

Theorem 9   [1]: A is a fuzzy subgroup of G if and only if $A(xy^{-1}) \geq \min(A(x), A(y))$.
Proof: Given $A(xy^{-1}) \geq \min(A(x), A(y))$.
Substituting $x = y$, we get $A(yy^{-1}) \geq \min(A(y), A(y)) = A(y) \Rightarrow A(e) \geq A(y)$.

Now $A\left(y^{-1}\right) = A\left(ey^{-1}\right) \geq \min(A(e), A(y)) = A(y)$ as $A(e) \geq A(y)$

From this, it follows that $A(xy) = A(x(y^{-1})^{-1}) \geq \min(A(x), A(y^{-1})) = \min(A(x), A(y))$.
Hence, A is a fuzzy subgroup of G.
Conversely as A is a subgroup, so $A(xy^{-1}) \geq \min(A(x), A(y^{-1}))$ implies $A(xy^{-1}) \geq \min(A(x), A(y))$.

Theorem 10   [1]: If A is a fuzzy subgroup and if $A(xy^{-1}) = A(e)$, then $A(x) = A(y)$.

Proof: We can write $A(x) = A(xy^{-1}y) \geq \min(A(xy^{-1}), A(y)) \geq \min(A(e), A(y)) = A(y)$. Again, $A(y) = A(yx^{-1}x) \geq \min(A(yx^{-1}), A(x)) \geq \min(A(e), A(y)) = A(x)$.

Thus, we get $A(x) \geq A(y)$ and $A(y) \geq A(x)$ implies $A(x) = A(y)$.

8.3   Fuzzy-level Subgroup

According to Zadeh, let A be a fuzzy subset of set X. Then, for $t \in [0,1]$, set $A_t = \{x \in X \mid A(x) \geq t\}$ is the level subset of A. If the set $A_t = \{x \in X \mid A(x) > t\}$, then A_t is the strong-level subset of A. From the level set, Das [8] introduced fuzzy-level subgroup and then it was analyzed by Mukherjee and Bhattacharya [7].

Definition 2   [7–9]: If A is a fuzzy subgroup of G, then for any $t \in [0,1]$ with $t \leq A(e)$, $A_t = \{x \in G \mid A(x) \geq t\}$ is called fuzzy-level subgroups of A. So, subgroups A_t are called level subgroups of A. If an image set of fuzzy subgroup A of a group G consists of $t_i = t_0, t_1, t_2, \ldots, t_n$, then the family of level subgroups is A_{t_i}, $\{A_{t_i} : 0 \leq i \leq n\}$ constitute a complete list of level subgroups of A.

If $t_0 > t_1 > t_2 > \cdots > t_n$, then a family of level subgroup of fuzzy subgroup A form a chain

$$A_{t_1} < A_{t_2} < A_{t_3} \cdots < A_{t_n} = G,$$

where $A(e) = t_0$.

Clearly, if A is a fuzzy subgroup of a group G, then for any x in G, $A(x) \leq A(e)$.

There are few theorems based on fuzzy-level subgroup.

Theorem 11   [8]: If G is a group and A is a fuzzy subgroup of G, then the level subset set A_t, $t \leq A(e)$ for $t \in [0,1]$ is a subgroup of G where e is an identity of G.

Proof: We know from the definition of level set that $A_t = \{x \in G \mid A(x) \geq t\}$ for $t \in [0,1]$.

Let $x,y \in A_t$, then $A(x) \geq t$ and $A(y) \geq t$.

As A is a fuzzy subgroup of G (given),

$$A(xy) \geq \min[A(x), A(y)] \Rightarrow A(xy) \geq t.$$

Hence, $xy \in A_t$.

Again from our assumption, $x \in A_t$ implies $A(x) \geq t$.

As A is a subgroup of G, so $A(x^{-1}) \geq A(x)$.

Substituting $A(x) \geq t$, we get $A(x^{-1}) \geq t$.
This implies $x^{-1} \in A_t$.
Thus, A_t is a subgroup of G.

Theorem 12 [8]: If G is a group and A is a fuzzy subset of G such that A_t is a subgroup of G, $\forall t \in [0,1]$, $t \leq A(e)$, then A is a fuzzy subgroup of G.
Proof: Let $x,y \in G$. Also, let $A(x) = t_1$ and $A(y) = t_2$.
So, $x \in A_{t_1}$ and $y \in A_{t_2}$.
Let us consider $t_1 > t_2$.
Then, $A_{t_1} \subseteq A_{t_2}$ that is A_{t_1} is the subset of A_{t_2}.
As A_{t_1} is the subset of A_{t_2}, so $x \in A_{t_2}$.
Thus, we get $x \in A_{t_2}$ and from the assumed condition, we have $y \in A_{t_2}$.
Now, as A_{t_2} is a subgroup of G, so $xy \in A_{t_2}$.
This implies $A(xy) \geq t_2 = \min[A(x),A(y)]$.
One condition of fuzzy subgroup is verified.
Second: Let $x \in G$, and let $A(x) = t$.
So, $x \in A_t$. As A_t is a subgroup, we have $x^{-1} \in A_t$ implies $A(x^{-1}) \geq t$.
Thus $A(x^{-1}) \geq A(x)$.
Hence, A is a subgroup of G.

Theorem 13 [8]: If G is a group and A is a fuzzy subgroup of G, then two level subgroups A_{t_1}, A_{t_2} of A with $t_1 < t_2$ are equal if and only if there is no $x \in G$ such that $t_1 \leq A(x) \leq t_2$.
Proof: Let us assume a hypothesis $A_{t_1} = A_{t_2}$. Suppose there exists $x \in G$ such that $t_1 \leq A(x) \leq t_2$, $(t_1 < t_2)$, then $A_{t_2} \subseteq A_{t_1}$. This implies that x belongs to A_{t_1} but not in A_{t_2} which contradicts our assumption.
Again, suppose there is no x in G such that $t_1 \leq A(x) \leq t_2$. Since $t_1 < t_2$, $A_{t_2} \subseteq A_{t_1}$. Let $x \in A_{t_1}$, then $A(x) \geq t_1$. As $A(x)$ does not lie between t_1 and t_2, so $A(x) \geq t_2$ implies $x \in A_{t_2}$.
Thus, $A_{t_1} \subseteq A_{t_2}$. Hence, $A_{t_1} = A_{t_2}$.

Theorem 14 [7]: If A is a fuzzy subgroupoid of a finite group G, then A is a fuzzy subgroup.
Proof: The condition that A is a fuzzy subgroupoid is:
$A(xy) \geq \min[A(x),A(y)]$ and we will prove that A is a fuzzy subgroup, i.e. $A(x) \geq A(x^{-1})$.
This can be proved using order of a group. Order or period of an element x in a group is the smallest integer n satisfying the relation $x^n = e$, where e is the identity element and x^n denotes the product of n copies of x.
Let $x \in G$. Since G is finite, x has finite order, say n. As per the definition of order of group, we know $x^n = e$, where e is the identity of group G.
Thus, we get $x^{n-1} = x^{-1}$.

Now, using the definition of fuzzy subgroupoid, we have

$$A\left(x^{-1}\right) = A\left(x^{n-1}\right) = A\left(x^{n-2}x\right) \geq \min\left[A\left(x^{n-2}\right), A(x)\right]. \tag{8.9}$$

Again, we can write $A(x^{n-2}) = A(x^{-2}) = A(x^{-1}x^{-1}) \geq \min[A(x^{-1}), A(x^{-1})] = A(x^{-1}) \Longrightarrow A(x^{n-2}) \geq A(x^{-1})$.

So, Eq. (8.9) may be written as:

$$A\left(x^{-1}\right) = A\left(x^{n-1}\right) \geq \min[A(x^{n-2}), A(x)] = A(x).$$

$$\Rightarrow A(x^{-1}) \geq A(x).$$

Interchanging x with x^{-1}, we get $A(x) \geq A(x^{-1})$.

Thus, A is a subgroup.

Theorem 15 [4]: If A is a subgroup of G, then for any integer n, $A(x^n) \geq A(x)$.

Proof: This can be done using the induction method.

For $n = 2$, $A(x^2) \geq A(x \cdot x) \geq \min(A(x), A(x)) = A(x)$.

So, $A(x^2) \geq A(x)$.

Likewise, for $n = 3$, we get $A(x^3) \geq A(x^2 \cdot x) \geq \min(A(x^2), A(x)) = A(x)$.

With integer $n > 3$, say k, we can make a hypothesis, $A(x^k) \geq A(x)$

Likewise, $A(x^{k+1}) \geq A(x^k \cdot x) \geq \min(A(x^k), A(x)) = A(x)$

For $n < 0$, we can write $A(x^{-1}) \geq A(x)$ and $A(x^{-2}) \geq A(x^{-1} \cdot x^{-1}) \geq \min(A(x^{-1}), A(x^{-1})) = A(x^{-1}) = A(x)$ (as A is a subgroup of G).

Thus, with $n < 0$, $A(x^k)^{-1} = A(x^{-k}) \geq A(x)$.

Hence, $A(x^n) \geq A(x)$.

8.4 Fuzzy Normal Subgroup

Definition 3 [7]: A fuzzy subgroup A of a group G that satisfies the conditions

i) $A(xy) = A(yx)$,

ii) $A\left(xyx^{-1}\right) = A(y)$

is called a fuzzy normal subgroup.

If A and B are two fuzzy subgroups of G, then the fuzzy subgroup A is called conjugate to a fuzzy subgroup B if there exists x in G such that for all g in G [7,10]

$$A(g) = B\left(x^{-1}gx\right), \forall x \in G.$$

Theorem 16 [7]: A fuzzy subgroup A of a group G is a fuzzy normal subgroup if and only if A is constant on the conjugate classes of G.

Proof: As A is a fuzzy normal subgroup, so $A(xy) = A(yx)$.

Then, we can write $A(y^{-1}xy) = A(xyy^{-1}) = A(x), \forall x, y \in G$.

Hence, $A(y^{-1}xy) = A(x)$ implies A is constant on conjugate class of G as per the definition of conjugate.

Conversely, let us consider A to be constant on the conjugate classes of G.

$$A(xy) = A\left(xyxx^{-1}\right)$$

$$= A\left[x(yx)x^{-1}\right] = A(yx), \forall x, y \in G \text{ (from the definition of conjugate).}$$

Thus, A is a normal fuzzy subgroup of G.

An alternative definition of normal fuzzy subgroup using "commutator" is given by Mukherjee [7].

Definition 4   [7]: If $x, y \in G$ and $x^{-1}y^{-1}xy$ is denoted by $[x,y]$, then $[x,y]$ is called commutator of x and y. If $xy = yx$, then obviously $[x,y] = e$. Thus, a group G is Abelian if $[x,y] = e$. That is, if x and y commute each other, then $[x,y] = e$.

Theorem 17   [10]: If A is a fuzzy subgroup of a group G, then A is a fuzzy normal subgroup implies $A[x,y]) = A(e)$, where $[x,y] = x^{-1}y^{-1}xy, \forall x, y \in G$.

Proof: We know from Theorem 16 that a fuzzy subgroup A of a group G is a fuzzy normal subgroup if and only if A is constant on the conjugate classes of G. Thus, we have that A is normal implies

$$A\left(x^{-1}y^{-1}x\right) = A\left(y^{-1}\right).$$

$$\Rightarrow A\left(x^{-1}y^{-1}xyy^{-1}\right) = A\left(y^{-1}\right)$$

$$\Rightarrow A\left([x,y]y^{-1}\right) = A\left(y^{-1}\right).$$

From Theorem 7 we know that if A is a fuzzy subgroup of a group G, then $A(xy) = A(y), \forall y \in G$, implies $A(x) = A(e)$.

Then, $A([x,y]) = A(e)$, considering $[x,y] = x$ and $y^{-1} = y$.

8.5   Fuzzy Subgroups Using T-norms

Anthony and Sherwood [6] generalized the definition of fuzzy subgroup that was introduced by Rosenfeld [1] by replacing the stronger condition minimum with a T-norm function $T: [0,1] \times [0,1] \rightarrow [0,1]$.

A T-norm is a function $T: [0,1] \times [0,1] \rightarrow [0,1]$ satisfying each of $x, y, z \in [0,1]$:

i) $T(x, 1) = x$ – identity

ii) $T(x, y) = T(y, x)$ – commutativity

iii) $T(x,y) \leq T(x,z)$, if $y \leq z$

iv) $T(x, T(y,z)) = T(T(x,y),z))$ – associativity.

The T-norms that are frequently encountered are as follows:

Min T-norm – $T_M(x,y) = \min(x,y) = \begin{cases} x, & \text{if } x \leq y \\ y & \text{if } y < x \end{cases}$,

Lukaseiwics T-norm – $T_L(x,y) = \max(x + y - 1,0)$,

Product T-norm – $T_P(x,y) = xy$,

Drastic T-norm –

$$T_D(a,b) = \begin{cases} 0, & \text{if } x,y \in [0,1]^2 \\ \min(x,y,) & \text{if } \max(x,y) = 1 \end{cases}$$

Schweizer and Sklar [11] introduced a "stronger than" relation and suggested that any T-norm T is stronger than drastic T-norm, T_D. T_M, the min T-norm is the strongest among all the T-norms. Any fuzzy subgroup with respect to min is a fuzzy subgroup with respect to any other T-norms. Since min is the strongest of all T-norms, so any fuzzy subgroup with respect to "min" is a fuzzy subgroup with respect to any other T-norm as well.

Definition 5   [3, 12]: Let $(G,*)$ be a group. A function $A : G \to [0,1]$ is a fuzzy subgroup of G with respect to T-norm is defined as follows:
For every $x, y \in G$

i) $A(xy) \geq T[A(x),A(y)]$.
   This inequality is obtained by replacing "min" term in T-norm in Eq. (8.10) by "T."

ii) $A(x^{-1}) = A(x)$,
iii) $A(e) = 1$ (e is an identity in G).

Further, A is called normal fuzzy subgroup under the T-norm T if $A(xy) = A(yx)$.

Just like fuzzy subgroup using minimum, there are also similar theorems of fuzzy subgroup using T-norms.

There are few theorems on fuzzy subgroup with respect to T-norm T.

Theorem 18   [3]: Let A be a fuzzy subgroup with respect to T-norm T. If $A(xy^{-1}) = 1$, then $A(x) = A(y)$.

Proof. As A is a fuzzy subgroup,

$$A(y) = A\left(y^{-1}\right) = A\left(x^{-1}xy^{-1}\right) \geq T\left(A\left(x^{-1}\right), A\left(xy^{-1}\right)\right)$$
$$\geq T(A(x), 1) = A(x).$$

Likewise, $A(x) = A(xy^{-1}y) \geq T(A(xy^{-1}), A(y)) = T(1, A(y)) = A(y)$.
Thus, $A(y) = A(x)$.

Theorem 19 [3]: Let A be a fuzzy subgroup and T be a T-norm. If $A(e) = 1$ and $A(xy^{-1}) \geq T(A(x), A(y))$ $\forall x, y \in G$, then A be a fuzzy subgroup of G with respect to T-norm T.

Proof: We can write

$$A\left(y^{-1}\right) = A\left(ey^{-1}\right) \geq T\left(A(e), A\left(y^{-1}\right)\right) = T(A(e), A(y))$$
$$= T(A(y), 1) = A(y).$$

Again, we see $A(y) = A(ey) \geq T(A(e), A(y)) = T(A(e), A(y^{-1})) = T(1, A(y^{-1})) = A(y^{-1})$. Hence, $A(y^{-1}) = A(y)$.

Moreover, $A(xy) = A(x(y^{-1})^{-1}) \geq T(A(x), A(y^{-1})) = T(A(x), A(y))$ as $A(xy^{-1}) \geq T(A(x), A(y))$.

Hence, A is a fuzzy subgroup of G with respect to T-norm T.

8.6 Product of Fuzzy Subgroups

Let G_1 and G_2 be two groups. If A_1 is a fuzzy subgroup of G_1 with respect to T-norm T and A_2 is a fuzzy subgroup of G_2 under T-norm T, then $A = A_1 \times A_2$ is a fuzzy subgroup of $G = G_1 \times G_2$ with respect to T-norm T.

To prove the theorem, an intermediate theorem related to associative property of T-norm, T, is required, which is defined below.

Theorem 20 Let $T: [0,1] \times [0,1] \to [0,1]$ be the T-norm. Then, for any $w, x, y, z \in [0.1]$, $T(T(w,x), T(y,z)) = T(T(w,y), T(x,z))$ [13].

Proof: Here, the associative property has been used, $T(T(x,y), z) = T(x, T(y,z))$. Considering $T(y,z) = m$, then LHS becomes

$$T(T(w,x), T(y,z)) = T(T(w,x), m) = T(w, T(x,m)) = T(w, T(x, T(y,z)))$$
$$= T(w, T(T(x,y), z)) = T(w, T(T(y,x), z)) = T(w, T(y, T(x,z))).$$

Thus, $T(T(w,x), T(y,z)) = T(w, T(y, T(x,z)))$.
Again, let us assume $T(x,z) = n$, then

$$T(w, T(y, T(x,z))) = T(w, T(y,n)) = T(T(w,y), n) = T(T(w,y), T(x,z))$$

Hence, $T(T(w, x), T(y, z)) = T(T(w, y), T(x, z))$.

Theorem 21 [13]: If A_1 is a fuzzy subgroup of G_1 with respect to T-norm T and A_2 is a fuzzy subgroup of G_2 under T-norm T, then $A = A_1 \times A_2$ is a fuzzy subgroup of $G = G_1 \times G_2$ with respect to T-norm T is defined by

$$A(x_1, x_2) = (A_1 \times A_2)(x_1, x_2) = T(A_1(x_1), A_2(x_2)).$$

Proof: Let $x = [x_1, x_2]$, $y = [y_1, y_2]$ be any elements of group $G = G_1 \times G_2$, then

$$A(xy^{-1}) = A(x_1 y_1^{-1}, x_2 y_2^{-1}) = T(A_1(x_1 y_1^{-1}), A_2(x_2 y_2^{-1}))$$

$$\geq T(T(A_1(x_1), A_1(y_1^{-1}))), T(A_2(x_2), A_2(y_2^{-1}))) \text{ using } A(xy) \geq T[A(x), A(y)]$$

$$= T(T(A_1(x_1), A_1(y_1))), T(A_2(x_2), A_2(y_2))) \text{ using } A(x^{-1}) = A(x)$$

$$= T(T(A_1(x_1), A_2(x_2))), T(A_1(y_1), (A_2(y_2))) \text{ using the theorem}$$

$$T(T(w,x), T(y,z)) = T(T(w,y), T(x,z))$$

$$= T(A(x_1, x_2), A(y_1, y_2)) = T(A(x), A(y))$$

Thus, $A(xy^{-1}) \geq T(A(x), A(y))$.

Moreover, $A(e) = A(e_1, e_2) = T(A_1 e_1, A_2 e_2) = T(1,1) = 1$.

Thus, A is a fuzzy subgroup of G with respect to T.

Definition 6 Sherwood [12] defined that for each $i = 1, 2, 3, \ldots, n$, let G_i be a fuzzy collection of groups and $G = G_1 \times G_2 \times \cdots \times G_n$ be the product group of G_i. Let A_i be the respective fuzzy subgroup of G_i with respect to T-norm T, then the T-product of A_i is a function

$$A_1 \times A_2 \times \cdots \times A_n : G_1 \times G_2 \times \cdots \times G_n \to [0,1]$$

given by:

$$A_1 \times A_2 \times \cdots \times A_n(x_1, x_2, x_3, \ldots, x_n) = T\{A_1(x_1), A_2(x_2), A_3(x_3), \ldots, A_n(x_n)\}.$$
$$(8.10)$$

Definition 7 [12, 14]: For each $i = 1, 2, 3, \ldots, n$, let A_i be the fuzzy subgroup of G under the minimum operation in a group G, then the membership function of the product $A = A_1 \times A_2 \times \cdots \times A_n(x)$ in $G = G_1 \times G_2 \times \cdots \times G_n$ is defined by:

$$A_1 \times A_2 \times \cdots \times A_n(x) = \min(A_1(x_1), A_2(x_2), A_3(x_3), \ldots, A_n(x_n)),$$

where $x = x_1, x_2, x_3, \ldots, x_n$.

There are few theorems on the product of fuzzy subgroups under two different T-norms.

If the two T-norms T_1 and T_2 are different and one T-norm is stronger than the other T-norm, then the following definition follows:

Definition 8 [1, 12]: A T-norm T_2 is said to be stronger than T_1 if $T_2(x, y) \geq T_1(x, y)$ for all $x, y \in [0,1]$. Let T_1 and T_2 be two T-norms and T_2 is stronger than T_1 $(T_2 \geq T_1)$ if

i) $T_2(x,y) \geq T_1(x,y) \forall x,y$ in $[0,1]$,

ii) For all a,b,c,d in $[0,1]$, T_2 dominates T_1 $(T_2 \gg T_1)$

iff $T_2(T_1(a,b), T_1(c,d)) \geq T_1(T_2(a,c), T_2(b,d))$. $\qquad$ (8.11)

Theorem 22 [15]: Let A_1 and A_2 be two fuzzy subgroups of group G with respect to T-norm T_1. Let T_2 be a T-norm that dominates T_1 $(T_2 > T_1)$. Then the T_2 product of A_1 and A_2 is a fuzzy subgroup of G with respect to T-norm T_1. Also, if A_1 and A_2 are normal, then the T_2 product of A_1 and A_2 is also normal.

Proof:

$$A_1 \times A_2(xy)^{-1} = T_2\left(A_1\left(xy^{-1}\right), A_2\left(xy^{-1}\right)\right)$$

$$\geq T_2\left(T_1\left(A_1(x), A_1\left(y^{-1}\right)\right), T_1\left(A_2(x), A_2\left(y^{-1}\right)\right)\right)$$

$$\geq T_2\left(T_1(A_1(x), A_1(y)), T_1(A_2(x), A_2(y))\right)$$

$$\geq T_1\left(T_2(A_1(x), A_2(x)), T_2(A_1(y), A_2(y))\right) \text{ [from definition 8 (ii)]}$$

From the theory of T product in Definition 5, we get

$$A_1 \times A_2(xy)^{-1} \geq T_1\left(T_2(A_1(x), A_2(x)), T_2(A_1(y), A_2(y))\right)$$

$$= T_1\left(A_1 \times A_2(x), A_1 \times A_2(y)\right)$$

Hence, $A_1 \times A_2(xy)^{-1} \geq T_1(A_1 \times A_2(x), A_1 \times A_2(y))$.

Moreover, $A_1 \times A_2(e) = T_2(A_1 e, A_2 e)$

$$= T_2(1,1) = 1,$$

Thus, $A_1 \times A_2$ is a fuzzy subgroup of G with respect to T-norm T_1 by Theorem 19.

Second: If A_1 and A_2 are normal

$$A_1 \times A_2(xy) = T_2(A_1(xy), A_2(xy))$$

$$= T_2(A_1(yx), A_2(yx)), \text{ as } A_1, A_2 \text{ are normal}$$

$$= A_1 \times A_2(yx)$$

Theorem 23 [12]: Let A_i be a fuzzy subgroup of G_i with respect to T-norm T_1. Let T_2 be a T-norm that dominates T_1 $(T_2 > T_1)$. Then, the T_2 product of $A_1 \times A_2 \times \cdots \times A_n$ is a fuzzy subgroup of $G_1 \times G_2 \times \cdots \times G_n$ under T_1.

This is similar to that of Theorem 23 but this is the product of $'n'$ subgroups. The proof can be done in a similar way, i.e. $A(xy^{-1}) \geq T(A(x), A(y))$ as that of Theorem 22. In this theorem, a different procedure is followed.

Proof: For any $(x_1, x_2, \ldots, x_n)$ and $(y_1, y_2, \ldots, y_n)$ in $(G_1 \times G_2 \times \cdots \times G_n)$ and T_2 is an increasing T-norm,

$$A_1 \times A_2 \times \cdots \times A_n((x_1, x_2, \ldots, x_n), (y_1, y_2, \ldots, y_n)) = A_1 \times A_2 \times \cdots \times A_n(x_1 y_1, x_2 y_2, \ldots, x_n y_n)$$

$$= T_2(A_1(x_1 y_1), A_2(x_2 y_2), \ldots, A_n(x_n y_n))$$

$$\geq T_2(T_1(A_1 x_1, A_1 y_1), T_1(A_2 x_2, A_2 y_2), \ldots, T_1(A_n x_n, A_n y_n)), (\text{as } A(xy) \geq T_1(A(x), A(y)))$$

$$\geq T_1(T_2(A_1 x_1, A_2 x_2, \ldots, A_n x_n), T_2(A_1 y_1, A_2 y_2, \ldots, A_n y_n)),$$

(as T_2 dominates T_1, the rule $T_2(T_1(a,b), T_1(c,d)) \geq T_1(T_2(a,c), T_2(b,d))$ follows)

From the definition of T-product in Definition 5, we get:

$$A_1 \times A_2 \times \cdots \times A_n((x_1, x_2, \ldots, x_n), (y_1, y_2, \ldots, y_n))$$

$$\geq T_1(T_2(A_1 x_1, A_2 x_2, \ldots, A_n x_n), T_2(A_1 y_1, A_2 y_2, \ldots, A_n y_n))$$

$$= T_1(A_1 \times A_2 \times \cdots \times A_n(x_1, x_2, \ldots, x_n), A_1 \times A_2 \times \cdots \times A_n(y_1, y_2, \ldots, y_n))$$

Hence,

i) $A_1 \times A_2 \times \cdots \times A_n((x_1, x_2, \ldots, x_n), (y_1, y_2, \ldots, y_n)) \geq T_1(A_1 \times A_2 \times \cdots \times A_n(x_1, x_2, \ldots, x_n), A_1 \times A_2 \times \cdots \times A_n(y_1, y_2, \ldots, y_n))$.

Thus, $A_1 \times A_2 \times \cdots \times A_n$ satisfies the first condition of Definition 5.

ii) $A_1 \times A_2 \times \cdots \times A_n((x_1, x_2, \ldots, x_n)^{-1}) = A_1 \times A_2 \times \cdots \times A_n(x_1^{-1}, x_2^{-1}, x_3^{-1}, \ldots, x_n^{-1})$

$$= T_2((A_1 x_1^{-1}), (A_2 x_2^{-1}), \ldots, (A_n x_n^{-1}))$$

$$= T_2(A_1 x_1, A_2 x_2, \ldots, A_n x_n), \text{ as } A(x^{-1}) = A(x)$$

$$= A_1 \times A_2 \times \cdots \times A_n(x_1, x_2, \ldots, x_n);$$

(second condition is verified)

Finally iii) $A_1 \times A_2 \times \cdots \times A_n(e_1 \times e_2 \times \cdots \times e_n) = T_2(A_1 e_1, A_2 e_2, \ldots, A_n e_n)$

$$= T_2(1, 1, \ldots, 1) = 1 \text{ as } A(e) = 1.$$

Thus, $A_1 \times A_2 \times \cdots \times A_n$ is a fuzzy subgroup of $G_1 \times G_2 \times \cdots \times G_n$ under T-norm T_1 as per Definition 5.

8.7 Summary

This chapter describes in detail the fuzzy subgroup and product of fuzzy group of theorems and propositions. Fuzzy normal subgroup, fuzzy-level subgroup, and fuzzy subgroup defined using T-norm are explained with propositions.

References

1 Rosenfeld, A. (1971). Fuzzy subgroups. *Journal of Mathematical Analysis and Application* 35: 512–517.

2 Negoite, C.V. and Ralescu, D.A. (1975). *Applications of Fuzzy Sets to System Analysis*. New York: Wiley.

3 Anthony, J.M. and Sherwood, H. (1979). Fuzzy groups redefined. *Journal of Mathematical Analysis and Applications* 69 (1): 124–130.

4 Akgul, M. (1988). Some properties of fuzzy subgroups. *Journal of Mathematical Analysis and Application* 133: 93–100.

5 Dixit, V.N., Kumar, R., and Azmal, N. (1990). Level subgroups and union of fuzzy subgroups. *Fuzzy Sets and Systems* 24: 79–86.

6 Anthony, J.M. and Sherwood, H. (1982). Characterization of fuzzy subgroups. *Fuzzy Sets and Systems* 7: 297–305.

7 Mukherjee, N.P. and Bhattacharya, P. (1984). Fuzzy normal subgroups and fuzzy cosets. *Information Science* 34: 225–239.

8 Das, S. (1981). Fuzzy groups and level subgroups. *Journal of Mathematical Analysis and Application* 84: 264–269.

9 Bhattacharya, P. (1987). Fuzzy subgroups: some characterization. *Journal of Mathematical Analysis and Application* 128: 241–252.

10 Bhattacharya, P. and Mukherjee, N.P. (1986). Fuzzy groups: some group theoretic analogs II. *Information Science* 39: 247–268.

11 Schweizer, B. and Sklar, A. (1960). Statistical metric spaces. *Pacific Journal of Mathematics* 10 (1): 313–334.

12 Sherwood. H (1983). Product of fuzzy subgroups. *Fuzzy Sets and Systems* 11: 79–89.

13 Abu Osman, M.T. (1984). On direct product of fuzzy subgroups. *Fuzzy Sets and Systems* 12: 87–91.

14 Chon, I. (2004). Fuzzy subgroups as products. *Fuzzy Sets and Systems* 141: 505–508.

15 Abu Osman, M.T. (1987). On some product of fuzzy subgroups. *Fuzzy Sets and Systems* 24: 79–86.

9

Application of Fuzzy/Intuitionistic Fuzzy Set in Image Processing

9.1 Introduction

This chapter discusses the application of fuzzy set and intuitionistic fuzzy set theory in medical image processing. Different types of fuzzy membership functions, fuzzy operators, fuzzy measures, fuzzy integrals, and entropy that are discussed in the previous chapters are used in processing these images. Processing includes enhancement, segmentation, retrieval, clustering, and edge detection. These are very much important for medical image diagnosis when detection of abnormal lesions/tumor/hemorrhage or counting blood cells or computing thinness of any vessels or any other detection is required. In this chapter, both fuzzy and intuitionistic fuzzy set theories are used so that the readers can visualize the difference of the image results. After going through the details of fuzzy and intuitionistic fuzzy mathematics, readers are now aware that intuitionistic fuzzy set considers two uncertainties – membership and nonmembership degrees apart from fuzzy set theory where only membership degree is used. We know that medical images are not uniformly illuminated, so the image boundaries/regions are not properly visible. So, these images contain more uncertainties as compared to other images and in that case intuitionistic fuzzy set may be of great use as it considers more uncertainties. In the following section we will see the use of fuzzy/intuitionistic fuzzy set in medical image processing where different types of operators, measures, and membership functions that are discussed earlier in different chapters are used.

9.2 Digital Images

There are two types of images – analog images and digital images. Analog images are those images that we see in television and photograph. It is generally

Fuzzy Set and Its Extension: The Intuitionistic Fuzzy Set, First Edition. Tamalika Chaira.
© 2019 John Wiley & Sons, Inc. Published 2019 by John Wiley & Sons, Inc.

continuous and not broken. For computer processing of images, analog images or raw images are converted into digital images through digitization. Digital images are made up of picture elements called pixels. Pixels are arranged in the form of a rectangular array. The size of the image depends on the dimension of the array. An image of size $M \times N$ has M rows and N columns and is thought to be a matrix. Each element in the matrix is called pixel and each pixel has its own intensity or brightness. Intensity values in digital images are defined by bits. A 8-bit intensity range has 256 (2^8) possible values, i.e. from 0 to 255. The brightness or intensity value ranges from 0 to 255. 0 means black and 255 means white. The values between 0 and 255 are the variation of intensity values ranging from black to white. These are called the gray levels. As the image is a matrix, each element or pixel of an image, say A, is represented by a_{ij}.

An image histogram is the frequency of occurrence of the pixels in an image, i.e. number of many times that pixel is present in the image. It is denoted as $h(g)$, g is the gray value of the image ranging from 0 to 255.

9.3 Image Enhancement

Preprocessing is required in almost all medical images, as the images are not uniformly illuminated. Enhancement increases the overall visual contrast of the image so that the image structures are more clear and distinguishable. It is applied to the images where the contrast between object and background is low, i.e. when object and background are not clearly differentiable. It makes gray region darker and bright region brighter. It is used to restore an image that has some kind of deterioration or to enhance certain features of the image. For visual analysis of medical images, images should be clear and have distinct structures for better diagnosis. So, if the quality of the image is improved, processing becomes easier. For this reason, medical image enhancement is extremely important and it becomes easier for specialists or doctors to spot any anomalies visually in the enhanced radiological or pathological images. It is very much useful where the intensities of important regions such as tissues, blood vessels, or fine structures in medical images are very low and it becomes very difficult to make out the structures with the human eye. Enhancement highlights the areas of low intensity, thus improves the readability.

Here, we will discuss few results on enhancement of medical images using fuzzy and intuitionistic fuzzy set theory.

9.3.1 Fuzzy Enhancement Method

This method uses an intensification (INT) operator to reduce the fuzziness and increase the image contrast. The basic requirement is to increase the dynamic range of the image. The principle of contrast stretching depends on the selection of a threshold T, which is user defined. A better contrast is obtained by darkening the levels that are below a certain threshold T and brightening the gray levels that are above T and this is done in a nonlinear fashion. If the contrast is low, fuzziness of the fuzzy set is more. Thus, enhancing the contrast would reduce the fuzziness present in the image [1]. This can be achieved by applying fuzzy INT operator on the membership function of the fuzzy set. The choice of fuzzy membership function is not unique and it depends on the type of application. If μ_{mn} be the membership values of a fuzzy set (image) at $(m,n)^{th}$ gray level, then the resulting fuzzy set that reduces the fuzziness will have a membership function μ'_{mn} which is given as:

$$\mu'_{mn} = 2 \cdot \left[\mu_{ij} \right]^2 \qquad \text{for } 0 \leq \mu_{ij} \leq 0.5$$

$$= 1 - 2 \cdot \left[1 - \mu_{ij} \right]^2 \quad \text{for } 0.5 \leq \mu_{ij} \leq 1$$

$$(9.1)$$

This expression is known as the INT operator. Here, the membership values are transformed in such a way that the membership values which are above 0.5 (default value) are transformed to much higher values and the membership values which are lower than 0.5 are transformed to much lower values in a nonlinear fashion to obtain a good contrast in an image μ'_{mn} is the modified image at (m,n)th gray level. Once the membership values are modified, modified gray values are then transformed to spatial domain using an inverse function:

$$g'_{mn} = G^{-1} \left(\mu'_{mn} \right)$$

9.3.2 Intuitionistic Fuzzy Enhancement Method

The enhancement method is suggested by Chaira [2]. The image is initially fuzzified as $\mu_A(g) = \dfrac{g - g_{min}}{g_{max} - g_{min}}$, where g is the gray value of the image. g_{min} and g_{max} are the minimum and maximum gray values of the image.

From Sugeno's fuzzy complement (described in Chapter 5), the intuitionistic fuzzy membership function is given as:

$$\mu_A^{IFS}(g) = 1 - \frac{1 - \mu_A(g)}{1 + \lambda \cdot \mu_A(g)} = \frac{(1 + \lambda) \cdot \mu_A(g)}{1 + \lambda \cdot \mu_A(g)}.$$

$$(9.2)$$

For computing the nonmembership function of an intuitionistic fuzzy image, λ in Eq. (9.2) is changed to $\lambda + 1$. This is done to obtain a better contrast-enhanced image but the equation should follow the condition $\mu_A^{IFS}(g) \leq 1 - \nu_A^{IFS}(g)$. The modified nonmembership function is given as:

$$
\nu_A^{IFS}(g) = \varphi\left(\mu_A^{IFS}(g)\right) = \frac{1 - \mu_A^{IFS}(g)}{1 + (\lambda + 1) \cdot \mu_A^{IFS}(g)} = \frac{1 - \mu_A(g)}{1 + 3\lambda \cdot \mu_A(g) + \lambda^2 \mu_A(g) + \mu_A(g)}.
$$

$$(9.3)$$

Hesitation degree is computed as: $\pi_A(g) = 1 - \mu_A^{IFS}(g) - \nu_A^{IFS}(g)$.

λ is computed using intuitionistic fuzzy entropy (IFE) which is given as:

$$
IE(A) = \frac{1}{N} \sum_{j=1}^{N} \sum_{i=1}^{N} \pi_A\left(g_{ij}\right) \cdot e^{1 - \pi_A\left(g_{ij}\right)},
$$

$$(9.4)$$

and the optimum value of λ is computed as:

$$
\lambda_{opt} = \max_\lambda\left(IE(A;\lambda)\right).
$$

$$(9.5)$$

With this λ_{opt} value, the membership degrees and the hesitation degrees are computed.

A modified intuitionistic fuzzy image is then formed using fuzzy hedge:

$$
\mu_{new}^{IFS}(g) = \left(\mu_A^{IFS}(g)\right)^{1.25}.
$$

Then, contrast stretching is applied on the intuitionistic image using INT operator using Eq. (9.1) as follows:

$$
\mu_{mn}^{enh} = 2 \cdot \left[\mu_{new}^{IFS}(g)\right]^2 \qquad \text{for } 0 \le \mu_{new}^{IFS}(g) \le 0.5
$$

$$
= 1 - 2 \cdot \left[1 - \mu_{new}^{IFS}(g)\right]^2 \quad \text{for } 0.5 \le \mu_{new}^{IFS}(g) \le 1.
$$

Results on CT-scanned brain image and a knee-patella image are shown. For both the methods, as the image is a medical image, global enhancement is not done, rather the image is divided into windows and the enhancement is carried out for each windows. As the images are real-time images, so the intensities are not uniform; global enhancement is not done.

It is observed that using the fuzzy method, the enhanced image is so bright that the blood vessels and other structures in the left portion of the knee image are not visible clearly. In the intuitionistic fuzzy image, the enhancement or contrast is better and all the portions are visible clearly (Figure 9.1).

9.4 Thresholding

Segmentation is a fundamental building block in image analysis. It is the first stage to analyze an image. It partitions the image into disjoint regions. Thresholding is a type of segmentation that is computationally fast and an inexpensive segmentation technique. It classifies the pixels into two categories – those pixels fall below the threshold and those that fall above the threshold. If the objects in an image are disjoint and their gray levels are clearly distinct from the background,

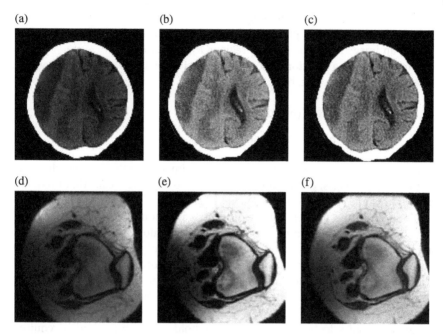

Figure 9.1 (a and d) CT scan brain image, (b and e) fuzzy enhanced image, and (c and f) intuitionistic fuzzy enhanced image (*Source:* Adapted from Refs. [2, 3]).

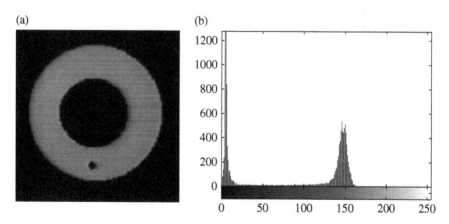

Figure 9.2 (a) Image and (b) image histogram.

thresholding is an appropriate type of segmentation. The gray-level histogram of the picture displays the peaks corresponding to the two gray levels – image object region and background. Figure 9.2 shows an image and its histogram. Histogram of an image is the frequency of occurrences of the gray values in the image. An appropriate threshold, T, is chosen that separates the image object regions from background. A histogram containing a single peak is called unimodal, two peaks is called bimodal, and multiple peaks is called multimodal. A complete segmentation of an image R is a finite set of regions $R_1, R_2, \ldots, R_N$.

Thresholding is a transformation of an input image A into a segmented output image B. If T be the threshold, then

$$b_{ij} = 1 \quad \text{for} \quad a_{ij} \geq T \, (\text{object class})$$

$$= 0 \quad \text{for} \quad a_{ij} \leq T \, (\text{background class})$$

There are various ways one can threshold an image. For the images, which are not real-time images, global thresholding can be done. It means that there is a single threshold throughout the image.

But in real-time images, e.g. medical images, finding a threshold is a very difficult task. As the illumination is not uniform, global threshold cannot be applied. In these images, there may present different objects of different gray levels and for these images, global threshold will not work. In that case, local or window-based threshold is used where the threshold varies throughout the image. Image is divided into subregions and threshold is computed for each region.

For threshold selection, an optimal threshold selection method is required where it uses a criterion function that yields a measure of separation between two regions. A criterion function may be Shannon's entropy, cross entropy, divergence, or any other measure.

Thresholding a medical image where both fuzzy and intuitionistic fuzzy is used is discussed briefly.

9.4.1 Intuitionistic Fuzzy Thresholding Method

This method is proposed by Chaira [4]. Let us take an image A. For a certain threshold t, that separates the object and background regions, the membership function of the object region is written as:

$\mu_A(a_{ij}) = 0.582 \left(e^{1 - |a_{ij} - m_0(t)|} - 1 \right)$, if $a_{ij} \leq t$ for object region, a_{ij} is the pixel value at the $(i, j)^{\text{th}}$ point of image A. This membership function is obtained from Eq. (1.7) using restricted equivalence function, described in Chapter 1.

Likewise, the membership function, $\mu_A(a_{ij})$, of the background region is written as:

$$\mu_A(a_{ij}) = 0.582 \left(e^{1 - |a_{ij} - m_B(t)|} - 1 \right), \quad \text{if} \quad a_{ij} > t \text{ for background} \tag{9.6}$$

with $m_0(t)$, $m_B(t)$ are the average gray levels of the background and object region that are computed as:

$$m_O(t) = \frac{\sum_{g=0}^{t} g \cdot h(g)}{\sum_{g=0}^{t} h(g)}, \quad m_B(t) = \frac{\sum_{g=t+1}^{L-1} g \cdot h(g)}{\sum_{g=t+1}^{L-1} h(g)},$$

g is the gray level that lies between 0 and 255. Here, $L = 256$.

To create an intuitionistic fuzzy image, Sugeno-type intuitionistic fuzzy generator is used to find the nonmembership function, which is written as:

$$\nu_A(a_{ij}) = \frac{1 - \mu_A(a_{ij})}{1 + \lambda \cdot \mu_A(a_{ij})}, \lambda > 0 \tag{9.7}$$

For thresholding an image, the image is initially filtered using a Gaussian filter of size 3 × 3. The filtered image is divided into several windows of size $(1/4)^*$. If smaller window size is selected, the threshold will pick very small particles and that will result in poor extraction and performance of the image.

Membership values and nonmembership values are computed using Eqs. (9.6 and 9.7) for each window. Hesitation degree is also computed using equation: $\pi_A(a_{ij}) = 1 - \mu_A(a_{ij}) - \nu_A(a_{ij})$. The value of λ in Eq. (9.7) is taken as $\lambda = 0.8$. For each threshold gray level, fuzzy divergence between an ideal and actual thresholded image is computed. Ideally, thresholded image (say B) is that image where the object/background regions exactly belong to their respective object/background regions and in that case $\mu_B(b_{ij}) = 1, \nu_B(b_{ij}) = 0$. The divergence measure between the ideal (B) and actual thresholded image (A) is computed as:

$$D_{IFS}(A,B) = \sum_i \sum_j \left[2 - \left(1 - \mu_A(a_{ij}) + \mu_B(b_{ij})\right) \cdot e^{\mu_A(a_{ij}) - \mu_B(b_{ij})} \right.$$
$$- \left(1 - \mu_B(b_{ij}) + \mu_A(a_{ij})\right) \cdot e^{\mu_B(b_{ij}) - \mu_A(a_{ij})} + 2 - \left(1 - \mu_A(a_{ij}) - \pi_A(a_{ij})\right)$$
$$+ \mu_B(b_{ij}) + \pi_B(b_{ij})\right) \cdot e^{\mu_A(a_{ij}) + \pi_A(a_{ij}) - \mu_B(b_{ij}) - \pi_B(b_{ij})} - \left(1 - \mu_B(b_{ij})\right)$$
$$\left. - \pi_B(b_{ij}) + \mu_A(a_{ij}) + \pi_A(a_{ij})\right) \cdot e^{\mu_B(b_{ij}) + \pi_B(b_{ij}) - \mu_A(a_{ij}) - \pi_A(a_{ij})} \right],$$

which reduces to

$$D_{IFS}(A,B) = \sum_i \sum_j \left[2 - \left(2 - \mu_A(a_{ij})\right) \cdot e^{\mu_A(a_{ij}) - 1} - \mu_A(a_{ij}) \cdot e^{1 - \mu_A(a_{ij})} \right.$$
$$+ 2 - \left(2 - \mu_A(a_{ij}) - \pi_A(a_{ij})\right) \cdot e^{\mu_A(a_{ij}) + \pi_A(a_{ij}) - 1} \tag{9.8}$$
$$\left. - \left(\mu_A(a_{ij}) + \pi_A(a_{ij})\right) \cdot e^{1 - \mu_A(a_{ij}) - \pi_A(a_{ij})} \right].$$

For all threshold gray levels, intuitionistic fuzzy divergence (IFD) is computed. The threshold level corresponding to the minimum divergence is selected as an optimum threshold.

This is computed for all the windows. The final threshold is the optimal threshold minus the $(1/4)^{\text{th}}$ of the standard deviation of the window.

$Final\_Th = Th^{opt}_{window} - (1/4)^* S \cdot D_{window}$, where $S \cdot D$ is the standard deviation.

9.4.2 Fuzzy Thresholding Method

The same procedure is followed using fuzzy set where only fuzzy membership function is required. For selecting the optimum threshold, fuzzy divergence is used. Fuzzy divergence between two images A and B is given as:

$$D_{FS}(A,B) = \sum_i \sum_j \left[2 - \left(1 - \mu_A(a_{ij}) + \mu_B(b_{ij})\right) \cdot e^{\mu_A(a_{ij}) - \mu_B(b_{ij})} \right. \tag{9.9}$$
$$\left. - \left(1 - \mu_B(b_{ij}) + \mu_A(a_{ij})\right) \cdot e^{\mu_B(b_{ij}) - \mu_A(a_{ij})} \right].$$

Substituting $\mu_B(b_{ij}) = 1$, for ideally segmented image in Eq. (9.9), divergence may be written as:

$$D_{FS}(A,B) = \sum_i \sum_j \left[2 - \left(2 - \mu_A(a_{ij})\right) \cdot e^{\mu_A(a_{ij}) - 1} - \mu_A(a_{ij}) \cdot e^{1 - \mu_A(a_{ij})} \right].$$
$$\tag{9.10}$$

Divergence is computed for all threshold gray levels and the threshold corresponding to the minimum divergence value is selected as the optimum threshold.

Results on two blood vessel/blood cell images are shown using fuzzy set and intuitionistic fuzzy set theory methods. It is observed that using fuzzy set theory, the blood vessels are not clearly segmented. Using intuitionistic fuzzy set, blood cells/vessels are clearly visible. If the thresholding/segmentation is done properly, physicians can properly detect the abnormal regions and diagnosis can be done better (Figure 9.3).

9.5 Edge Detection

Edge-based segmentation is a type of segmentation that determines the boundaries of the image regions such as organ structures/abnormalities in medical images. It gives information about edge boundaries. Edges simplify the analysis of the images by reducing the data to be processed. Edges are present in many directions and these are detected using some edge-detecting operators. Edge-detecting operators should be very efficient so that they are capable of being tuned to any desired scale. Edge detectors should consider all the edge directions to detect all the edges in images. The most common edge-detection operators are Sobel, Prewitt, Roberts, and Canny's edge detectors. An image may contain noise, so whether the pixel is an edge pixel or noise pixel, it depends on the gray value of the pixel and its surrounding pixels. Smoothing is necessary to remove the noise present in the image and Gaussian smoothing is the most common filter used for smoothing. There are many methods for edge detection but in this section we will discuss fuzzy and intuitionistic fuzzy edge-detection methods.

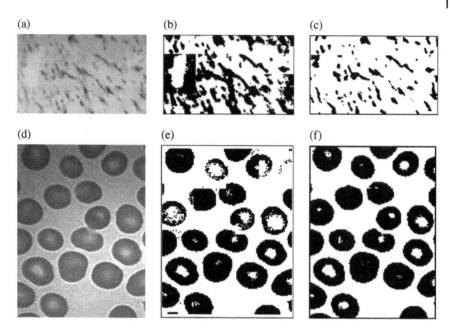

Figure 9.3 (a and d) Original image, (b and e) thresholded image using fuzzy method, and (c and f) thresholded image using intuitionistic fuzzy method (*Source:* Adapted from [4]).

9.5.1 Fuzzy Edge-detection Method

In this method [5], a set of 16 fuzzy templates representing different edge profiles each of size $p \times p$ (here $p = 3$) are used.

$$
\begin{bmatrix} 0 & b & a \\ 0 & b & a \\ 0 & b & a \end{bmatrix}
\begin{bmatrix} a & a & a \\ 0 & 0 & 0 \\ b & b & b \end{bmatrix}
\begin{bmatrix} a & a & b \\ a & b & 0 \\ b & 0 & 0 \end{bmatrix}
\begin{bmatrix} b & b & b \\ 0 & 0 & 0 \\ a & a & a \end{bmatrix}
\begin{bmatrix} b & a & a \\ 0 & b & a \\ 0 & 0 & b \end{bmatrix}
\begin{bmatrix} b & a & 0 \\ b & a & 0 \\ b & a & 0 \end{bmatrix}
\begin{bmatrix} a & 0 & b \\ a & 0 & b \\ a & 0 & b \end{bmatrix}
\begin{bmatrix} 0 & 0 & 0 \\ b & b & b \\ a & a & a \end{bmatrix}
$$

$$
\begin{bmatrix} a & a & a \\ b & b & b \\ 0 & 0 & 0 \end{bmatrix}
\begin{bmatrix} a & b & 0 \\ a & b & 0 \\ a & b & 0 \end{bmatrix}
\begin{bmatrix} 0 & 0 & 0 \\ a & a & a \\ b & b & b \end{bmatrix}
\begin{bmatrix} 0 & a & b \\ 0 & a & b \\ 0 & a & b \end{bmatrix}
\begin{bmatrix} b & b & b \\ a & a & a \\ 0 & 0 & 0 \end{bmatrix}
\begin{bmatrix} b & 0 & a \\ b & 0 & a \\ b & 0 & a \end{bmatrix}
\begin{bmatrix} b & 0 & 0 \\ a & b & 0 \\ a & a & b \end{bmatrix}
\begin{bmatrix} 0 & 0 & b \\ 0 & b & a \\ b & a & a \end{bmatrix}
$$

The templates are the edges, which are considered as images. 'a', 'b', and '0' represent the pixels of the edge templates. The chosen values of $a = 0.4$, $b = 0.9$ are arbitrary. The size of the templates is less than the size of the image. The image is initially normalized. The center of each template is placed at each pixel position (i,j) over the image. Fuzzy divergence measure $Div(i,j)$ at each pixel position (i,j) in the image, where the template is centered, is computed between each of the elements of the image window and the template and minimum value are selected. This is followed for all the 16 fuzzy templates and the maximum value among the 16 divergence values is selected.

$$Div(i,j) = \max_N[\min_r(Div(A,B))] \tag{9.11}$$

where N = number of templates, r = elements in the square template, i.e. $3^2 = 9$. Then fuzzy divergence measure between each of the elements a_{ij} of image A and b_{ij} of image B is computed as:

$$IFD(a_{ij}, b_{ij}) = 2 - \left(1 - \mu_A(a_{ij}) + \mu_B(b_{ij})\right) \cdot e^{\mu_A(a_{ij}) - \mu_B(b_{ij})}$$
$$- \left(1 - \mu_B(b_{ij}) + \mu_A(a_{ij})\right) \cdot e^{\mu_B(b_{ij}) - \mu_A(a_{ij})}$$

In the fuzzy divergence computation, image A represents the chosen window in the test image and the image B is the template. $\mu_A(a_{ij})$ is the membership value of the (i,j)th pixel of the normalized image A and $\mu_B(b_{ij})$ is the value of the template image B. Max–min value from Eq. (9.11) is computed for all pixel positions and a fuzzy divergence matrix is formed. The divergence matrix is thresholded and thinned to obtain an edge-detected image.

9.5.2 Intuitionistic Fuzzy Edge Detection

Edge detection using intuitionistic fuzzy set is the same as that of fuzzy set, the only difference is that the nonmembership and hesitation terms are also considered [6].

In this method, a set of 16 fuzzy templates each of size 3 × 3, representing the edge profiles of different types, are used.

The IFD measure at each pixel position (i,j) in the image, where the template was centered, $IFD(i,j)$, is computed between the image window and the template using the max–min relationship:

$$IFD(i,j) = \max_N[\min_r(IFD(A,B))] \tag{9.12}$$

The IFD between A and B, $IFD(A,B)$, is computed between each of the elements a_{ij} and b_{ij} of image window A and that of template B which is given as:

$$IFD(a_{ij}, b_{ij}) = 2 - \left(1 - \mu_A(a_{ij}) + \mu_B(b_{ij})\right) \cdot e^{\mu_A(a_{ij}) - \mu_B(b_{ij})} - \left(1 - \mu_B(b_{ij})\right)$$
$$+ \mu_A(a_{ij})) \cdot e^{\mu_B(b_{ij}) - \mu_A(a_{ij})} + 2 - \left(1 - \mu_A(a_{ij}) + \mu_B(b_{ij}) - \pi_A(a_{ij})\right)$$
$$+ \pi_B(b_{ij})) \cdot e^{\mu_A(a_{ij}) - \mu_B(b_{ij}) + \pi_A(a_{ij}) - \pi_B(b_{ij})} - \left(1 + \mu_A(a_{ij})\right)$$
$$+ \pi_A(a_{ij}) - \mu_B(b_{ij}) - - \pi_B(b_{ij})) \cdot e^{\mu_B(b_{ij}) - \mu_A(a_{ij}) + \pi_B(b_{ij}) - \pi_A(a_{ij})}$$

For hesitation degree computation, $\pi_A(a_{ij}) = c \cdot (1 - \mu_A(a_{ij}))$, where c is a constant. The value of c should be such that $\mu_A(a_{ij}) + \nu_A(a_{ij}) + \pi_A(a_{ij}) = 1$ holds. $IFD(a_{ij}, b_{ij})$ is the IFD between each element of the template b_{ij} and image window a_{ij}. It is computed for all pixel positions of the image and an IFD matrix, which is of same size as that of image, is formed. This IFD matrix is thresholded and thinned to get an edge-detected image. The edge-detected image is a binary image. Results of two CT scan brain images are shown. For clear visibility of the edges, the binary edge image is superimposed on the original images so that the mapping of the edges on the image structures will be seen clearly. Edge detection in medical images helps the physicians to look at the boundaries of the different abnormal regions of an image. Suppose, in an ultrasound or CT scanned images of tumor or clot/hemorrhage or cyst or calcification, edge detection will clearly detect the region boundaries or in chronic obstructive pulmonary disease (COPD), edge detection will show the reduction in the air passage boundary and other abnormalities. Results of two CT scan brain images are shown in Figure 9.4 using both fuzzy and intuitionistic fuzzy methods. It is observed that in the intuitionistic fuzzy method, edges are detected clearly.

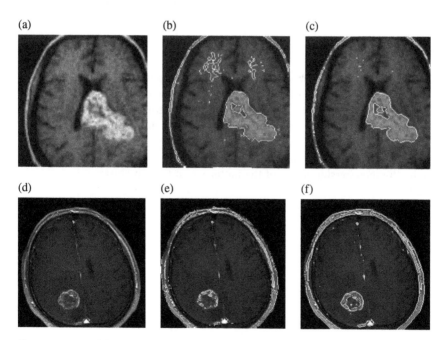

Figure 9.4 (a and d) Original image, (b and e) edge-detected image using fuzzy method, and (c and f) edge-detected image using intuitionistic fuzzy method (*Source:* Adapted from [5, 6]).

9.6 Clustering

Clustering is also a type of segmentation where different regions in an image are grouped together according to the similarity of pixels. There are non-fuzzy K means and fuzzy c means (FCM) and also intuitionistic fuzzy c means clustering (IFCM) methods but the traditional method, i.e. K-means, puts each data into exactly one cluster. But in most of the cases where overlapped datasets are present and in that case some data may be allocated to multiple clusters and K-means clustering may not analyze the dataset clearly. To achieve better clustering, FCM is used. The first fuzzy method to segment the regions of an image is the FCM, introduced by Bezdek et al. [7]. In many real-time images, especially the medical images, there is no sharp boundary between the regions and this problem can be alleviated by assigning a membership value to each data in every cluster such that each data has some kind of similarity in every cluster with membership values lie between 0 and 1. 1 signifies full presence and 0 signifies no presence. It signifies the similarity or closeness of data among the clusters. Fuzzy c means partitions the data in such a way that a data can belong to many clusters and the belongingness is represented in terms of membership grades – partial membership grades among the clusters.

9.6.1 Fuzzy c Means Clustering (FCM)

Given a set of data points, it classifies the set of data points $X = \{x_1, x_2, x_3, \ldots, x_n\}$ into c homogeneous groups or clusters and these clusters are represented as fuzzy sets $(F = F_1, F_2, F_3, \ldots, F_c)$. The algorithm requires a priori knowledge of the number of clusters in the image to be partitioned.

Let $X = \{x_1, x_2, x_3, \ldots, x_n\}$ be a set of data points where each data point x_i, $i = 1, 2, 3, \ldots, n$ is partitioned in c number of clusters. Let $U = [u_{ki}]$ be a $c \times n$ matrix and u_{ki} be the membership grade of the pattern x_i to the kth cluster.

$$U = \begin{bmatrix} c_1 & u_{11} & u_{12} & \cdots & u_{1n} \\ c_2 & u_{21} & & & \cdot \\ & & & & \cdot \\ & \vdots & & & \cdot \\ c_c & u_{c1} & \cdot & \cdot & \cdot & u_{cn} \end{bmatrix}$$

The membership distribution has the following properties:

$$u_{ki} \in [0,1], \quad \forall i,k \quad i = 1, 2, 3, \ldots, n \text{ and } k = \text{no.of classes}$$

$$0 < \sum_{i=1}^{n} u_{ki} < n, \quad 1 \leq k \leq c$$

$$\sum_{k=1}^{c} u_{ki} = 1, \quad 1 \le i \le n.$$

The objective is to obtain a "c" partition by minimizing the criterion function using Lagrangian multiplier method:

$$J_m(U, V : X) = \sum_{i=1}^{n} \sum_{k=1}^{c} u_{ki}^m \|x_i - v_k\|^2 \qquad (9.13)$$

The membership matrix U is initialized as:

$$u_{ki} = \frac{1}{\sum_{j=1}^{c} \left(\frac{\|x_i - v_k\|^2}{\|x_i - v_j\|^2} \right)^{1/(m-1)}}, \quad \forall k = 1, 2, 3, \ldots, c, \quad i = 1, 2, 3, \ldots, n \qquad (9.14)$$

and cluster center is computed as:

$$v_k = \frac{\sum_{i=1}^{n} u_{ki}^m \cdot x_i}{\sum_{i=1}^{n} u_{ki}^m} \qquad (9.15)$$

u_{ki} is the membership of the data x_i to the kth fuzzy cluster with centroid v_k. m is user defined and generally it is taken as 2. $\|.\|$ is a norm and it may be the Euclidean distance or any distance measure that is used to find the similarity between the cluster center and the data points.

This iteration will stop when $\max_{k,i}\left\{ \left| u_{ki}^{t+1} - u_{ki}^{t} \right| \right\} < \varepsilon$, where ε is a tolerance level which lies in between 0 and 1. t and $t + 1$ are the successive iterations.

9.6.2 Intuitionistic Fuzzy Clustering

Chaira [8] suggested an intuitionistic fuzzy clustering algorithm for the first time and in the construction of the algorithm, objective criterion function is modified by incorporating an IFE in the criterion function.

Initially, an intuitionistic fuzzy image is constructed using Yager's intuitionistic fuzzy generator which is given as: $A_{IFS} = \left\{ x, \mu_A(x_i), [1 - \mu_A(x_i)]^{\frac{1}{\alpha}} \right\}$ and the hesitation degree is computed as:

$$\pi_A(x_i) = 1 - \mu_A(x_i) - [1 - \mu_A(x_i)]^{\frac{1}{\alpha}}. \qquad (9.16)$$

A second objective function called IFE given by Chaira [8] is introduced in the conventional clustering algorithm that maximizes the good points in the class. It is given as:

$$J_2 = \sum_{k=1}^{c} \pi_k^* \cdot e^{1 - \pi_k^*},$$

where $\pi_k^* = \frac{1}{N}\sum_{i=1}^{n}\pi_{ki}$, $i \in [1,N]$, π_{ki} is the hesitation degree of the ith element in cluster $'k'$ which is computed using Eq. (9.16).

So, the criterion function becomes:

$$J_2 = \sum_{i=1}^{n}\sum_{k=1}^{c} u_{ki}^{*m}\|x_i - v_k\|^2 + \sum_{k=1}^{c}\pi_k^* \cdot e^{1-\pi_k^*}, m = 2, \tag{9.17}$$

where $\|.\|$ is a norm that is the Euclidean distance (or any other distance measure) between v_k (cluster center) of each region and x_i (points in the pattern). Intuitionistic fuzzy membership values (u_{ki}^*) of ith data in k^{th} class are obtained as:

$$u_{ki}^* = u_{ki} + \pi_{ki}. \tag{9.18}$$

Just like FCM algorithm, the cluster center using intuitionistic fuzzy membership value is given as:

$$v_k^* = \frac{\sum_{i=1}^{n} u_{ki}^{*m} \cdot x_i}{\sum_{i=1}^{n} u_{ki}^{*m}} \tag{9.19}$$

Cluster center is updated and simultaneously the membership matrix is updated.

Like FCM, at each iteration, the cluster center and the membership matrix are updated and the algorithm stops when the updated membership matrix and the previous matrix, i.e.

$\max_{i,k}\left|U_{ki}^{*new} - U_{ki}^{*prev}\right| < \epsilon$, ϵ is a user-defined value and is selected as $\epsilon = 0.03$.

Results of CT scanned brain images are shown using fuzzy c means and intuitionistic FCM algorithm in Figure 9.5.

In the clustering algorithm, three features are considered and these are pixel gray value, pixel mean, and standard deviation. A small square window of size 3 × 3 is moved throughout the image to calculate the mean and the standard deviation. In the intuitionistic fuzzy clustering algorithm, $\alpha \geq 0.5$, in Eq. (9.16) is used.

9.6.3 Kernel Clustering

Kernel clustering can also be done where a kernel function is introduced in the inner product of the clustering function. The inner product in the clustering algorithm is replaced by a nonlinear mapping function called kernel function [9] that transforms the input data to a high-dimensional feature space where the data are expected to be more separable, and may result in better performance by working in the new space. Direct computation in high-dimensional feature space consumes more time and so Mercer kernels are used. Kernel function may be hyper tangent or radial basis or Gaussian. Kernel clustering performs better than FCM.

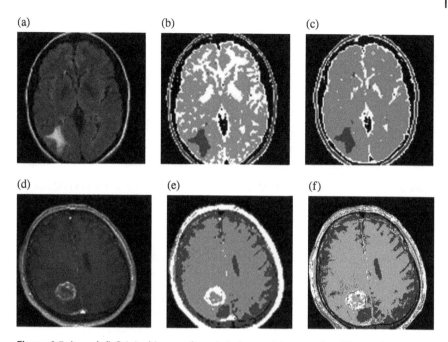

Figure 9.5 (a and d) Original image, (b and e) clustered image using FCM, and (c and f) clustered image using IFCM (*Source:* Adapted from [8]).

A brief note on kernel is given along with an example. As mentioned earlier, kernel function may be Gaussian, hyper tangent, or radial basis. Here, we will show a result using hyper tangent kernel function. While using kernel function, distance function in clustering algorithm in Eq. (9.12) is replaced by a kernel function.

For hyper tangent kernel: $K(x_i, v_k) = 1 - \tanh\left(\dfrac{-\|x_i - v_k\|^2}{\sigma^2}\right)$.

So, from Eq. (9.13)

$$\|\varphi(x_i) - \varphi(v_k)\|^2 = \langle \varphi(x_i), \varphi(x_i) \rangle + \langle \varphi(v_k), \varphi(v_k) \rangle - 2\langle \varphi(x_i), \varphi(v_k) \rangle$$
$$= K(x_i, x_i) + K(v_k, v_k) - 2K(x_i, v_k),$$

where $i = 1, 2, 3, \ldots, n$ and $k = 1, 2, 3, \ldots, c$.
For the hyper tangent kernel, $K(x_i, x_i) = K(v_k, v_k) = 1$,
so, $\|\varphi(x_i) - \varphi(v_k)\|^2 = 2(1 - K(x_i, v_k))$.
The final objective function becomes for IFCM is:

$$J_m(U, V : X) = 2 \sum_{i=1}^{n} \sum_{k=1}^{c} u_{ki}^m (1 - K(x_i, v_k)).$$

(a)   (b)   (c)

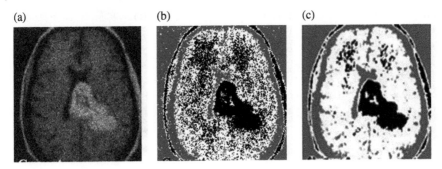

Figure 9.6 (a) Original image, (b) IFCM, and (c) IFCM with hyper tangent kernel (*Source:* Adapted from [10]).

The membership matrix is : $u_{ki} = \dfrac{\left(1-K\left(x_i,v_k\right)\right)^{-\frac{1}{m-1}}}{\sum_j^c \left(1-K\left(x_i,v_j\right)\right)^{-\frac{1}{m-1}}}.$ 　(9.20)

The intuitionistic fuzzy membership matrix is created using Eq. (9.18).

Cluster center, after incorporating intuitionistic property, using hyper tangent kernel is written as:

$$v_k^* = \frac{\sum_{i=1}^{n} u_{ki}^{*m} K(x_i,v_k)\left(1+\tan h\left(\frac{-\|x_i-v_k\|^2}{\sigma^2}\right)\right)x_i}{\sum_{i=1}^{n} u_{ki}^{*m} K(x_i,v_k)\left(1+\tan h\left(\frac{-\|x_i-v_k\|^2}{\sigma^2}\right)\right)}.$$ 　(9.21)

Similar procedure as intuitionistic FCM algorithm is followed. A result on a CT scan brain image is shown [10] in Figure 9.6 and it is observed that on mixing noise to the original image, the algorithm can pick out the hemorrhage region properly.

9.7 Mathematical Morphology

Mathematical morphology is used to retrieve structural properties of an image. Mathematical morphology is a set-theoretic method for extraction of shape from a scene. It interacts with a matching pattern through well-defined local operators such as erosion, dilation, and others to obtain geometric information of different objects in the image. The matching pattern is called structuring element. By varying the size and shape of the structuring element that may be rhombus, circle, square, it is possible to transform an original image to an image using the image structure of certain shape and size. It provides an approach to analyze the geometric characteristics of images and is a tool for extracting different image components that is useful in representation and description of image regions, boundaries, shapes, or skeletons. Initially, mathematical morphology was dealt with binary images and many binary operators were introduced such as erosion, dilation, opening, closing, and skeletonization. Later, it was generalized to gray-scale image and then to fuzzy images. Many operators such as union, intersection, difference, conjunctor–implicator, T-norms, and T-conorms are used in morphological image processing.

If $a(x,y)$ is a gray-scale image and $b(x,y)$ is a flat structuring element, gray-scale dilation is defined as the maximum value of the image within neighborhoods represented by the structuring element, b, i.e. the maximum value of the image in the window outlined by the structuring element b, when the origin of b is (x,y).
It is written as:

$$[a \oplus b](x,y) = \max_{p,q \in b}[A(x-p,y-q)].$$

Similar to gray-scale dilation, gray-scale erosion is defined as the minimum value of the image in the window outlined by the structuring element, b.
It is written as:

$$[a \ominus b](x,y) = \min_{p,q \in b}[a(x+p,y+q)].$$

Dilated gray-scale image with structuring element is computed as the maximum of the sum of the gray levels of $b(x,y)$ in B with each points of $a(x,y)$ [11]:

$$d(x,y) = \max_{p,q \in b}[A(x-p,y-q) + b(x,y)].$$

Eroded gray-scale image is computed as the minimum of the difference of the gray levels of $b(x,y)$ in B with each of $a(x,y)$ [11]:

$$e(x,y) = \min_{p,q \in b}(a(x+p,y+q) - b(x,y)).$$

Fuzzy mathematical morphology – Considering images as fuzzy sets, different definitions of erosion and dilation are given by many authors where A is the image, B is the structuring element, which are defined over space S, at any point x are as follows:

i) De Baets and Kerre [12] and Bloch and Maitre [13]

$$D(A,B) = \sup_{y \in S}\min[A(y),B(y-x)],$$
$$E(A,B) = \inf_{y \in S}\max[A(y),1-B(y-x)].$$

ii) Bloch and Maitre [13]

$$D(A,B) = \sup_{y \in S}[A(y)B(y-x)],$$
$$E(A,B) = \inf_{y \in S}[A(y)B(y-x)+1-B(y-x)].$$

iii) Sinha and Dougherty [14]

$$D(A,B) = \sup_{y \in S} \max[0, A(y) + B(y - x) - 1],$$
$$E(A,B) = \inf_{y \in S} \min[1, 1 + A(y) - B(y - x)].$$

9.7.1 Fuzzy Approach

For fuzzy cases, the image is initially fuzzified and then erosion and dilation are carried out [14]. Erosion removes the structures of certain shapes and sizes and it shrinks the objects whereas dilation dilates the objects and closes the holes and gaps of certain shapes and sizes, using a structuring element. The dilated image with flat structuring element computes the maximum value, so dilation brightens the image. The erosion is opposite or complement of dilation.

From the definition of dilation and erosion by De Baets and Kerre [12]:

$$D(A,B) = \sup_{y \in S} i[B(y - x), A(y)] \tag{9.22}$$

$$E(A,B) = \inf_{y \in S} u[c \cdot B(y - x), A(y)] \tag{9.23}$$

where i is T-norm, u is a T-conorm, and c is a complement. A is an image and B is a structuring element. Structuring element is placed on the image and is moved throughout the image. At each location, dilation/erosion operation is performed.

Using Lukasiewicz T-norm and T-conorm:
T-norm – $T(A,B) = \max(0, A + B - 1)$,
T-conorm – $T^*(A,B) = \min(1, A + B)$.
So, the dilation and erosion in Eqs. (9.22 and 9.23) become:

$$D(A,B) = \sup_{y \in S}[\max(0, A(y) + B(y - x) + 1)], \tag{9.24}$$

$$E(A,B) = \inf_{y \in S}[\min(1, A(y) + c \cdot B(y - x))]$$
$$= \inf_{y \in S}[\min(1, 1 + A(y) - B(y - x))]. \tag{9.25}$$

Initially, the image is fuzzified and then dilation and erosion are performed. If we take the difference of dilation and erosion, we can also get an edge-detected image.

9.7.2 Intuitionistic Fuzzy Approach

As explained earlier, an intuitionistic fuzzy image is created and the nonmembership degree is computed using Sugeno's intuitionistic fuzzy generator [15].

From Sugeno's fuzzy complement or intuitionistic fuzzy generator, intuitionistic fuzzy membership function at (m,n)th point is computed as:

$$\mu'_{mn} = 1 - \frac{1 - \mu_{mn}}{1 + \lambda \cdot \mu_{mn}} = \frac{(1 + \lambda) \cdot \mu_{mn}}{1 + \lambda \cdot \mu_{mn}}.$$

Nonmembership function of an intuitionistic fuzzy image is computed using Sugeno-type negation as:

$$\nu'_{mn} = \varphi\left(\mu'_{mn}\right) = \frac{1 - \mu'_{mn}}{1 + \lambda \cdot \mu'_{mn}} = \frac{1 - \mu_{mn}}{1 + 2\lambda \cdot \mu_{mn} + \lambda^2 \, \mu'_{mn}},$$

where $\mu_{mn} = \dfrac{g_{mn} - g_{\min}}{g_{\max} - g_{\min}}$ is the membership function of the fuzzified image. $g_{\max}$ and $g_{\min}$ are the maximum and minimum gray values of the image. g_{mn} is the gray level at $(m,n)^{\text{th}}$ point.

λ is computed using IFE:

$$IE(A) = \sum_{n=1}^{N} \sum_{m=1}^{M} \pi_A\left(x_{mn}\right) \cdot e^{1 - \pi_A\left(x_{mn}\right)},$$

where $\pi_{mn} = 1 - \nu'_{mn} - \mu'_{mn}$ is the hesitation degree.

The optimum value of λ is $\lambda_{opt} = \max_\lambda \left(IE(A, \lambda)\right)$.

Then, using dilation and erosion in Eqs. (9.24 and 9.25), an edge-detected image is obtained. The structuring element used is:

$$\begin{bmatrix} 0.8 & 0.2 & 0.8 \\ 0.2 & 0.8 & 0.2 \\ 0.8 & 0.2 & 0.8 \end{bmatrix}$$

An example of a CT scanned brain image where both fuzzy and intuitionistic fuzzy edge detection is shown by using dilation and erosion is shown in Figure 9.7.

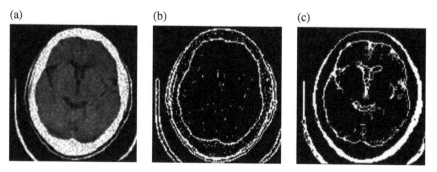

(a)   (b)   (c)

Figure 9.7 (a) Noisy CT scan brain image, (b) edge image using fuzzy approach, and (c) edge image using an intuitionistic fuzzy approach (*Source:* Adapted from [15]).

9.8 Summary

This chapter discusses the application of fuzzy and intuitionistic fuzzy set in image processing. It explains the method to use different types of operators, measures, and other related mathematical operations in image processing. Applications of image enhancement, image thresholding/segmentation, clustering, edge detection, and morphology are discussed along with the results.

References

1 Pal, S.K. and King, R.A. (1981). Image enhancement using smoothing with fuzzy sets. *IEEE Transactions on Systems Man and Cybernetics SMC* 11 (7): 494–501.

2 Chaira, T. (2012). Construction of intuitionistic fuzzy contrast enhanced medical images. *IEEE International Conference on Human Computer Interaction* (27–29 December), IIT Kharagpur.

3 Chaira, T. (2015). *Medical Image Processing: Advanced Fuzzy Set Theoretic Techniques*. Boca Raton, FL: CRC Press.

4 Chaira, T. (2010). Intuitionistic fuzzy segmentation of medical images. *IEEE Transaction of Biomedical Engineering* 57 (6): 1430–1436.

5 Chaira, T. (2003). Image segmentation and color retrieval: a fuzzy and intuitionistic fuzzy set theoretic approach. PhD thesis, IIT Kharagpur, India.

6 Chaira, T. and Ray, A.K. (2007). A new measure using intuitionistic fuzzy set theory and its application to edge detection. *Applied Soft Computing* 8: 919–927.

7 Bezdek, J.C., Hall, L.O., and Clark, L.P. (1993). Review of MR segmentation technique in pattern recognition. *Medical Physics* 10 (20): 33–48.

8 Chaira, T. (2011). A novel intuitionistic fuzzy c means clustering algorithm and its application to medical images. *Applied Soft Computing* 11 (2): 1711–1717.

9 Shawe-Taylor, J. and Cristianini, N. (2004). *Kernel Methods for Pattern Analysis*. Cambridge: Cambridge University Press.

10 Chaira, T. and Panwar, A. (2013). An Atanassov's intuitionistic fuzzy kernel based clustering for medical image segmentation. *International Journal of Computational Intelligence Systems* 7 (2): 360–370.

11 Dong, P. (1997). Implementation of mathematical morphological operations for spatial data processing. *Computers & Geosciences* 23 (1): 103–107.

12 De Baets, B. and Kerre, E. (1993). An Introduction to fuzzy mathematical morphology. *Proceedings of the North America Fuzzy Information Processing Society (NAFIPS'93)* (22–25 August), Allentown, PA.

13 Bloch, I. and Maitre, H. (1995). Fuzzy mathematical morphology: a comparative study. *Pattern Recognition* 28 (9): 1341–1387.

14 Sinha, D. and Dougherty, E.R. (1996). Fuzzy mathematical morphology. *Journal of Visual Communication and Representation* 3 (3): 286–303.

15 Chaira, T. (2015). Fuzzy mathematical morphology using Hamacher T operators and its application to images. *Journal of Intelligence and Fuzzy Systems* 28 (5): 2269–2277.

10

Type-2 Fuzzy Set

10.1 Introduction

The concept of type-2 fuzzy set was introduced by Zadeh [1] as an extension of type 1 or ordinary fuzzy set. Such sets are fuzzy sets whose membership grades are type-1 fuzzy set. Membership grades in type-1 fuzzy set are not precise. A type-2 fuzzy set is a fuzzy set that models the uncertainty in membership values in type-1 fuzzy sets. The membership function of type-2 fuzzy sets provides an additional degree of freedom that makes it possible to model the uncertainty in the membership grades in type-1 fuzzy set. For example, uncertainty of "the membership degree of a young man is 0.9" is on a type-1 fuzzy set. Now, if we say "membership grade of the man in the category 'possibly young' is 0.9," then it is a type-2 fuzzy set. So, the membership grades of type-1 fuzzy set are crisp while the membership grades of type-2 fuzzy set are fuzzy.

Type-2 fuzzy sets are useful in those situations, where it is difficult to determine an exact membership function for a fuzzy set. According to Mendel [2, 3], type-2 fuzzy set provides us with more degrees of freedom, so type-2 fuzzy sets have the potential to outperform type-1 fuzzy sets, especially when we are in uncertain environments.

There are different sources of uncertainties in type-1 fuzzy sets which are (i) inaccurate measurements, (ii) disagreement of the membership values with the accurate membership values of the data, and (iii) uncertainty in the location, or shape or other parameters. According to Mendel and Bob John [3], type-1 fuzzy sets cannot directly model such uncertainties because their membership functions are totally crisp. But, type-2 fuzzy sets are able to model such uncertainties as their membership functions are fuzzy.

Fuzzy Set and Its Extension: The Intuitionistic Fuzzy Set, First Edition. Tamalika Chaira.
© 2019 John Wiley & Sons, Inc. Published 2019 by John Wiley & Sons, Inc.

10.2 Type-2 Fuzzy Set

Consider a blurry type-1 fuzzy membership function that means that the membership function of type-1 fuzzy set, A, is blurred and all of its points are shifted either left or right of the triangle. The shifting value is not necessary to be equal. Consider a point $x = x'$ in Figure 10.1. In the figure, the inner triangle is the type-1 fuzzy membership function and when it is blurred, the region is shown by two hyphenated nonlinear lines filled with small triangles. It is seen that at the specific value of $x = x'$, there is a single membership value, say u' in type-1 fuzzy membership function, but in the blurry type-1 membership function, there is no longer a single membership value for the membership function u', rather, assumes the values whenever x' intersects the blurry region. The values need not be the same rather they take different values or amplitudes to all the points. So, there will a third dimension that will show the amplitude of the points for all $x \in X$. That is, corresponding to each primary membership, there is a secondary membership which also lies in [0,1] that defines the possibilities for the primary membership as shown in Figure 10.1.

A fuzzy set A in a set X is characterized by a membership function, $\mu_A(x)$, that lies in the interval [0,1] which is denoted as:

$$\mu_A : X \rightarrow [0,1],$$

where μ_A is the value of membership of element x in A.

Fuzzy set is represented as:

$$A = \frac{\mu_A(x_1)}{x_1} + \frac{\mu_A(x_2)}{x_2} + \frac{\mu_A(x_3)}{x_3} + \cdots + \frac{\mu_A(x_n)}{x_n}.$$

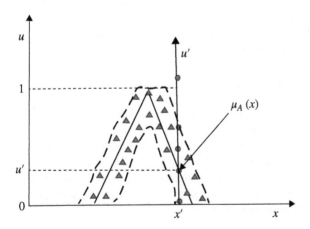

Figure 10.1 Blurred (region filled with triangles) type-1 fuzzy membership function showing the discretization at $x = x'$. (*Source:* Adapted from [3].)

For representing type-2 fuzzy set, we consider a type-2 fuzzy set, $\tilde{A}$, whose membership grade of x in $\tilde{A}$ is $\mu_{\tilde{A}}(x)$. The elements of $\mu_{\tilde{A}}(x)$ are called primary memberships and the memberships of the primary memberships in $\mu_{\tilde{A}}(x)$ are called secondary memberships of x in $\tilde{A}$.

Here, we present definitions of type-2 fuzzy set by different authors:

i) *Definition* 1 by Mendel and John [3] – Type-2 fuzzy set, characterized by a fuzzy membership function, $\mu_{\tilde{A}}(x,u)$, $0 \le \mu_{\tilde{A}}(x,u) \le 1$, is represented as:

$$\tilde{A} = \left\{ (x,u), \mu_{\tilde{A}}(x,u) \right\} \forall x \in X, u \in J_x \subseteq [0,1].$$

Type-2 fuzzy set $\tilde{A}$ can also be represented as:

$$\tilde{A} = \int_{x \in X} \int_{u \in J_x} \frac{\mu_{\tilde{A}}(x,u)}{(x,u)}, J_x \subseteq [0,1]. \tag{10.1}$$

For discrete cases, integration is replaced by summation. $\iint$ represents the union over all x and u. $\mu_{\tilde{A}}(x,u)$ is the secondary membership grade of x in $\tilde{A}$. J_x is the primary membership of x which is the domain of secondary membership function, and u is the element of primary membership J_x.

ii) *Definition* 2 by Mendel and John [3] – At each value of x, say $x = x'$, a 2-D plane is visualized, whose axes are u and $\mu_{\tilde{A}}(x',u)$, which is called a vertical slice of $\mu_{\tilde{A}}(x,u)$ as shown in Figure 10.2. Vertical slice of $\mu_{\tilde{A}}(x,u)$ is a secondary membership function, which is represented as:

$$\mu_{\tilde{A}}(x = x',u) = \mu_{\tilde{A}}(x') = \int_{u \in J_{x'}} \frac{p_{x'}(u)}{u}, \text{for } x \in X, u \in J_{x'} \subseteq [0,1],$$

in which $0 \le p_{x'}(u) \le 1$. $p_{x'}(u)$ is the secondary membership grade, which is the amplitude of the secondary membership function.

As $\forall x' \in X$, the prime notation on $\mu_{\tilde{A}}(x')$ is dropped for simplicity and is written as $\mu_{\tilde{A}}(x)$ to signify the secondary membership function throughout the chapter.

The possibilities for primary membership are represented as (for discrete cases):

$$\mu_{\tilde{A}}(x) = \frac{p_x(u_1)}{u_1} + \frac{p_x(u_2)}{u_2} + \frac{p_x(u_3)}{u_3} + \cdots + \frac{p_x(u_n)}{u_n} \tag{10.2}$$

where $\mu_{\tilde{A}} : X \rightarrow [0,1]^{[0,1]}$, and $\mu_{\tilde{A}}(x)$ is the fuzzy grade and being a fuzzy set in the interval [0,1], $u \in J_x$. p_x is the membership function of fuzzy grade $\mu_{\tilde{A}}(x)$ and is a fuzzy set in the unit interval [0,1] or $p_x(u)$ represents the membership grades of u_i in J_x or the secondary memberships (grades) of x.

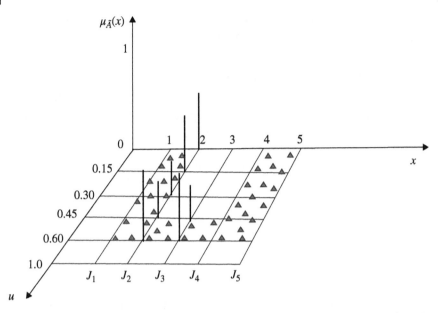

Figure 10.2 (a) Type-2 membership function showing the vertical slice at $x = 2$ and $x = 3$ (triangular spotted region is the blurred region). (*Source:* Adapted from [3].)

This can be explained using Figure 10.2. Let us assume $X = \{1,2,3,4,5\}$, $U = \{0.0,0.15,0.30,0.45,0.60\}$, x and u are discrete. There are five vertical slices associated with type-2 membership function. Vertical slices at $x = 2$ and $x = 3$ are shown in Figure 10.2. It is observed that the primary memberships are: $J_1 = J_2 = J_4 = J_5 = \{0,0.15,0.3,0.45,0.6\}$ and $J_3 = \{0.45,0.6\}$.

At $x = 2$, the secondary membership function is:

$$\mu_{\tilde{A}}(2) = \frac{0.4}{0} + \frac{0.3}{0.15} + \frac{0.2}{0.30} + \frac{0.2}{0.45} + \frac{0.5}{0.60},$$

where 0.4, 0.3, 0.2, 0.2, and 0.5 are the amplitudes or possibilities of the points 0, 0.15, 0.3, 0.45, and 0.6, respectively. Likewise, at $x = 3$, $\mu_{\tilde{A}}(3) = \frac{0.3}{0.45} + \frac{0.5}{0.60}$.

It is shown in Figure 10.2.

If we consider a type-2 fuzzy set:

$$\tilde{A} = \frac{(0.3/0.1)}{x_1} + \frac{(0.2/0.4)}{x_1} + \frac{(0.5/0.7)}{x_1} + \frac{(0.4/0.6)}{x_2} + \frac{(0.2/0.3)}{x_2},$$

its vertical slice representation can also be given as in Figure 10.3.

When all the uncertainties disappear, then type-2 membership function reduces to type-1 membership function.

Figure 10.3 Vertical slice of $\tilde{A}$. (*Source:* Adapted from [3].)

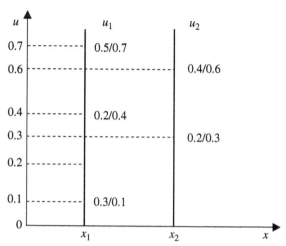

A fuzzy grade is called convex if

$$p_x(u_2) \geq p_x(u_1) \wedge p_x(u_3),$$

where $u_1 \leq u_2 \leq u_3$ and $u_1, u_2, u_3 \in [0,1]$.
A fuzzy grade is called normal if

$$\bigvee_u p_x(u) = 1.$$

Otherwise, it is subnormal.
If a fuzzy grade is normal and convex, then the fuzzy grade is called normal convex fuzzy grade.
α level set of fuzzy grade $\mu_{\tilde{A}}$ is a set where $p_x(u) \geq \alpha$. It is written as:

$$\mu_{\tilde{A}}{}^{\alpha} = \{u \mid p_x(u) \geq \alpha\}, 0 < \alpha \leq 1$$

10.3 Operations on Type-2 Fuzzy Set

Mizumoto and Tanaka [4] studied set theoretic operations, properties of membership grades of type-2 fuzzy sets, and also the operations of algebraic product and algebraic sum of type-2 fuzzy sets [5–7]. Operations of type-2 fuzzy sets are explained using Zadeh's extension principle.

Just like in ordinary fuzzy set, if $\{x_1, x_2, x_3, \ldots, x_n\}$ are the elements and $\mu_A(x_1)$, $\mu_A(x_2)\ldots\mu_A(x_n)$ are the membership grades, then the fuzzy set A is represented as:

$$A = \frac{\mu_A(x_1)}{x_1} + \frac{\mu_A(x_2)}{x_2} + \frac{\mu_A(x_3)}{x_3} + \cdots + \frac{\mu_A(x_n)}{x_n}$$

$$= \sum_i \frac{\mu_A(x_i)}{x_i}.$$

Likewise, for type-2 fuzzy set, let $\tilde{A}$ and $\tilde{B}$ be the two type-2 fuzzy sets and $\mu_{\tilde{A}}(x)$ and $\mu_{\tilde{B}}(x)$ be two fuzzy grades of these two fuzzy sets. Then, the type-2 fuzzy sets $\tilde{A}$ and $\tilde{B}$ are represented as:

$$\mu_{\tilde{A}}(x) = \frac{p_x(u_1)}{u_1} + \frac{p_x(u_2)}{u_2} + \frac{p_x(u_3)}{u_3} + \cdots + \frac{p_x(u_m)}{u_m}$$

$$= \sum_{i=1}^{m} \frac{p_x(u_i)}{u_i}, u_i \in J_x$$

$$\mu_{\tilde{B}}(x) = \frac{q_x(w_1)}{w_1} + \frac{q_x(w_2)}{w_2} + \frac{q_x(w_3)}{w_3} + \cdots + \frac{q_x(w_n)}{w_n}$$

$$= \sum_{j=1}^{m} \frac{q_x(w_i)}{w}, w_i \in J_x,$$

where $\mu_{\tilde{A}}(x)$, $\mu_{\tilde{B}}(x)$ are the fuzzy grades and $p_x(u_i)$, $q_x(w_j) \in [0,1]$ are the membership functions of the fuzzy grades $\mu_{\tilde{A}}(x)$, $\mu_{\tilde{B}}(x)$ for u_i and w_j, respectively, and u_i and w_j are the elements of primary membership J_x.

The following operations holds for type-2 fuzzy set [4]:

If $\mu_{\tilde{A}}$ and $\mu_{\tilde{B}}$ are two fuzzy grades, then

i) Union:

$$\tilde{A} \cup \tilde{B} \rightarrow \mu_{\tilde{A} \cup \tilde{B}}(x) = \mu_{\tilde{A}}(x) \sqcup \mu_{\tilde{B}}(x)$$

$$= \sum_{i=1}^{m} \left(\frac{p_x(u_i)}{u_i}\right) \sqcup \sum_{j=1}^{n} \left(\frac{q_x(w_j)}{w_j}\right) \qquad (10.3)$$

$$= \sum_{i,j} \left(\frac{p_x(u_i) \wedge q_x(w_j)}{u_i \vee w_j}\right)$$

ii) Intersection:

$$\tilde{A} \cap \tilde{B} \rightarrow \mu_{\tilde{A} \cap \tilde{B}}(x) = \mu_{\tilde{A}}(x) \sqcap \mu_{\tilde{B}}(x)$$

$$= \sum_{i=1}^{m} \left(\frac{p_x(u_i)}{u_i}\right) \sqcap \sum_{j=1}^{n} \left(\frac{q_x(w_j)}{w_j}\right) \qquad (10.4)$$

$$= \sum_{i,j} \left(\frac{p_x(u_i) \wedge q_x(w_j)}{u_i \wedge w_j}\right)$$

iii) Complement:

$$\tilde{\tilde{A}} \rightarrow \mu_{\tilde{A}}(x) = \sum_i \frac{p_x(u_i)}{(1-u_i)} \tag{10.5}$$

Operation $\sqcup$ is the join operation and $\sqcap$ is the meet operation.

iv) Algebraic product:

$$\tilde{A}\tilde{B} = \mu_{\tilde{A}}(x) \cdot \mu_{\tilde{B}}(x) = \sum_{i,j} \left(\frac{p_x(u_i) \wedge q_x(w_j)}{u_i w_j} \right) \tag{10.6}$$

v) Algebraic sum:

$$\tilde{A} \dotplus \tilde{B} = \mu_{\tilde{A}}(x) + \mu_{\tilde{B}}(x) = \sum_{i,j} \left(\frac{p_x(u_i) \wedge q_x(w_j)}{u_i + w_j - u_i w_j} \right) \tag{10.7}$$

Example 1 Let $J = \{0, 0.1, 0.2, 0.3, 0.4, 0.5, 0.6, 0.7, 0.8, 0.9, 1\}$ and the fuzzy grades be

$$\mu_{\tilde{A}}(x) = \frac{0.5}{0} + \frac{0.8}{0.2} + \frac{1}{0.3} + \frac{0.6}{0.5}$$

$$\mu_{\tilde{B}}(x) = \frac{0.8}{0} + \frac{1}{0.2} + \frac{0.7}{0.3} + \frac{0.2}{0.5}$$

Union:

$$\mu_{\tilde{A}}(x) \sqcup \mu_{\tilde{B}}(x) = \left(\frac{0.5}{0} + \frac{0.8}{0.2} + \frac{1}{0.3} + \frac{0.6}{0.5} \right) \sqcup \left(\frac{0.8}{0} + \frac{1}{0.2} + \frac{0.7}{0.3} + \frac{0.2}{0.5} \right)$$

$$= \frac{0.5 \wedge 0.8}{0 \vee 0} + \frac{0.5 \wedge 1}{0 \vee 0.2} + \frac{0.5 \wedge 0.7}{0 \vee 0.3} + \frac{0.5 \wedge 0.2}{0 \vee 0.5} + \frac{0.8 \wedge 0.8}{0.2 \vee 0} + \frac{0.8 \wedge 1}{0.2 \vee 0.2}$$

$$+ \frac{0.8 \wedge 0.7}{0.2 \vee 0.3} + \frac{0.8 \wedge 0.2}{0.2 \vee 0.5} + \frac{1 \wedge 0.8}{0.3 \vee 0} + \frac{1 \wedge 1}{0.3 \vee 0.2} + \frac{1 \wedge 0.7}{0.3 \vee 0.3}$$

$$+ \frac{1 \wedge 0.2}{0.3 \vee 0.5} + \frac{0.6 \wedge 0.8}{0.5 \vee 0} + \frac{0.6 \wedge 1}{0.5 \vee 0.2} + \frac{0.6 \wedge 0.7}{0.5 \vee 0.3} + \frac{0.6 \wedge 0.2}{0.5 \vee 0.5}$$

$$= \frac{0.5}{0} + \frac{0.5}{0.2} + \frac{0.5}{0.3} + \frac{0.2}{0.5} + \frac{0.8}{0.2} + \frac{0.8}{0.2} + \frac{0.7}{0.3} + \frac{0.2}{0.5} + \frac{0.8}{0.3}$$

$$+ \frac{1}{0.3} + \frac{0.7}{0.3} + \frac{0.2}{0.5} + \frac{0.6}{0.5} + \frac{0.6}{0.5} + \frac{0.6}{0.5} + \frac{0.2}{0.5}$$

$$= \frac{0.5}{0} + \frac{\max(0.5, 0.8, 0.8)}{0.2} + \frac{\max(0.5, 0.7, 0.8, 1, 0.7)}{0.3}$$

$$+ \frac{\max(0.2, 0.2, 0.2, 0.6, 0.6, 0.6, 0.2)}{0.5} \text{(using extension principle)}$$

$$= \frac{0.5}{0} + \frac{0.8}{0.2} + \frac{1}{0.3} + \frac{0.6}{0.5}$$

Intersection:

$$\mu_{\tilde{A}}(x) \sqcap \mu_{\tilde{B}}(x) = \left(\frac{0.5}{0} + \frac{0.8}{0.2} + \frac{1}{0.3} + \frac{0.6}{0.5}\right) \sqcap \left(\frac{0.8}{0} + \frac{1}{0.2} + \frac{0.7}{0.3} + \frac{0.2}{0.5}\right)$$

$$= \frac{0.5 \wedge 0.8}{0 \wedge 0} + \frac{0.5 \wedge 1}{0 \wedge 0.2} + \frac{0.5 \wedge 0.7}{0 \wedge 0.3} + \frac{0.5 \wedge 0.2}{0 \wedge 0.5} + \frac{0.8 \wedge 0.8}{0.2 \wedge 0} + \frac{0.8 \wedge 1}{0.2 \wedge 0.2}$$

$$+ \frac{0.8 \wedge 0.7}{0.2 \wedge 0.3} + \frac{0.8 \wedge 0.2}{0.2 \wedge 0.5} + \frac{1 \wedge 0.8}{0.3 \wedge 0} + \frac{1 \wedge 1}{0.3 \wedge 0.2} + \frac{1 \wedge 0.7}{0.3 \wedge 0.3}$$

$$+ \frac{1 \wedge 0.2}{0.3 \wedge 0.5} + \frac{0.6 \wedge 0.8}{0.5 \wedge 0} + \frac{0.6 \wedge 1}{0.5 \wedge 0.2} + \frac{0.6 \wedge 0.7}{0.5 \wedge 0.3} + \frac{0.6 \wedge 0.2}{0.5 \wedge 0.5}$$

$$= \frac{0.5}{0} + \frac{0.5}{0} + \frac{0.5}{0} + \frac{0.2}{0} + \frac{0.8}{0} + \frac{0.8}{0.2} + \frac{0.7}{0.2} + \frac{0.2}{0.2} + \frac{0.8}{0}$$

$$+ \frac{1}{0.2} + \frac{0.7}{0.3} + \frac{0.2}{0.3} + \frac{0.6}{0} + \frac{0.6}{0.2} + \frac{0.6}{0.3} + \frac{0.2}{0.5}$$

$$= \frac{\max(0.5, 0.5, 0.5, 0.2, 0.8, 0.8, 0.6)}{0} + \frac{\max(0.8, 0.7, 0.2, 1, 0.6)}{0.2}$$

$$+ \frac{\max(0.7, 0.2, 0.6)}{0.3} + \frac{0.2}{0.5} \text{ (using extension principle)}$$

$$= \frac{0.8}{0} + \frac{1}{0.2} + \frac{0.7}{0.3} + \frac{0.2}{0.5}$$

Complement: $\mu_{\bar{\tilde{A}}}(x) = \dfrac{0.5}{1} + \dfrac{0.8}{0.8} + \dfrac{1}{0.7} + \dfrac{0.3}{0.5}$

Algebraic product:

$$\tilde{A}\tilde{B} = \mu_{\tilde{A}}(x) \cdot \mu_{\tilde{B}}(x) = \left(\frac{0.5}{0} + \frac{0.8}{0.2} + \frac{1}{0.3} + \frac{0.6}{0.5}\right) \cdot \left(\frac{0.8}{0} + \frac{1}{0.2} + \frac{0.7}{0.3} + \frac{0.2}{0.5}\right)$$

$$= \sum_{i,j} \left(\frac{p(u_i) \wedge q(w_j)}{u_i w_j}\right)$$

$$= \frac{0.5 \wedge 0.8}{0 \times 0} + \frac{0.5 \wedge 1}{0 \times 0.2} + \frac{0.5 \wedge 0.7}{0 \times 0.3} + \frac{0.5 \wedge 0.2}{0 \times 0.5} + \frac{0.8 \wedge 0.8}{0.2 \times 0}$$

$$+ \frac{0.8 \wedge 1}{0.2 \times 0.2} + \frac{0.8 \wedge 0.7}{0.2 \times 0.3} + \frac{0.8 \wedge 0.2}{0.2 \times 0.5} + \frac{1 \wedge 0.8}{0.3 \times 0} + \frac{1 \wedge 1}{0.3 \times 0.2}$$

$$+ \frac{1 \wedge 0.7}{0.3 \times 0.3} + \frac{1 \wedge 0.2}{0.3 \times 0.5} + \frac{0.6 \wedge 0.8}{0.5 \times 0} + \frac{0.6 \wedge 1}{0.5 \times 0.2}$$

$$+ \frac{0.6 \wedge 0.7}{0.5 \times 0.3} + \frac{0.6 \wedge 0.2}{0.5 \times 0.5}$$

$$= \frac{0.5}{0} + \frac{0.5}{0} + \frac{0.5}{0} + \frac{0.2}{0} + \frac{0.8}{0} + \frac{0.8}{0.04} + \frac{0.7}{0.06} + \frac{0.2}{0.1} + \frac{0.8}{0}$$

$$+ \frac{1}{0.06} + \frac{0.7}{0.09} + \frac{0.2}{0.15} + \frac{0.6}{0} + \frac{0.6}{0.1} + \frac{0.6}{0.15} + \frac{0.2}{0.25}$$

$$= \frac{0.8}{0} + \frac{0.8}{0.04} + \frac{1}{0.06} + \frac{0.6}{0.1} + \frac{0.7}{0.09} + \frac{0.6}{0.15} + \frac{0.2}{0.25}.$$

In a similar way, algebraic sum is also done.

10.4 Inclusion Measure and Similarity Measure

Similar to fuzzy sets, similarity measures between type-2 fuzzy sets are also defined. There are few authors who suggested similarity and inclusion measures between type-2 fuzzy sets. Let X be a universe of discourse and let $\mu_{\tilde{A}}(x,u) : (x,u) \rightarrow [0,1]$ $\forall x \in X$, $u \in J_x$ be the type-2 fuzzy membership function of type-2 fuzzy set $\tilde{A}$.

Let $\tilde{A}$ and $\tilde{B}$ be two type-2 fuzzy sets and their secondary membership functions be $g_X(x,u) = \mu_{\tilde{A}}(x,u)$ and $h_X(x,u) = \mu_{\tilde{B}}(x,u)$, respectively. According to Mizumoto and Tanaka [6], if $\tilde{A} \subset \tilde{B}$, then $\mu_{\tilde{A}}(x) \le \mu_{\tilde{B}}(x)$.

For any $\tilde{A}, \tilde{B} \in \Phi(x)$, if $\tilde{A} \subseteq \tilde{B}$, then $0 \le g_X(u) \le h_X(u) \le 1 \, \forall x \in X$, $u \in J_x$.

Inclusion measure – Just like fuzzy set, inclusion measure indicates the degree to which a fuzzy set $\tilde{A}$ is contained in another fuzzy set $\tilde{B}$. A real function I: $\Phi(x) \times \Phi(x) \rightarrow [0,1]$ is called the inclusion measure if it satisfies the following properties [6]:

i) $I(\tilde{A}, \tilde{A}) = 1$
ii) $\tilde{A} \subseteq \tilde{B} \Longleftrightarrow I(\tilde{A}, \tilde{A}) = 1$
iii) For any $\tilde{A}, \tilde{B}, \tilde{C} \in \Phi(x)$, if $\tilde{A} \subseteq \tilde{B} \subseteq \tilde{C}$, then $I(\tilde{C}, \tilde{A}) \le I(\tilde{B}, \tilde{A}), I(\tilde{C}, \tilde{A})$ $\le I(\tilde{C}, \tilde{B})$.

$$I(\tilde{A}, \tilde{B}) = \frac{1}{\int_{x \in X} dx} \int_{x \in X} \frac{\int_{u \in J_x} \{\min(u \cdot g_X(u), u \cdot h_X(u))\} du}{\int_{u \in J_x} \{u \cdot g_X(u)\} du} dx \qquad (10.8)$$

For discrete cases, integral is replaced by summation.
If $\tilde{B} = \tilde{A}$, then

$$I(\tilde{A}, \tilde{A}) = \frac{1}{\int_{x \in X} dx} \int_{x \in X} \frac{\int_{u \in J_x} \{\min(u \cdot g_X(u), u \cdot g_X(u))\} du}{\int_{u \in J_x} \{u \cdot g_X(u)\} du} dx$$

$$= \frac{1}{\int_{x \in X} dx} \int_{x \in X} dx = 1.$$

Again, if $\tilde{A} \subseteq \tilde{B}$, then $0 \le g_X(u) \le h_X(u) \le 1$, the term $\{\min(u \cdot g_X(u), u \cdot h_X(u))\}$ becomes $\{u \cdot g_X(u)\}$, then the inclusion measure,

$$I(\tilde{A}, \tilde{B}) = \frac{1}{\int_{x \in X} dx} \int_{x \in X} \frac{\int_{u \in J_x} \{\min(u \cdot g_X(u), u \cdot h_X(u))\} du}{\int_{u \in J_x} \{u \cdot g_X(u)\} du} dx$$

$$= \frac{1}{\int_{x \in X} dx} \int_{x \in X} \frac{\int_{u \in J_x} \{u \cdot g_X(u)\} du}{\int_{u \in J_x} \{u \cdot g_X(u)\} du} dx$$

$$= 1.$$

10.4.1 Similarity Measure

Just like fuzzy set, similarity measure indicates the degree to which a type-2 fuzzy set $\tilde{A}$ is close or similar to another type-2 fuzzy set $\tilde{B}$. A real function $S : \Phi(x) \times \Phi(x) \to [0,1]$ is called a similarity measure if it satisfies the following properties [8]:

i) $S(\tilde{A}, \tilde{B}) = S(\tilde{B}, \tilde{A})$,

ii) $S(\tilde{C}, \tilde{C}^c) = 0$, $\forall C \in P(X)$, $P(X)$ is the power set of X, $\tilde{C}^c$ is the complement of $\tilde{C}$,

iii) $S(\tilde{D}, \tilde{D}) = \max_{A,B \in \Phi(x)} S(\tilde{A}, \tilde{B})$, $\forall \tilde{D} \in \Phi(x)$,

iv) For any $\tilde{A}, \tilde{B}, \tilde{C} \in \Phi(x)$, if $\tilde{A} \subseteq \tilde{B} \subseteq \tilde{C}$, then $S(\tilde{A}, \tilde{B}) \geq S(\tilde{A}, \tilde{C})$ and $S(\tilde{B}, \tilde{C})$ $\geq S(\tilde{A}, \tilde{C})$.

Similarity measure between two fuzzy sets $\tilde{A}$ and $\tilde{B}$ is computed as:

$$S(\tilde{A}, \tilde{B}) = \frac{1}{\int_{x \in X} dx} \int_{x \in X} \frac{\int_{u \in J_x} \{\min\{u \cdot g_X(u), u \cdot h_X(u)\} du}{\int_{u \in J_x} \{\max\{u \cdot g_X(u), u \cdot h_X(u)\} du} dx.$$

If $\tilde{A} = \tilde{B}$, then

$$S(\tilde{A}, \tilde{A}) = \frac{1}{\int_{x \in X} dx} \int_{x \in X} \frac{\int_{u \in J_x} \{u \cdot g_X(u)\} du}{\int_{u \in J_x} \{u \cdot g_X(u)\} du} dx \tag{10.9}$$

$$= 1.$$

This implies that for two similar sets, similarity value is 1.
Singh [9] suggested another similarity measure as:

$$S(\tilde{A}, \tilde{B}) = \frac{1}{\int_{x \in X} dx} \int_{x \in X} \left\{ \frac{\int_{u \in J_x} \{u \cdot g_X(u) \cdot u \cdot h_X(u)\} du}{\sqrt{\int_{u \in J_x} \{u \cdot g_X(u)\}^2 du} \cdot \sqrt{\int_{u \in J_x} \{u \cdot h_X(u)\}^2 du}} dx \right\}. \tag{10.10}$$

If $\tilde{A} = \tilde{B}$, then

$$S(\tilde{A}, \tilde{A}) = \frac{1}{\int_{x \in X} dx} \int_{x \in X} \frac{\int_{u \in J_x} \{u \cdot g_X(u)\}^2 du}{\int_{u \in J_x} \{u \cdot g_X(u)\}^2 du} dx = 1.$$

Similarity measure can also be computed from inclusion measure [7] as:

$$S(\tilde{A}, \tilde{B}) = \min\left(I(\tilde{A}, \tilde{B}), I(\tilde{B}, \tilde{A})\right) = S(\tilde{B}, \tilde{A}).$$

Example 2 Consider two patterns denoted using two type-2 fuzzy sets: $\tilde{A} = \{(x_i,\mu_{\tilde{A}}(x_i)) = \{x_1\,x_2,x_3\}\}$, $\tilde{B} = \{(x_i,\mu_B(x_i)) = \{x_1,x_2,x_3\}\}$, where Pattern A:

$$\mu_{\tilde{A}}(x_1) = \{(0.2,0.07),(0.3,0.1),(0.5,0.3),(0.8,0.5)\},$$

$$\mu_{\tilde{A}}(x_2) = \{(0.1,0.3),(0.3,0.6),(0.6,1.0),(0.7,0.7)\},$$

$$\mu_{\tilde{A}}(x_3) = \{(0.2,0.1),(0.4,0.4),(0.5,0.7),(0.9,0.3)\}.$$

Pattern B:

$$\mu_{\tilde{B}}(x_1) = \{(0.2,0.05),(0.3,0.2),(0.5,0.4),(0.8,0.7)\},$$

$$\mu_{\tilde{B}}(x_2) = \{(0.1,0.2),(0.3,0.5),(0.6,0.8),(0.7,0.5)\},$$

$$\mu_{\tilde{B}}(x_3) = \{(0.2,0.08),(0.4,0.3),(0.5,0.6),(0.9,0.2)\}.$$

Consider an unknown pattern, $\tilde{C}$, with

$$\mu_{\tilde{C}}(x_1) = \{(0.2,0.2),(0.3,0.4),(0.5,0.5),(0.8,0.7)\},$$

$$\mu_{\tilde{C}}(x_2) = \{(0.1,0.1),(0.3,0.3),(0.6,0.5),(0.7,0.3)\},$$

$$\mu_{\tilde{C}}(x_3) = \{(0.2,0.1),(0.4,0.2),(0.5,0.5),(0.9,0.6)\}.$$

Find a pattern, $\tilde{A}$ or $\tilde{B}$, that matches the unknown pattern $\tilde{C}$.

Solution

From the definition of similarity measure, we know

$$S(\tilde{A},\tilde{C}) = \frac{1}{3}\left\{ \begin{array}{l} \frac{\min(0.2\times0.07,\ 0.2\times0.2)+\min(0.3\times0.1,0.3\times0.4)+\min(0.5\times0.3,0.5\times0.5)+\min(0.8\times0.5,0.8\times0.7)}{\max(0.2\times0.07,\ 0.2\times0.2)+\max(0.3\times0.1,0.3\times0.4)+\max(0.5\times0.3,0.5\times0.5)+\max(0.8\times0.5,0.8\times0.7)} + \\[6pt] \frac{\min(0.1\times0.3,0.1\times0.1)+\min(0.3\times0.6,0.3\times0.3)+\min(0.6\times1.0,0.6\times0.5)+\min(0.7\times0.7,0.7\times0.3)}{\max(0.2\times0.3,0.1\times0.1)+\max(0.3\times0.6,0.3\times0.3)+\max(0.6\times1.0,0.6\times0.5)+\max(0.7\times0.7,0.7\times0.3)} + \\[6pt] \frac{\min(0.2\times0.1,0.2\times0.1)+\min(0.4\times0.4,0.4\times0.2)+\min(0.5\times0.7,0.5\times0.5)+\min(0.9\times0.3,0.9\times0.6)}{\max(0.2\times0.1,0.2\times0.1)+\max(0.4\times0.4,0.4\times0.2)+\max(0.5\times0.7,0.5\times0.5)+\max(0.9\times0.3,0.9\times0.6)} \end{array} \right\}$$

$$= \frac{1}{3}\left\{\frac{0.014+0.03+0.15+0.4}{0.04+0.12+0.25+0.56}+\frac{0.01+0.09+0.3+0.21}{0.03+0.18+0.6+0.49}+\frac{0.02+0.08+0.25+0.27}{0.02+0.16+0.35+0.54}\right\}$$

$$= \frac{1}{3}\left\{\frac{0.594}{0.97}+\frac{0.61}{1.3}+\frac{0.62}{1.07}\right\} = \frac{0.612+0.47+0.5794}{3} = 0.55.$$

$$S(\tilde{B},\tilde{C}) = \frac{1}{3}\left\{ \begin{array}{l} \frac{\min(0.2\times0.05,0.2\times0.2)+\min(0.3\times0.2,0.3\times0.4)+\min(0.5\times0.4,0.5\times0.5)+\min(0.8\times0.7,0.8\times0.7)}{\max(0.2\times0.05,0.1\times0.2)+\max(0.3\times0.2,0.3\times0.4)+\max(0.5\times0.4,0.4\times0.5)+\max(0.8\times0.7,0.8\times0.7)} + \\[6pt] \frac{\min(0.1\times0.2,0.1\times0.1)+\min(0.3\times0.5,0.3\times0.3)+\min(0.6\times0.8,0.6\times0.5)+\min(0.7\times0.5,0.7\times0.3)}{\max(0.1\times0.2,0.1\times0.1)+\max(0.3\times0.5,0.3\times0.3)+\max(0.6\times0.8,0.6\times0.5)+\max(0.7\times0.5,0.7\times0.3)} + \\[6pt] \frac{\min(0.2\times0.08,0.2\times0.1)+\min(0.4\times0.3,0.4\times0.2)+\min(0.5\times0.6,0.5\times0.5)+\min(0.9\times0.2,0.9\times0.6)}{\max(0.2\times0.08,0.2\times0.1)+\max(0.4\times0.3,0.4\times0.2)+\max(0.5\times0.6,0.5\times0.5)+\max(0.9\times0.2,0.9\times0.6)} \end{array} \right\}$$

$$= \frac{1}{3}\left\{\frac{0.01+0.06+0.2+0.56}{0.04+0.12+0.25+0.56}+\frac{0.01+0.09+0.3+0.21}{0.02+0.15+0.48+0.35}+\frac{0.016+0.08+0.25+0.18}{0.02+0.12+0.30+0.54}\right\}$$

$$= \frac{1}{3}\left\{\frac{0.83}{0.97}+\frac{0.61}{1.0}+\frac{0.526}{0.98}\right\} = \frac{0.85+0.61+0.537}{3} = 0.67.$$

Since $S(\tilde{B},\tilde{C}) > S(\tilde{A},\tilde{C})$, this means that pattern $\tilde{C}$ is much similar to pattern $\tilde{B}$.

10.5 Interval Type-2 Fuzzy Set

Till now we discussed general type-2 fuzzy set and we see that the computational complexity of using type-2 fuzzy set is more. People mostly use interval type-2 fuzzy set where computations are manageable. Interval-valued fuzzy set is a particular type of fuzzy set. A type-2 fuzzy set is written as:

$$\tilde{A} = \left\{ (x,u), \mu_{\tilde{A}}(x,u) \right\} \forall x \in X, u \in J_x \subseteq [0,1]$$

or it can also be represented as:

$$\tilde{A} = \int_{x \in X} \int_{u \in J_x} \frac{\mu_{\tilde{A}}(x,u)}{(x,u)}, J_x \subseteq [0,1],$$

where $\mu_{\tilde{A}}(x,u)$ is the secondary membership function and J_x is the primary membership of x. The notations are already explained in Section 10.2. When all $\mu_{\tilde{A}}(x,u) = 1$, then type-2 fuzzy set, $\tilde{A}$, becomes an interval type-2 fuzzy set. This means that the amplitude of the secondary membership function or the secondary grade is equal to 1. We know, as described earlier, that due to the uncertainty in the primary membership grades of type-2 fuzzy set, $\tilde{A}$, the set $\tilde{A}$ consists of a bounded region, which is called footprint of uncertainty. It is the union of all primary membership functions.

$$FOU(\tilde{A}) = \bigcup_{x \in X} J_x.$$

Thus, the secondary grade of an interval type-2 fuzzy set carries no information as the amplitude of the secondary membership grades is equal to 1. The upper membership function and lower membership function of type-2 fuzzy set, $\tilde{A}$, are the two type-1 fuzzy membership functions that bound the FOU. The upper membership function is the upper bound of the footprint of uncertainty while the lower membership function is the lower bound of the footprint of uncertainty.

The upper membership function is denoted as $\bar{\mu}_{\tilde{A}}(x)$ and the lower membership function is denoted as $\underline{\mu}_{\tilde{A}}(x)$.

So, J_x can be represented as [10]:

$$J_x : \left\{ (x,u) : u \in \left[\bar{\mu}_{\tilde{A}}(x), \underline{\mu}_{\tilde{A}}(x) \right] \right\} \text{ and}$$

$$FOU(\tilde{A}) = \bigcup_{x \in X} \left[\bar{\mu}_{\tilde{A}}(x), \underline{\mu}_{\tilde{A}}(x) \right].$$

10.6 Application of Interval Type-2 Fuzzy Set in Image Segmentation

Selection of upper and lower membership functions of type-2 fuzzy set plays a significant role for uncertainty measure. A more practical form of representing type-2 fuzzy set is:

$$\tilde{A} = \{x, \mu_U(x), \mu_L(x) \mid x \in X\},$$

where $\mu_U(x) \le \mu(x) \le \mu_U(x) \in [0,1]$.

The upper and lower membership functions are defined in terms of linguistic hedges [11]. These are written as:

$$\mu^{upper} = [\mu(x)]^{1/\alpha}$$

$$\mu^{lower} = [\mu(x)]^{\alpha}, \tag{10.11}$$

where $\alpha \in [1, \infty]$.

Tizhoosh used type-2 fuzzy set for image segmentation where he used the value of $\alpha \in (1,2]$. The general concept of image thresholding is already explained in Chapter 9. A measure of ultrafuzziness is used to find the optimal threshold T. It aims at capturing/eliminating the uncertainties within fuzzy systems using type-2 fuzzy sets.

Consider an image A, which is fuzzified with any user-defined membership function as explained in Chapter 9. Type-2 fuzzy image is formed, with lower and upper membership functions from Eq. (10.11) using $\alpha = 2$. For each threshold gray level, the measure of ultrafuzziness is computed and the maximum value is computed. The threshold gray level that corresponds to the maximum ultrafuzziness value corresponds the optimum threshold of the image. Ultrafuzziness is defined as:

$\xi(A) = \dfrac{1}{MN} \sum_{g=0}^{L-1} h(g) [\mu_U(g) - \mu_L(g)]$, where g is the gray level of the image that lies between $[0,255]$ for gray images, $M \times N$ is the size of the image, and $h(g)$ is the frequency of gray level. A result is shown in the blood vessel image (Figure 10.4).

Another example on image enhancement is given where the upper and lower membership functions are combined together using Chaira T-conorm [12].

Original image is initially fuzzified with membership function

$\mu(g) = \dfrac{g - g_{min}}{g_{max} - g_{min}}$, where g is the gray level of the image.

Then, using type-2 fuzzy set, two membership levels, $\mu^{upper}(g)$, $\mu^{lower}(g)$, are computed with $\alpha = 0.75$.

A new membership function is computed using fuzzy T-conorm by Chaira, described in Chapter 5. The new membership function using fuzzy T-conorm by Chaira is computed as:

$$\mu^{enh}(g) = \frac{\mu^{upper}(g) + \mu^{lower}(g) + \lambda \cdot \mu^{upper}(g) \cdot \mu^{lower}(g)}{(1 + \lambda) \cdot \mu^{upper}(g) \cdot \mu^{lower}(g) + 1}. \tag{10.12}$$

This is obtained using Eq. (5.28), $T_C^*(x, y) = \frac{x + y + \gamma xy}{(1 + \gamma)xy + 1}$.

$\mu^{upper}(g)$, $\mu^{upper}(g)$ are the upper and lower membership functions of the type-2 fuzzy set.

$\lambda = im\_avg$, where im_avg is the average of the image. The new image with the new membership function so formed is the enhanced image. A result on image enhancement is shown (Figure 10.5).

(a)   (b)

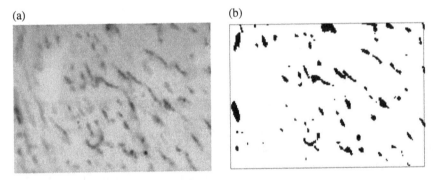

Figure 10.4 (a) Blood vessel image. (b) Tizhoosh type-2 fuzzy method.

(a)   (b)

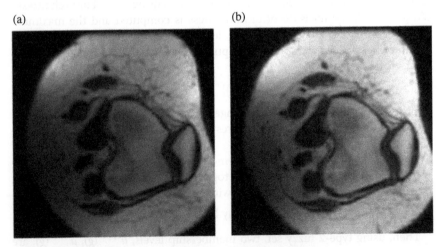

Figure 10.5 (a) Knee-patella image. (b) Enhanced image with Chaira T-conorm.

10.7 Summary

This chapter introduces type-2 fuzzy set theory, which is a fuzzy fuzzy set. Type-2 fuzzy operations such as union, intersection, addition, and multiplication along with similarity measures and inclusion measures are explained with examples. Interval-value type-2 fuzzy set is also explained where computational complexity is less and also application on medical image enhancement and thresholding is given.

References

1 Zadeh, L.A. (1975). The concept of a linguistic variable and its application to approximate reasoning-1. *Information Science* 8: 199–249.

2 Mendel, J.M. (2007). Type-2 fuzzy sets and systems: an overview. *IEEE Computational Intelligence Magazine* 2 (1): 20–29.

3 Mendel, J.M. and John, R.I.B. (2002). Type-2 fuzzy sets made simple. *IEEE Transactions on Fuzzy Systems* 10 (2): 117–127.

4 Mizumoto, M. and Tanaka, K. (1976). Some properties of fuzzy sets of type 2. *Information and Control* 31: 312–340.

5 Karnik, N.K. and Mendel, J.M. (2001). Operations on type-2 fuzzy sets. *Fuzzy Sets and Systems* 122: 327–348.

6 Mizumoto, M. and Tanaka, K. (1981). Fuzzy sets of type 2 under algebraic product and algebraic sum. *Fuzzy Sets and Systems* 5: 277–290.

7 Nieminen, J. (1977). On the algebraic structure of fuzzy sets of type-2. *Kybernetica* 13 (4): 261–273.

8 Yang, M.S. and Lin, D.C. (2009). On similarity and inclusion measures between type-2 fuzzy sets with an application to clustering. *Computers and Mathematics with Applications* 57: 896–907.

9 Singh, P. Similarity measure for type-2 fuzzy sets with an application to students' evaluation. *Computer Applications in Engineering Education* 23 (5): 694–702.

10 Dhar, S. and Kundu, M.K. (2018). A novel method for image thresholding using interval 2 type-2 fuzzy set and Bat algorithm. *Applied Soft Computing* 63: 154–166.

11 Tizhoosh, H.R. (2005). Image thresholding using type II fuzzy sets. *Pattern Recognition* 38: 2363–2372.

12 Chaira, T. (2014). An improved medical image enhancement scheme using Type II fuzzy set. *Applied Soft Computing* 25 (C): 293–308.

Beyond Your Doubts

1 Consider a fuzzy set A defined in the interval $x = [0,10]$ of integers by the membership function $\mu(x) = \dfrac{x}{x+1}$. Find α-cut corresponding to $\alpha = 0.4$.

2 For two fuzzy sets: $A = \{(2,0.4), (4,0.5), (6,0.1)\}$, $B = \{(3,0.3), (7,0.5)\}$, compute the product $A \times B$.

3 Consider two fuzzy relations:
$$\mathcal{R}_1 = \begin{bmatrix} 0.2 & 0.4 & 0.6 \\ 0.6 & 0.2 & 0.7 \\ 0.1 & 0.5 & 0.3 \end{bmatrix} \text{ and } \mathcal{R}_2 = \begin{bmatrix} 0.2 & 0.1 & 0.2 \\ 0.6 & 0.1 & 0.3 \\ 0.8 & 0.3 & 0.7 \end{bmatrix}.$$
Compute the following: (i) $\mathcal{R}_1 \cup \mathcal{R}_2$, (ii) $\mathcal{R}_1 \cap \mathcal{R}_2$, and (iii) $\mathcal{R}_1 \cdot \mathcal{R}_2$.

4 For the relation, $\mathcal{R} = \begin{bmatrix} 0.2 & 0.4 & 0.6 \\ 0.6 & 0.2 & 0.7 \\ 0.1 & 0.5 & 0.3 \end{bmatrix}$,
compute first projection, second projection, and total projection.

5 For the two fuzzy relations:
$$\mathcal{R}_1 = \begin{bmatrix} 0.2 & 0.4 & 0.6 \\ 0.6 & 0.2 & 0.7 \\ 0.1 & 0.5 & 0.3 \end{bmatrix}, \mathcal{R}_2 = \begin{bmatrix} 0.2 & 0.1 & 0.2 \\ 0.6 & 0.1 & 0.3 \\ 0.8 & 0.3 & 0.7 \end{bmatrix},$$
compute max-average, max-prod, and max-min composition.

Fuzzy Set and Its Extension: The Intuitionistic Fuzzy Set, First Edition. Tamalika Chaira.
© 2019 John Wiley & Sons, Inc. Published 2019 by John Wiley & Sons, Inc.

6 Consider a fuzzy relation $\mathcal{R} = \begin{bmatrix} 0.1 & 0.4 & 0.6 \\ 0.6 & 0.3 & 0.0 \\ 0.1 & 0.4 & 0.3 \end{bmatrix}$.

Find the transitive closure of the relation.

7 Consider a relation $\mathcal{R} = \begin{bmatrix} 1.0 & 0.7 & 0.3 & 0.7 \\ 0.7 & 1.0 & 0.5 & 0.0 \\ 0.3 & 0.5 & 1.0 & 0.0 \\ 0.7 & 0.0 & 0.0 & 1.0 \end{bmatrix}$ which is symmetric and

reflexive.

Show, if the relation satisfies the transitive relation. If not, then how this relation is reformed to an equivalence relation.

8 Consider two intuitionistic fuzzy sets:

$A = \{(1,0.2,0.6),(2,0.4,0.3),(4,0.7,0.2),(6,1.0,0.6),(7,0.9,0.0),(10,0.4,0.5)\}$

$B = \{(1,0.3,0.6),(2,0.3,0.4),(4,0.5,0.3),(6,1.0,0.0),(7,0.7,0.2),(10,0.3,0.5)\}$

Compute the following: (i) $A \cup B$, (ii) $A \cap B$, and (iii) $A^{1/2}$.

9 Consider two intuitionistic fuzzy relations R and S

$R = \begin{bmatrix} (0.3,0.4) & (0.4,0.3) & (0.6,0.3) \\ (0.7,0.1) & (0.5,0.2) & (0.2,0.8) \\ (0.2,0.8) & (0.2,0.7) & (1.0,0.0) \end{bmatrix}$, $S = \begin{bmatrix} (0.3,0.5) & (0.5,0.4) & (0.2,0.6) \\ (0.7,0.3) & (0.4,0.3) & (0.2,0.7) \\ (0.1,0.8) & (0.3,0.6) & (0.2,0.7) \end{bmatrix}$

Compute $R \cup S$, $R \cap S$.

Also, show whether the relations R and S are transitive.

10 Consider a fuzzy set:

$A = \{(1,0.2),(2,0.4),(3,0.6),(4,0.7),(6,1.0),(7,0.9),(8,0.7),(10,0.4)\}$

Compute α level and strong α level at $\alpha = 0.6$.

11 Compute the sets of α-cut for the membership function:

$$\mu_A(x) = \begin{vmatrix} 1-(1-x)^2, & \text{if } 1 \le x \le 4 \\ 0 & \text{otherwise} \end{vmatrix}$$

12 Consider a trapezoidal fuzzy membership function:

$$\mu_A(x) = \begin{vmatrix} \dfrac{x-1}{2}, & 1 \le x \le 3 \\[2mm] 1, & 3 \le x \le 5 \\[2mm] \dfrac{x-7}{-2}, & 5 \le x \le 7 \\[2mm] 0, & x > 7 \end{vmatrix}$$

Compute the α-cut interval of fuzzy number. Also, compute α-cut at $\alpha = 0.6$. Show the diagram depicting the $\alpha = 0.6$ level.

13 Let $A = (1,2,4)$ and $B = (2,4,5)$ be triangular fuzzy numbers.
Compute the α-cut of A and B. Also, compute the extended addition, extended subtraction, extended multiplication and show that multiplication of two fuzzy numbers is not a triangular fuzzy number.

14 Consider f, a function from X to Y and A is a fuzzy set on X,
$$A = \frac{0.3}{-4} + \frac{0.7}{-2} + \frac{1}{1} + \frac{0.6}{2} + \frac{0.5}{4}.$$
The function is $f(x) = 2x^2 - 1$. Using the extension principle, find the image of A.

15 Consider two fuzzy numbers $A = [1,3,5]$ and $B = [4,6,8]$, compute the following: (i) $A(+)B$, (ii) $A(-)B$, (iii) A^α, (iv) B^α, (v) $[A + B]^\alpha$, and (vi) $[A - B]^\alpha$.

16 Consider f to be a function from $(x,y) = x + y$. For two fuzzy subsets
$$A = \frac{0.2}{3} + \frac{0.5}{4} + \frac{1}{5} + \frac{0.8}{6},$$
$$B = \frac{0.2}{2} + \frac{0.4}{4} + \frac{1}{6} + \frac{0.8}{8},$$
compute the membership degree of fuzzy subset $f(A,B)$ through the function $f(x,y) = x + y$. Also, find the membership degree at $\mu_c(z)$ at $z = 10$, where $z = f(x,y)$.

17 Consider triangular fuzzy numbers A and B, $A = (-3,2,4)$, $B = (-1,0,6)$. Find the α-level intervals A^α, B^α.

18 For two triangular fuzzy numbers $A = [3,5,7]$ and $B = [5,8,10]$, compute $[A + B]^\alpha$, $[A - B]^\alpha$, $[A \cdot B]^\alpha$, $\left(\dfrac{A}{B}\right)^\alpha$.

19 If two fuzzy sets A and B are given with membership functions, $\mu_A(x) = \{0.3, 0.4, 0.7, 0.6, 0.4, 0.2\}$, $\mu_A(x) = \{0.2, 0.5, 0.7, 0.6, 0.3, 0.2\}$, compute $\mu(A \cap B)$, $\mu(A \cup B)$. Also, find the complement of $\mu(A \cap B)$.

20 Consider two fuzzy sets:

$A = \{(x_1, 0.2), (x_2, 0.6), (x_3, 0.8), (x_4, 0.6)(x_5, 0.3)\}$

$B = \{(x_1, 0.4), (x_2, 0.5), (x_3, 0.8), (x_4, 0.5)(x_5, 0.2)\}$,

compute $A - B$, $A + B$, $A \oplus B$, $A \ominus B$.

21 Compute ordered weighing averaging operator of four variables $F(0.5, 0.8, 0.3, 0.4)$ whose associated weighting vector is $\begin{bmatrix} 0.3 \\ 0.2 \\ 0.1 \\ 0.4 \end{bmatrix}$.

22 Consider argument variables along with order-induced variables as: $\langle u_j, a_j \rangle = \langle 0.7, 0.5 \rangle, \langle 0.5, 0.8 \rangle, \langle 0.4, 0.2 \rangle, \langle 0.2, 0.4 \rangle$. The associated weighting vector is $\begin{bmatrix} 0.4 \\ 0.3 \\ 0.2 \\ 0.1 \end{bmatrix}$.

Compute induced-ordered weighing operator at $\lambda = 1$.

23 Let us consider five intuitionistic fuzzy values – $p_1 = (0.3, 0.5)$, $p_2 = (0.4, 0.4)$, $p_3 = (0.5, 0.3)$ with weight vector $w = (0.3, 0.2, 0.1, 0.4)^T$ and $\lambda = 1$, $p_j (j = 1, 2, 3, 4)$.

Compute GIFOWA.

24 Let us consider five intuitionistic fuzzy values
$p_1 = (0.2, 0.7)$, $p_2 = (0.5, 0.2)$, $p_3 = (0.6, 0.3)$, $p_4 = (0.2, 0.6)$, $p_5 = (0.5, 0.3)$ and the weight vector of $p_j (j = 1, 2, 3, 4)$, $\omega = (0.3, 0.25, 0.1, 0.2, 0.4)^T$ and $\lambda = 1$, compute induced fuzzy hybrid operator.

25 Consider a set $X = \{l,m,n\}$. Fuzzy densities values are:

$$g_\lambda(\{l\}) = 0.6, g_\lambda(\{m\}) = 0.4, g_\lambda(\{n\}) = 0.2.$$

Compute λ and also compute the fuzzy measures $g_\lambda(\{l,m\})$, $g_\lambda(\{l,n\})$, $g_\lambda(\{m,n\})$.

26 Consider a set $X = \{x_1, x_2, x_3\}$ and the function $f(x)$ or the range is defined as:
$f(x_1) = 0.4$, $f(x_2) = 0.3$, $f(x_3) = 0.2$. The fuzzy densities g are given as follows:

$$g(\{x_1\}) = 0.2, g(\{x_2\}) = 0.3, g(\{x_3\}) = 0.5.$$

Compute Sugeno integral.

27 Consider a set $X = \{x_1, x_2, x_3\}$ and the function $f(x)$ or the range is defined as:
$f(x_1) = 0.5$, $f(x_2) = 0.3$, $f(x_3) = 0.2$. The fuzzy density values are given as:

$$g(\{x_1\}) = 0.2, g(\{x_2\}) = 0.5, g(\{x_3\}) = 0.6.$$

Find the Choquet integral for λ-fuzzy measure.

28 In continuation with problem 27, let μ be a fuzzy measure in space $A = (A_1, A_2, A_3\}$ and assume the fuzzy measure of attributes A_i $(i = 1,2,3,4\} = \{0.3, 0.4, 0.6\}$ and attribute sets of A_i as

$$\mu(\{A_1\}) = 0.2, \mu(\{A_2\}) = 0.5, \mu(\{A_3\}) = 0.6.$$

(a) Compute $g(\{A_1, A_2\})$, $g(\{A_2, A_3\})$, $g(\{A_1, A_3\})$.
(b) Find Choquet ordered aggregation.
(c) Assuming we have four pairs $A_1 = \langle 6,0 \rangle$, $A_2 = \langle 7,0.2 \rangle$, $A_3 = \langle 2,0.9 \rangle$, $A_4 = \langle 3,0.6 \rangle$ where the first component is the order-inducing variable, then compute induced-ordered Choquet aggregation.

29 Consider a possibility distribution of the proposition "X is an integer close to 10"

$$\pi_X = \{(6,0.4),(7,0.6),(8,0.6),(9,0.7),(10,1),(11,0.9),(12,0.7),(13,0.5),(14,0.3)\}$$

and a crisp set $A = \{7,9,12\}$. Find the possibility measure of A.

30 Consider two fuzzy matrices:

$$A = \begin{bmatrix} 0.4 & 0.1 & 0.7 & 0.9 \\ 0.0 & 0.6 & 0.4 & 0.8 \\ 0.5 & 0.3 & 0.1 & 0.9 \\ 1.0 & 0.4 & 0.7 & 0.1 \end{bmatrix}, B = \begin{bmatrix} 0.3 & 0.2 & 0.6 & 1.0 \\ 0.4 & 0.7 & 0.5 & 0.0 \\ 0.7 & 0.3 & 0.2 & 0.3 \\ 1.0 & 0.2 & 0.5 & 0.9 \end{bmatrix}.$$

Compute (i) $A \vee B, A \wedge B$, fuzzy matrix multiplication – AB and BA using max–min operator. Also, show $AB = BA$.

31 Find the determinant of a fuzzy matrix,

$$A = \begin{bmatrix} 0.3 & 0.5 & 0.6 \\ 0.6 & 0.3 & 0.9 \\ 0.1 & 0.3 & 0.2 \end{bmatrix}.$$

32 Consider two intuitionistic fuzzy matrices, A and B

$$A = \begin{bmatrix} (0.4,0.5) & (0.6,0.2) & (0.4,0.4) \\ (0.8,0.2) & (0.6,0.2) & (0.8,0.1) \\ (0.6,0.2) & (0.7,0.2) & (0.3,0.4) \end{bmatrix}, B = \begin{bmatrix} (0.7,0.3) & (0.6,0.3) & (0.5,0.3) \\ (0.6,0.3) & (0.7,0.2) & (0.5,0.2) \\ (0.5,0.3) & (0.8,0.2) & (0.7,0.2) \end{bmatrix}$$

Compute $A + B, AB$.

33 For the fuzzy matrices in Problem 31, compute $A^c, B^c, A^c + B^c$. Also, compute $(A + B)^c$.
See if $A^c + B^c = (A + B)^c$.

34 Consider $J = \{0,0.1,0.2,0.3,0.4,0.7,0.8,0.9,1\}$ and the fuzzy grades of type-2 fuzzy set, $\tilde{A}$, be

$$\mu_{\tilde{A}}(x) = \frac{0.4}{0} + \frac{05}{0.1} + \frac{0.7}{0.3} + \frac{0.4}{0.5},$$

$$\mu_{\tilde{B}}(x) = \frac{0.7}{0} + \frac{0.6}{0.2} + \frac{0.5}{0.4} + \frac{0.2}{0.6}.$$

Compute (a) Union – $\mu_{\tilde{A}}(x) \sqcup \mu_{\tilde{B}}(x)$, (b) Intersection – $\mu_{\tilde{A}}(x) \sqcap \mu_{\tilde{B}}(x)$.

Index

Fuzzy Set and Its Extension: The Intuitionistic Fuzzy Set, First Edition. Tamalika Chaira.
© 2019 John Wiley & Sons, Inc. Published 2019 by John Wiley & Sons, Inc.